Safety Science: Methods to Prevent Incidents and Worker Health Damage at the Workplace

Authored By

Eduardo Calixto

Production Engineering Department - LATEC
Federal Fluminense University
Brazil

DEDICATION

I dedicate this book to professor Dr. Gilson Brito Alves Lima from Federal University Fluminense for all support and sponsor my academic and professional career and friendship.

THANKS

Special thanks for David Thompson from RAMsoft UK, my colleagues Julio Elsio, Marcolan from VALE for all support in Safety and Occupational Health System implementation and management and also for my colleagues Carlos Daniel, Wilson Alves, Cid Atusi, Darlene Barbosa and Geraldo Alves from Petrobras Refineries for all support and learning of implementing and support the risk management process in Brazilian Oil and Gas refineries. Thanks for my wife Isabel Katrin Calixto for all support.

CONTENTS

FOREWORD

Occupational Risk management should be a continuous process and be in constant development. It is applied to the Organization's strategy and on the implementation of that strategy. It shall examine methodically all the risks inherent in past, present and in particular, in the future activities of an organization.

It should be integrated in the culture of the organization with an effective policy and be part of a program led by top management. It should translate the strategy into tactical and operational objectives, assigning responsibilities in risk management throughout the organization, as part of job descriptions. This practice supports the accountability, performance evaluation and respective reward, promoting in this way the operational efficiency at all levels of the organization.

In general, no event starts big! Risk management is a central element in any strategy management of any organization. It is the process used by organizations to methodically analyze risks inherent in their activities, aiming at achieving a sustained advantage in each individual activity and in the set of all activities.

The central point of a good risk management is the identification and treatment of risks. Its main objective is to add value in a sustained manner to all activities of the organization. It coordinates interpretation of the potential positive and negative aspects of all the factors that may affect the organization. It increases the likelihood of success and reduces both the probability of failure and the uncertainty of obtaining the overall objectives of the organization.

The evident and timely contribution of this work is based on the context presented above. Safety Science: Methods to Prevent incident and health damage in workplace is a book written focused on two main contexts: the corporate and the academic.

Firstly, because it is recognized that this is one of the aspects, whose importance in the management of organizations, has been increasing considerably in recent times; secondly, because it is apparent that it has not had the attention it deserved by our schools in general.

In the dialectic of the academic context, it will serve as the basis for the formation of the professional competence of higher education students in engineering courses, management and the like. The majority of these courses includes at least one subject related to health and safety management, whose program and content, do not deviate significantly from the general plan of the book.

Secondly, in the corporate context, a strong academic background and extensive experience of the author, as doctor in Engineering Sciences, engineer and consultant on several medium and large international companies, allowed the combination of the scientific rigor with pragmatism in the explanation of the various theories here exposed. The book will also be useful for managers, including those engaged in or intend to come to the position of general manager and wish to make a recycling or deeper their general knowledge.

As usual in training books, this one was written based on the four fundamental management functions – planning, organization, direction and control – and was divided into 3 parts: qualitative techniques, quantitative techniques, instruments and management standards, which include nine chapters.

The first chapter seeks to introduce the reader into the context of occupational risk, focusing on the physical, biological and chemical hazards aspects and also addresses the ergonomic factors, as a necessary knowledge to the understand the global environment of the organization.

On the assumption that the study of the planning would be incomplete without a reference to the way to implement it, some chapters were added about techniques and tools in support of the decision-making process. Chapter two to seven present the main qualitative and quantitative tools in the context of risk management, while managerial and technical knowledge necessary, without however, moving the central axis from the State-of-the-art risk management.

In order to address the analysis of the human factors related to the culture of organizations and the "engaging" feature in the practice of risk management, a set of methodologies is presented in chapter seventh. It describes the techniques inherent to the Human Factor, among them: Technique for Human Error Rate Prediction (THERP), Operator Action Tree (OAT), Accident Sequence Evaluation Program (ASEP), Social technical Analysis of Human Reliability (STAHR), Standardized Plant Analysis Risk Human Reliability (SPAR-H), Human Error Assessment Reduction technique (HEART) and Bayesian Belief Network analysis (BBN).

The main safety standard that support the structure of current management systems throughout the world are presented in the eight chapters. Among these: OHSAS 18001, ISO 31000, a safety case applied to oil and gas industry, EN 51026 (risk management applied to railway industry), a safety case applied to nuclear industry, a key program asset integrity, IEC 61508 (safety integrity level standard). The final chapter addresses Safety and Occupational Health Management.

The final message from the author is that no event starts big. The assumption paradigm being proposed is that risk management should be a central element in the strategy management of any organization and should be regarded as the process through which organizations analyze methodically risks inherent to their activities, with the aim of achieving a sustained advantage in each individual activity and in the set of all activities.

Gilson Brito Alves Lima
Technology Laboratory Safety Engineering
Post Graduation Coordinator
Fluminense Federal University
Brasil

PREFACE

This book aim to discuss the mains methods to prevent incident and employee's health on workplace. Despite a huge effort from different organizations all over the world is still a challenge to prevent incident and health damages in workplace. The challenge faced are related more to human rather than technological issues. The organizational culture, leadership as well as organizational learning has an important hole in safety and occupational health effectiveness. Indeed those factors are the safety management pillars and also the main problem on most of organization in different industry.

Concerning the methods, different approaches are described in this book to prevent incident and health damages like risk assessment, emergency response, incident analysis methods, human factor, safety standards, and safety management.

In general terms, such methods are not well applied in many cases due to many reasons like lack of awareness about the importance of risk mitigation, lack of dedicated time to perform better analysis and lack of investment on safety.

Unfortunately, many organizational leader do not address their attention to safety into organization which might explain why such major still accident happen nowadays and many health damages are still in high level in many organizations all over the world.

Therefore, it's important to be aware about the best approaches to address to different types of problems on safety as well as to establish a consistent and effective Safety occupational health management which is supported for such methods. By this way, the organizations will be able to achieve high performance and maintain it in long term.

Safety is not a matter of indexes, standards and reactive actions. Safety is a preventive culture which is reflected by preventive actions and attitude from all employees in different organizations levels which allow such organizations achieve high performance in long term.

CONFLICT OF INTEREST

The author confirms that this book contents have no conflict of interest of any part. All references are cited in the book and the text was written for the author.

ACKNOWLEDGEMENTS

I would like to please the high level specialist who contributed with their technical keen remarks and comments about the book chapters. That's enable me to improve the quality of this book. Thanks a lot for your support.

David Thompsom
Maintenance & Reliability Consultant
University of Salford
Manchester
UK

Deshai Botheju
Safety Engineer Specialist
University of Moratuwa
Sri Lanka

Karl Butler
Aerospace Engineer, ILS/RAMS Specialist
University of Manchester
UK

Paulo Maia
Risk Management Specialist
Instituto Superior Tecnico
Lisbon, Portugal

&

Eduardo Calixto
Production Engineering Department - LATEC
Federal Fluminense University
Brazil
E-mail: Eduardo.calixto@hotmail.com

CHAPTER 1

Occupational Risk

Abstract: The occupational risk analysis encompasses different types of risk faced by employee's in the workplace, such as physical, mechanical, chemical, biological and ergonomic. Depends on the type of industry and activities one type of hazard is more frequent than others. During the last decades, many procedures, laws and methods has been applied to reduce the health damage caused by such hazards. Even though, it's still necessary a high effort to mitigate some of these hazards such as drowning, machinery, ergonomic, obesity and stress. In general terms, fishing, farming, and building are the activities related to the higher number of accidents or health damage. By the other way round, some industry such as Chemical, Oil & Gas and Nuclear are responsible for the worst accident in terms of consequence for employee's health, environment and society. Therefore, the first step to avoid or mitigate these occupational hazards is to be aware about them in the workplace. Moreover, it's necessary to implement a systematic occupational risk management with the best methods to mitigate and communicate such risk. This chapter aims to describe the occupational hazards in the workplace as well as the best methods to identify, classify, assess and mitigate such hazards.

Keywords: Occupational hazard, physical hazard, mechanical hazard, chemical hazard, biological hazards, ergonomic hazard, social and organizational factor.

1.1. INTRODUCTION

The occupational safety and health regulation are a relatively new. Despite all effort since the industrial revolution to improve workers' health in workplace the modern concerns about health and safety starts at 1950 when the International Labor Organization (ILO) and the World Health Organization (WHO) have shared a common definition of occupational health.

In 1995 such organizations review and adopt a common occupational health definition such as:

"Occupational health should aim at: the promotion and maintenance of the highest degree of physical, mental and social well-being of workers in all occupations; the prevention amongst workers of departures from health caused by their working conditions; the protection of workers in their employment from risks resulting from factors adverse to health; the placing and maintenance of the worker in an occupational environment adapted to his physiological and psychological capabilities; and, to summarize, the adaptation of work to man and of each man to his job".

Eduardo Calixto

Based on such definition the occupational health is summarized in three objectives such as:

- To maintain the employee's health and safety in the workplace.
- To implement continuous workplace conditions improvement.
- To promote the organizational´s health and safety culture.

Such concern starts from enterprise concepts in project phase by choice of technologies that minimize risk of bad effect on employee and society health. Depend on situation, companies have not too much idea about projects, concepts and risk decision because such technologies are not under their competency domain, but even though they're also responsible for choosing such technologies and understand the risk involved from project concepts to decommission.

New products and processes are always a big challenge in terms of Occupational safety and health because such occupational risks are unknown. In this situation, huge effort must to be done in the design phase in order to minimize occupational risk and produce robust product against unsafe failures and conditions.

Indeed, the law is only the first parameter to take into account, but moral and company's images are also associated to Occupational safety and health. In some cases, finances drivers as reduce products cost and improve profits can lead to design not robust products and process which cause serious damages to employee's health. Actually, all organizations must be aware about their product and process unsafe conditions and implement preventive action to avoid health damages on their employee's and society.

Actually, the cost of damage employee's health can be measured having directly and indirectly cost associated based on expected compensation for loss of employee's or society welfare caused by an accident. That will be discussed in the next chapter and definitely it is an important topic that supports many decisions in favor to invest in prevention. In some cases, the direct and indirect cost is lower than the preventive cost to avoid the accident, but when the existing value is taken into account like company image and reputation, the preventive cost is mostly of time lower than accident cost.

One of the biggest challenge of OSH is interactions between different subjects such as, occupational medicine, occupational hygiene, safety engineering, chemistry, ergonomics and psychology. That requires that professional from different subjects work together with their specific knowledge to aggregate and complement the other subject's knowledge of interactions to have a better solution. In addition, it is necessary that different professional has the minimum knowledge about others subject to have a better interface and systemic problem comprehension. But the previous issue is what we understand for occupational health.

The safety culture is the hardest challenge for all companies, no matter industry, technology level, process complexity and number of employee's. Hardly will be achieved 100% of employee's pro occupational and safety and that is harder when leaders take place safety in favor of production, reduce costs, short project time, reduce lead time. Once such priorities are established, harder is to implement or even maintain occupational and safety culture that is reflected in prevention attitudes in all activities. The companies recognition and valuation for occupational and safety professional is also an important issue that must be taken into account. It is not usual to see many companies with executives and CEO who came from occupational a safety subjects. Hardly such professional achieve high manager hierarchy level of their organization. Such fact reflect in some way that such occupational and safety professional are not the most important for many organizations.

The effect of fatal accident can be seen in different industries and activities as reported by the United States Bureau of Labor Statistics as shown in Table **1** based on statistic from 2006.

Table 1: Number of Fatal work injuries by job occupation in 2006 (US) (Source: U.S. Bureau of Labor Statistics, Census of Fatal Occupational Injuries program)

Job	**Number of fatalities**
Truck drivers	957
Farmers and ranchers	292
Miners	156
Pilots	104
Roofers	81
Timber cutter	66
Fishermen	53
Power-line workers	38
Waste collectors	37
Structural metal workers	36
All occupations	**5840**

Once comparing such number with previous years, it's possible to see the reduction of fatal accident. Such reduction reveal all effort made in legislation as well as occupational health management. The Fig. **1** shows the total number of fatal work injuries from 1992 to 2006 in the United States.

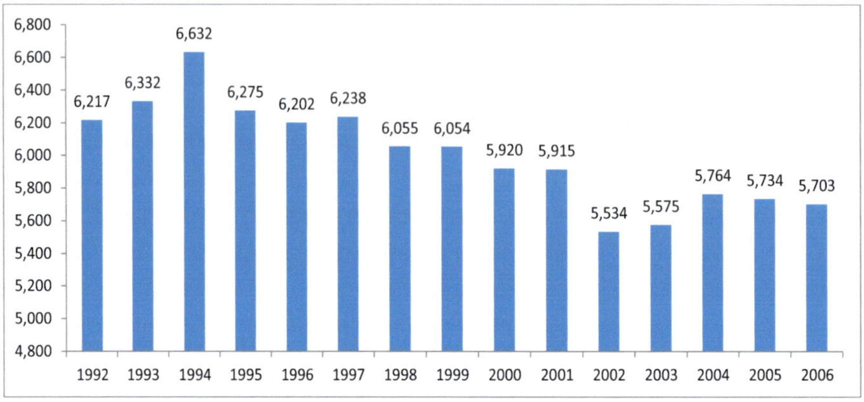

Figure 1: Number of fatal work injuries (US: 1992-2006) (Source: U.S. Bureau of Labor Statistics, Census of Fatal Occupational Injuries program).

Regarding occupational risk is important to understand the difference between chronic risk and acute risk. The Occupational Acute risk is related to punctual health damaged caused by accidents and may have low or high consequences for employee's health. An example of Occupational Acute risk is falling accident risk in the construction industry, the risk of drowning in fishermen activities and all activities with fatal accidents that is a reality in other industry as well like nuclear, Petrochemical, Oil and Gas, metallurgic. The Occupational Chronic risk is related to health damaged that has resulted from exposure to harm that effect employee's heath after a long period of time, having low or high consequence on employee's health. Such effects can be occurring in times, days or years. An example of such risk is disease develop with exposure to bad work condition in agriculture industry, exposure a toxic products in mine, bad ergonomic conditions in offices all over the world.

In the next sections different types of occupational risk like physical, chemical, biological and ergonomics will be explained and discussed with examples.

1.2. PHYSICAL AND MECHANICAL HAZARD IN WORKPLACE

Physical hazards are related to health damage caused by machinery, tools, or physical energy, such as noise, vibration, radiation and extreme temperatures.

Physical hazards can cause injuries and their effects can happen immediately after an accident, such as exposure like fractures and lacerations or can be delayed for many years, such as a gradual loss of hearing from noise exposure.

In general terms Physical hazards can be caused by:

- Slips, trips and falls
- Electricity
- Tools and machinery
- Heat stress
- Cold stress
- Noise
- Vibration
- Radiation
- Fire

In general, the slips and trips mostly happen to loss of contact of the foot and the floor. That is a very common occurrence when the floor has cleaned and people walking under this wet floor or moment after this floor be cleaned because it is still wet. Another common reason is to have not appropriate shoes like tennis shoes or high shoes in operational areas where there is an irregular floor or there are chemical products, dust or sand spread out on the floor.

In order to prevent slips is advisable the following actions:

- Wear appropriate shoes for industrial and office area.
- Introduce high resistance nonslip surfaces.
- Clean and safe surfaces which are appropriate for the intended tasks performed on them.

The fall is a type of physical hazard which may cause serious consequences with severe injuries or even death when over than 1 meter in case of an accident. The most common cause of falls is unsafe scaffold and ladder that are used in construction and maintenance activities in high places. Many of such accident would be avoided by correct use of the protections that are required in such situations.

It is advisable to prevent trips the following actions:

- Follow the ladder and scaffold manufacturer's instructions for use.
- Do not use a stepladder as a support.
- Inspect latters and scaffold before use it.
- To use only stable flat surface to install scaffold.
- Do not put the additional height at ladder or scaffolds.
- Avoid to positioning ladder and scaffold in place where it could be hit by a person or vehicle.

The noise is a physical hazard which has a lower hazard perception by most of employee's due to its long term effects.

The noise in the workplace can be classified as continuous and impulse noise.

The continuous noise refers to the exposure level of noise in the workplace for eight hours. We can assume as reference the following values:

- In case of continuous noise, regarding eight hour workday, we consider that is necessary to use personal hearing protectors when the exposure values go higher than 80 decibels (dB).

Impulse noise is a sudden loud noise over 135 dB. We can assume as reference the following values:

- The upper exposure action value of 137 dB and lower exposure value of 135 dB that requires hearing protectors.
- The limit exposure value is 140 dB that must be reduced below the limit value.

- In order to have a precise impulse noise assessment in work place a range of measurements are needed.

Regarding temperature, we need to take into account that temperature takes influence in employee's welfare as well as performance at the workplace. It´s advisable that such temperature be regulated based on specific value limits. The following temperature recommendations for different kinds of work considering the physical conditions required to task are:

- Light workload: 19 ºC – 25 ºC.
- Moderately workload: 17 ºC – 21 ºC.
- Heavy workload: 12 ºC – 17 ºC.

It is also important the ventilation system to help in such workplace temperature achievement. Indeed, due to new layout or huge modifications that mostly happen in offices, the temperature is deregulated in many workplaces. Consequently, there is some place extremely cold and other are extremely hot and depends on seasons, it may cause a real discomfort and reduce employee productivity.

Extreme temperatures are dangerous to employee's health. Heat stress may cause health damages such as dehydration, exhaustion, cramps, heart stroke as well as burns when employee's work near to hot surfaces.

By the other way round, extremely cold temperature is also dangerous to employee's health. The extreme cold conditions can cause hypothermia, frostbite, trench foot, or chilblains.

The other physic risk that must always take into account and prevent accident injuries is electricity**.** The electrical can cause injuries such as fatal electrocution, electric shock and burns.

In order to prevent accidents with electricity at workplace some actions must be taken place like:

- Inspect all tools with electrical connections and replace them whenever it´s been necessary.
- Make sure that tools with electrical connections are properly grounded.

- Make sure that the correct size fuse is being used.
- Make sure that nonconductive materials is being used when working around exposed, energized electrical equipment and wires.
- Be aware of wet location to avoid electrical shock.
- Be aware about the breakers and boxes location.
- Do not touch a person or equipment energized after electrical contact.
- Avoid to use extension cords.

The Vibrating machinery can cause work related illness and injury. The daily exposure limit value for whole body vibration is 1.15 m/s2.

The lighting at workstations is also an important physical hazard. In case of activities which requires high precision it´s necessary local lighting at workstations. When employee's work in the workplace with deficient lighting, the risk of accident or fatigue increases.

The walkways and outdoor must have sufficient general lighting in order to avoid accidents.

Other specific physical hazards, not usual in industrial area, but in specific laboratories are:

- Ionizing radiation like X-rays, gamma-rays and radon.
- Ultraviolet radiation.
- Laser radiation
- Infrared radiation
- Microwaves radiation.
- Electromagnetic fields.

In general terms, the physical hazard risk mitigation can be achieved by safety methods and procedures as well as an effective occupational risk program considering each workplace features.

In addition to all preventive measures is always advisable to take place Personal Protection Equipment as well as Collective Protection Equipment.

The employer responsibility is to provide such protection and train employee's to use them effectively. In fact, different work stations and activities requires specific personal protection such as eye protection, safety footwear, overalls and Earplugs, earmuffs, Industrial safety helmets and gloves. Indeed, to minimize the physical hazard effects, it´s important to identify the hazard property in order to define the best protection for each specific workplace activity.

The future physical risk is of course all related risk above and a new one that poses a huge challenge in terms on occupational risk analysis and prevention that are robots.

Nowadays, when the robot is working in automatic mode, the robot work zone is separated from human work zone in order to avoid accidents. In addition, the robot is turned off whenever maintenance task is required and the employee's must work on the robot work zone. The robot work zone has different protections, such as barriers, fences, controlled doors, light curtains and laser scanners are examples of protections against people entrance. In most of cases, whenever such barriers are violated robot stops by emergency mode (Tadeusz Missala, 2012).

Figure 2: Robot in pallet handling (Source: Tadeusz Missala, 2012).

The main robot applications developed are: painting, cutting (gas and plasma), welding (gas and electric arc), some kinds of automated assembly, packing, positioning on platforms, trucks and palettes, other material and object handling. These applications are realized till now and will be realized in the future. The Fig. **2** shows an example of robot application in industry.

Based on scientific fiction in future robots may carry out more complex tasks and have a more intensive relation with human in the workplace and in usual life as well. Despite new risk that new technology brings to the worlds there is also a good application like virtual reality to assess the occupational risk in the workplace.

The virtual reality process has simplified and speeded up the design work to a large extent. Computer simulations allow for solving many design problems, the identification of which was previously, *i.e.*, not having these techniques at one's disposal, possible only after building the physical model of a machine. The most advanced, rapidly developing computer technique consisting in the virtual reality approach. Recently, the technique is commonly applied also in supporting the design process of machines and manufacturing systems. Usually, virtual simulations are carried out to provide the necessary information necessary for the analysis of technical properties, technological potential and working parameters, respectively, of the designed devices. Very often the techniques are employed for modeling the human factor and performing the ergonomic analyses. When creating s new design the machine designer should take into consideration the issue of machine operator's safety. One of the methods for reduction of the risk due to machine operation consists in the application of protective devices (Dźwiarek M., Jankowski J. 2012).

The research aims at the application of virtual reality technique for solving the safety of Machinery issues, combining those ways two fields of science. In conducting the research one should take into consideration the knowledge on both the virtual prototyping as well as safety of machinery. The virtual technique, *i.e.*, the investigations in which advanced computer simulation methods are used, creates a new research field that develops rapidly.

The constructed VR models of a manufacturing system are nowadays employed for tests. Once selecting a system to be tested a diversity of dangerous zones appearing is decisive, so as a number of protective devices of different types employed would be as high as possible. The Fig. **3** shows an example of Virtual reality applied to assess physics occupational risk in the presence of robots in a workstation.

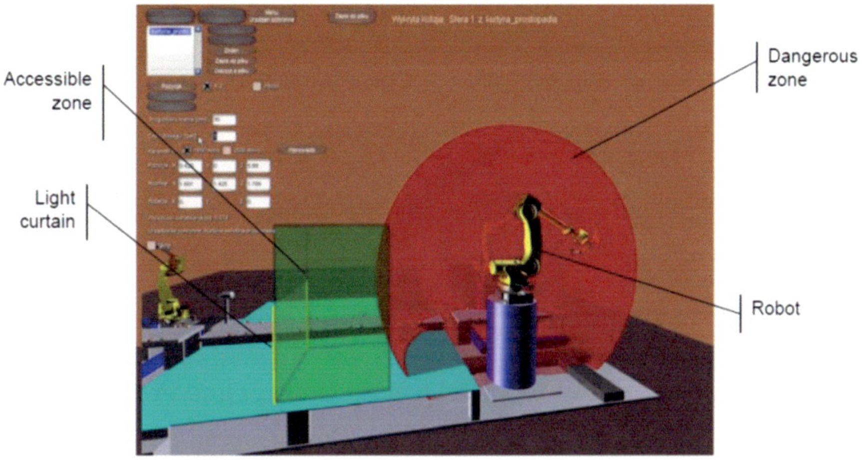

Figure 3: Virtual models of dangerous and accessible zones (Source: Dźwiarek M., Jankowski J. *et al* 2012).

This virtual reality applications are the result of the research tasks carried out all over the world. A good example of virtual reality research is carried out by the Polish government with the main objective of improvement of safety and working conditions considering robot in the workplace.

As a result of the research conducted the following results have been obtained:

- Virtual models of protective devices to be implemented to object databases that can be used in developing virtual models of machines and manufacturing systems as well simulating technological processes in terms virtual modeling.

- Software allows for visualization of dangerous zones when creating a virtual Prototype of a machine or manufacturing system.

This king of application must be more and more applied in the next future to assess not only physical hazards but other types of occupational hazards.

1.3. BIOLOGICAL HAZARDS

Biological hazards refer to hazard posed in organisms that may cause damage to employee's health. The biological hazard sources are different depends on workplace, such as parasites, viruses, bacteria, fungi and protein. In general, there are two major routes of entry for these biological hazards in our body, which are the respiratory system and skin. Such contamination occurs by contact with human body fluid with infected object or people.

The employee´s health damaged caused by biological hazards are infections, allergy and poisoning.

Whenever employers deal with biological hazards the first step is identifying such hazards in the workplace, assess and classify them. The following step is to communicate such dangerous consequence of such biological hazard to workforces and regarding workplace characteristics an each task which is exposure to different biological hazard, preventive actions and controls must be take place. An example of Biological Hazard Classification can be seen in Table **2** based on The United States' Centers for Disease Control and Prevention (CDC).

Thereby, based on specific biological hazard classification, such hazards can be identified for each specific task and workplace in order to give specific instruction for each employee to prevent for biological hazard consequences. It is important to take into account different routines, duration of exposure as well as protection against hazards. In order to assess biological hazard a questionnaire may ne apply and the type of task, a workplace, a type of biological hazard, source of biological as well as preventive actions and required protection must take place. The biological hazard workplaces are:

⇒ Hospital and medical laboratory.

⇒ Healthcare facilities.

⇒ Cleaning services in hospital.

⇒ Hospital hygiene and cleaning services.

⇒ Agriculture, fishery, veterinary services.

⇒ Process plants that use plant or animal based raw materials.

⇒ Indoor workplace areas with central air-conditioning.

Whenever such workplace is assessed the key success factor is to find out how the infectious agent are successful to contaminate the employee's and that answers

Table 2: Biological Hazard Classification (Source: The United States' Centers for Disease Control and Prevention)

Category	Decription
Biohazard Level 1	Bacteria and viruses including *Bacillus subtilis, canine hepatitis, Escherichia coli, varicella* (*chicken pox*), as well as some cell cultures and non-infectious bacteria. At this level precautions against the biohazardous materials in question are minimal, most likely involving gloves and some sort of facial protection.
Biohazard Level 2	Bacteria and viruses that cause only mild disease to humans, or are difficult to contract *via* aerosol in a lab setting, such as hepatitis A, B and C, influenza A, Lyme disease, salmonella, mumps, measles, scrapie, dengue fever, and HIV. "Routine diagnostic work with clinical specimens can be done safely at Biosafety Level 2, using Biosafety Level 2 practices and procedures. Research work (including co-cultivation, virus replication studies, or manipulations involving concentrated virus) can be done in a BSL-2 (P2) facility, using BSL-3 practices and procedures.
Biohazard Level 3	Bacteria and viruses that can cause severe to fatal disease in humans, but for which vaccines or other treatments exist, such as anthrax, West Nile virus, Venezuelan equine encephalitis, SARS virus, tuberculosis, typhus, Rift Valley fever, Rocky Mountain spotted fever, yellow fever, and malaria. Among parasites *Plasmodium falciparum*, which causes *Malaria*, and *Trypanosoma cruzi*, which causes trypanosomiasis, also come under this level.
Biohazard Level 4	Viruses and bacteria that cause severe to fatal disease in humans, and for which vaccines or other treatments are not available, such as Bolivian and Argentine hemorrhagic fevers, Marburg virus, Ebola virus, hantaviruses, Lassa fever virus, Crimean-Congo hemorrhagic fever, and other hemorrhagic diseases. Variola virus (smallpox) is an agent that is worked with at BSL-4 despite the existence of a vaccine. When dealing with biological hazards at this level the use of a positive pressure personnel suit, with a segregated air supply, is mandatory. The entrance and exit of a Level Four Biolab will contain multiple showers, a vacuum room, an ultraviolet light room, autonomous detection system, and other safety precautions designed to destroy all traces of the Biohazard. Multiple airlocks are employed and are electronically secured to prevent both doors opening at the same time. All air and water service going to and coming from a Biosafety Level 4 (P4) lab will undergo similar decontamination procedures to eliminate the possibility of an accidental release.

depends to have a clears understand of "Classic Chain of infection" which take place in one specific workplace. The Classic Chain of infection takes into account six factors as shown in Fig. **4**. The first one is the infectious agent that is all kinds of organism which causes people health damage like those who are classified in

Table **2** above. The second one is a reservoir that is a source which enable the infective agent to growth and multiplicity for example people, material and equipment. The third one is a portal to exist and that means the way that the infection agent can leave from reservoir for example skin or blood. The fourth one is a model of transmitting that means the channel which infection agent goes from reservoir to another place for example direct or indirect contact with people and objects. The fifth one is portals to Host which means the site where the infection agent can gain access to the host for example mucous membrane and skin. The last one is the Susceptible Host that is the invidious who is infected by infection agent.

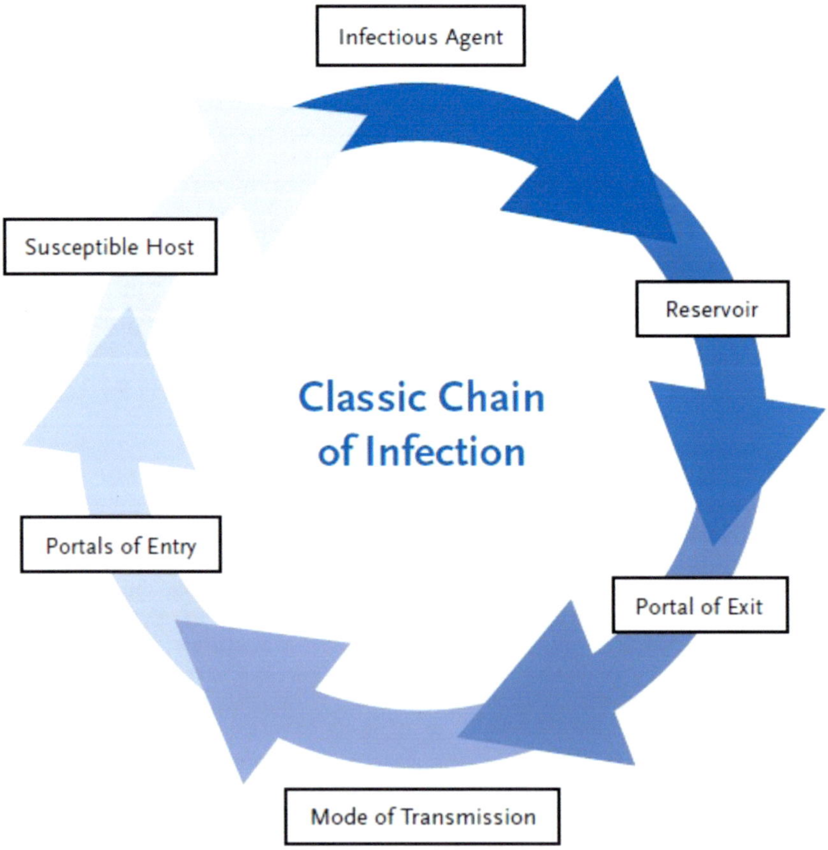

Figure 4: Classic Chain of infection (Source: www.employment.alberta.ca).

In order to prevent such contamination, it's necessary to carry out preventive action, such as improvement in ventilation system, isolation of contamination sources and use ultraviolet lamps which helps to contain the spread of contamination. In addition the correct use of personal equipment is essential to reduce the chance of contamination.

Once the biological hazard is assessed as well as preventive and control measure, is necessary to communicate and trainee employee's to ensure that such preventive actions and control will be take place property.

Whenever applying biological hazard prevention and control action we need to have in mind that three main factors like Biological agents, host and environment have high influence in the success or fail in such actions like shows Fig. **5**.

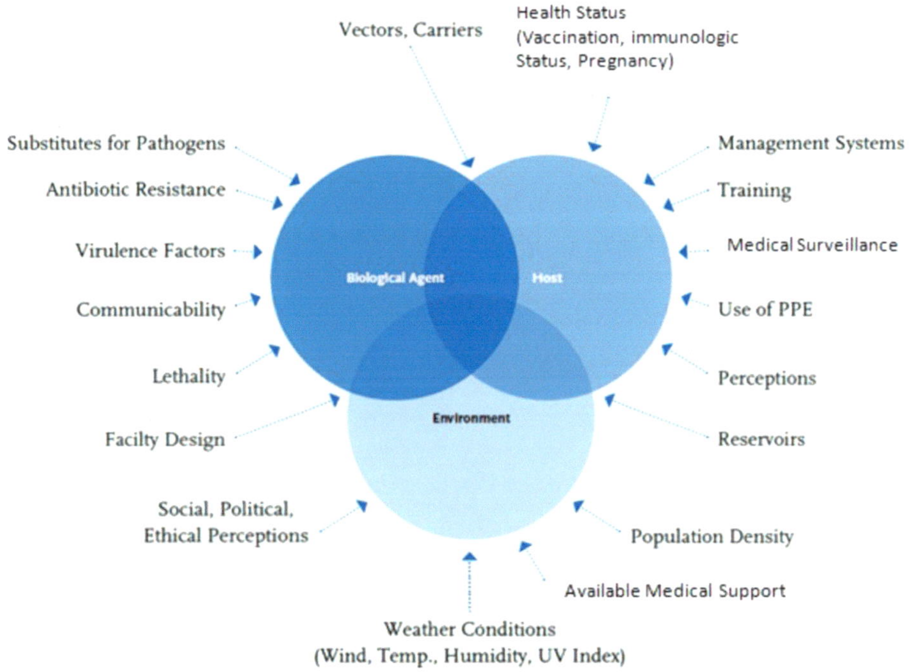

Figure 5: Classic Chain of infection (Source: www.employment.alberta.ca).

The personal protective equipment includes masks, gloves, protective clothing, eye shields, face shields and shoe covers.

In 2014, the known EVD (Ebola Virus Disease) has spread out an epidemic throughout some Africa countries and also achieved some victims in Europe and US since the first outbreak EVD infection in march 2014. Despite all effort to control such epidemic with more than 20.000 reported cases, up to 27 December,

7.857 people had been reported as having died from the disease in six countries; Liberia, Guinea, Sierra Leone, Nigeria, the US and Mali (www.bbc.co.uk/news/world-africa). Among the victims, the health professionals starts to be infected that might be caused by lack of precision in following procedures in personal and collective protective equipment or even inefficiency of such personal protective equipment. In fact, the personal protective equipment are not 100% reliable against the infections, but, it´s extremely important to use the correct one and follow the procedure when such equipment are removed to avoid contaminations. The Fig. **6** shows the difference of the previous guidelines for the new one.

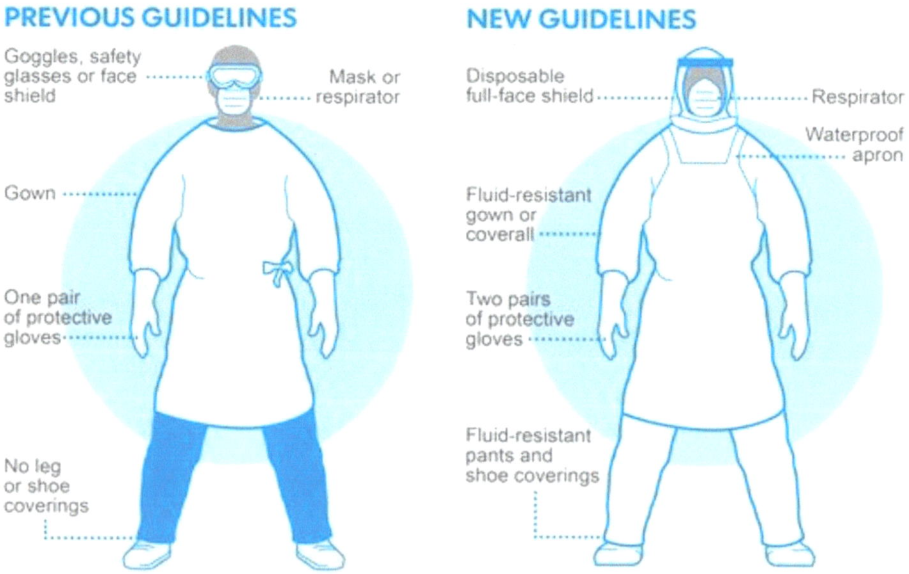

Figure 6: EVD personal protection equipment (Source: www.USAtoday.com).

1.4. CHEMICAL HAZARDS

The chemical hazard can be defined as an element or mixture of elements or synthetic substances that are considered harmful to employee's or other person's exposure.

The chemical hazards can pose in toxic, flammable, explosive and reactive form depends on the way, such substance is stock as well as the physical conditions in case of liberation and spill. In order to assess such hazard effects in employee's health in case of accident is necessary to take into account not only the properties but also the exposure level. Such analysis was carried out once in first chapter

(Risk Analysis) when we discussed about consequence analysis and defined the individual and societal risk.

Mostly chemical hazard is classified and identified in workplace to advise employee's about such hazards. Thereby, the main object of classification is to identify all the chemical substances in order to educate employee's about the risk when handling or use such products. In addition, employee's must be aware about the emergency response action concerning such products.

The chemical hazard classification is based on characteristics such as flammability and toxicity. Indeed, such products are classified is different "hazard categories" based on the Severity of the hazard. The chemical hazard classification is represented by symbols in the workplace as shows in Fig. **7** in order to advise employee's about hazards.

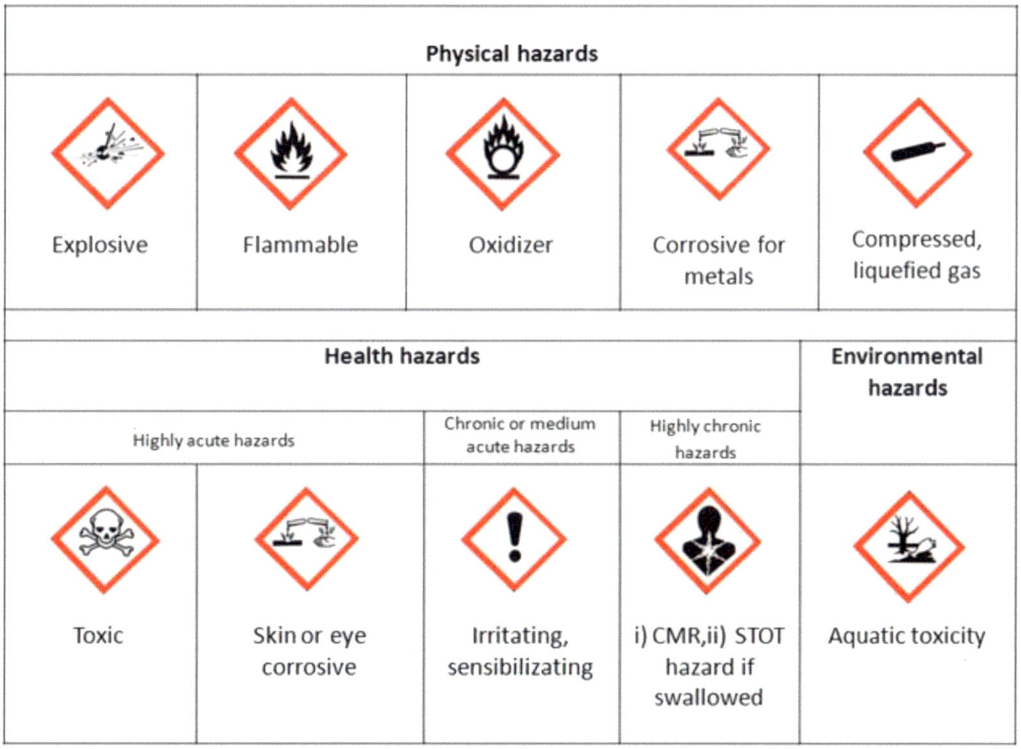

Figure 7: Classic Chain of infection (Source: www.employment.alberta.ca).

The symbol flammable is related to fire or burning. In fact, the fuel can be storages or disposal in process in solid, liquid or vapor form. Actually, only in vapor form occurs the combustion. A combustion will only occur the fuel, oxidizer (generally oxygen in air) and an ignition source are combined in a proper quantity. Those three elements are also known as the fire triangle as represented by Fig. **8**.

Figure 8: Fire Triangle.

Another combustion process is the explosion that can be a detonation or deflagration. In the first case, the shock wave travels at supersonic velocity, that means speed greater than sound and pressures in a wave is much higher than in deflagrations case. In deflagration case, this velocity is significantly lower. Thus detonations are more destructive than deflagrations. A deflagration may turn into a detonation, particularly when travelling down a long pipe for example.

The distinction between fires and explosions is the rate of energy release. In fires accident, the energy release is slower. On the other hand, in explosions accident cases, the energy release happens very fast.

The important concepts to understand why fire and explosion phenomenon happen are flash point, auto-ignition point and explosion limit. In the first case, the "flash point (of a liquid is the lowest temperature at which it gives off enough vapor to form an ignitable mixture with air. Liquids with low flash point are more flammable than liquids with a higher flash point.

In the second case, "auto-Ignition (is the lower temperature that one combustive is able to maintain the combustion without the presence of an ignition source. The third one is the "explosion limit), which is in fact, combustion limit, which means the interval of concentration necessary to have combustion. The mixture will not burn when the composition is lower than the lower explosive limit (LEL) or upper explosive limit (UEL). The Table **3** below shows some example of such three important parameters for some specific chemical products.

Table 3: Chemical combustion main characteristic

Chemical	**Flash Point (°C)**	**Flammable Limits in Air(%)**		**Auto-ignition Temperature (°C)**
		Lower (LEL/LFL)	Upper **(UEL/UFL)**	
Ammonia	NA	16	25	651
Benzene	-11.1	1.2	7.1	498
Butane	-60	1.8	8.4	287
Isobutane	-82.7	1.8	8.4	462
Methane	NA	5.0	15.4	537
Propane	-104.4	2.2	9.5	450

Due to have high severity consequences for employee's health, plants damages as well as society damage, fire and explosion must to be watch out constantly and some preventive measures must be taken places like:

- A non-flammable liquid with a higher flash point must take place a flammable liquid whenever possible.
- The flammable liquid must be storage is a safe area which doesn't affect employee's, facilities and community I case of an accident.
- The facilities with flammable liquid must have an efficient ventilation system.
- The quantity of storing flammable liquid must be kept as low as possible in order to reduce the consequence in case of an accident.
- Instrument and alarms must be installed to advise operators about unsafe conditions
- Layer of protection must be installed to prevent the incident.

- The preventive maintenance program which include inspection and test must take into account the layer of protection and safety devices.

- A proper emergency response program must to take into account all asset with storage or process chemical products.

The toxicity effect is also the other danger which chemical products by inhalation, skin absorption and ingestion may affect employee health. Such effect depends on the level of exposure and to avoid it is necessary to take place control measures as listed above and also,

The effects of exposure to chemicals may be classified as:

- Acute effect, is an immediate illness, irritation and even death caused by a short time exposure.

- Chronic effect, is a late disease caused by prolonged or repeated exposure to low concentrations of toxic substances.

- Reversible, is a temporary health effect caused by exposure, but ceases in a short period of time.

- Irreversible, is a permanent health effect caused by exposure which will not ceases even if the exposure be eliminated.

The chemical product are classified by effect on human health as defined in Table **4**.

Table 4: Class of Toxic Chemical (Source: Guideline on prevention and control of chemical hazard)

Class of Toxic Chemicals	Description	Examples
Asphyxiant (A chemical that interferes with the ability of living tissue to absorb oxygen.)	Simple asphyxiant in the presence of a gas reduces the oxygen to very low levels.	Nitrogen, acetylene, carbon dioxide, methane.
	Chemical asphyxiant interferes with the body's ability to transport and utilize oxygen.	Carbon monoxide, hydrogen cyanide, hydrogen sulfide.

Table 4: contd…

“Corrosive	A chemical that destroys or damages living tissue on contact.	Strong acids and alkalis such as phenol, sulphuric acid, sodium hydroxide.
Irritant	A chemical that produces local irritation or inflammation of the skin, eyes, nose or tissues of the respiratory system.	Nitrogen oxides, sulphur dioxide, chlorine, ammonia, formaldehyde.
Neurotoxic	A chemical that produces toxic effects on the nervous system.	Manganese, tetraethyl lead, hexane, mercury, carbon disulphide, methyl alcohol.
Carcinogen	A chemical that causes cancer.	Acrylonitrile, asbestos,arsenic, benzopyrene, vinyl chloride,benzidine, naphthylamine”.
“Hepatoxic	A chemical that causes damage to the liver.	Carbon tetrachloride, chloroform, trichloroethylene, perchloroethylene, vinyl chloride, nitrosamines.
Mutagen	A chemical that causes permanent damage to DNA in a cell. DNA is deoxyribonucleic acid, a molecule that carries genetic information to control the proper growth and function of cells.	Chloroprene.
Narcotic	A chemical that depresses the central Nervous system which may lead to coma and death.	Acetone, xylene, chloroform, isopropyl alcohol, ethyl ether.
Nephrotoxic	A chemical that causes damage to the kidneys.	Mercury, cadimum, lead, halogenated hydrocarbons.
Sensitizer	A chemical that causes or induces an allergic reaction. The effects will depend on Individual susceptibility to the chemical itself.	Toluene di-isocyanate, maleic anhydride, nickel or chromium compounds
Teratogen	A chemical that, if present in the blood stream of a woman and transported to the developing fetus will result in structural or congenital abnormalities in the child.	Lead, methyl mercury, formamides”.

In addition to chemical hazard classification and identification, is also necessary to have files with all characteristic, storage conditions as well as a preventive action to keep such product safe and emergency procedures to be taken place in the case of an accident. In fact, different companies take place different procedures to precede such chemical product information in files, that means electronic or paper. A good example of information about chemical product is taking place in many companies in the US which use the software CAMEO developed by EPA (Environment Protection Agency) as shown in Fig. **9**.

The Fig. **9** shows an example of important information for chemical product which encompasses the chemical inventory as well as technical data, facilities data, location, contact phones are available to local authority and in company site.

An effective risk analysis together with information about chemical product and integration between local authorities and companies is the best approach to secure the preventive actions as well as an effective response in case of emergencies. The integration between company and local authority as well as data integration is very necessary. The Fig. **10** shows an example of chemical integrated information. In this case, a unique data bank like CAMEO integrates with other software like ALOHA which provide results from risk consequence analysis and Marplot which provide the result of vulnerable areas in electronic map.

1.5. ERGONOMIC FACTORS

Ergonomics is a science that tries to concern all factors which take influence on human performance and health in order to fit workplace and task features to employee's characteristics as well as fit employee's in the workplace and task features regarding employee's physic, psychological, cognitive and social conditions.

Whenever is regarded ergonomic factors influence in an employee's performance it must to be taken into account physical, cognitive, psychological and sociologic factors.

The physical factors are related to employee's physical characteristics like height, weight, physical limitations, vision, audition as well as fitness, skill associated with tasks that enable employee's to carry on such tasks without any health damage.

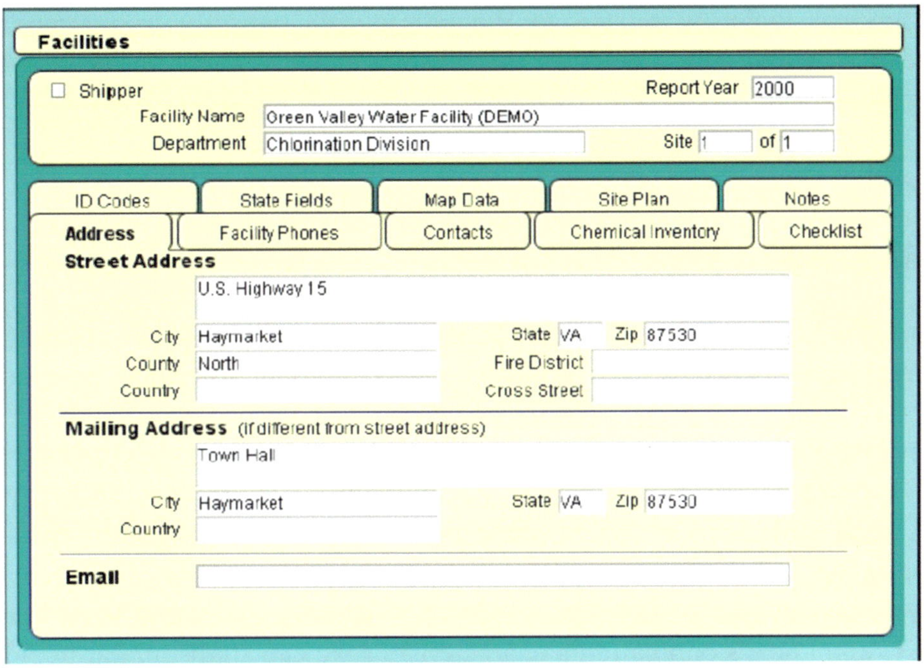

Figure 9: CAMEO template (Source: Manual CAMEO software (EPA, 2010)).

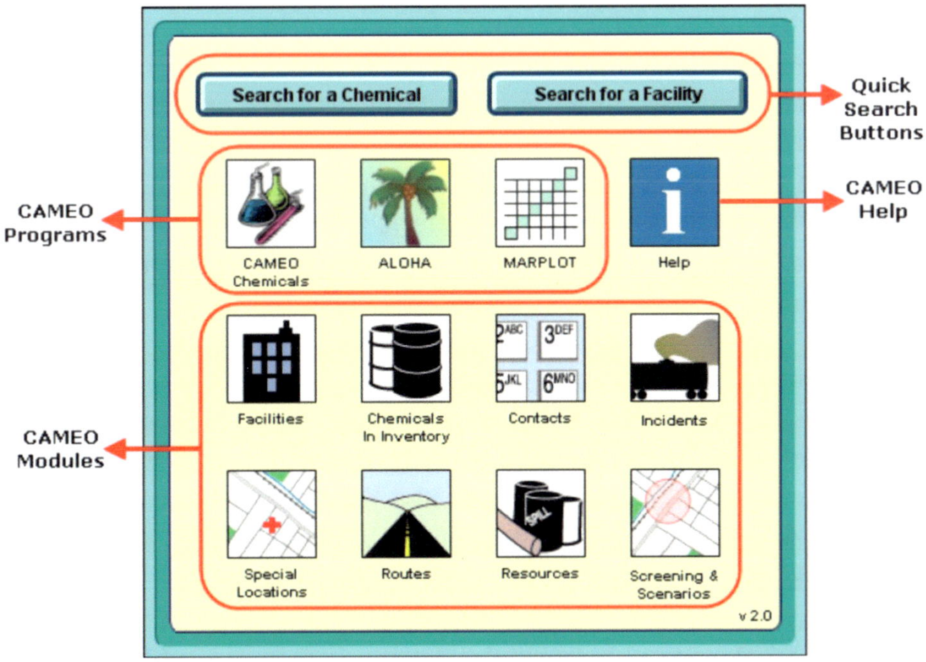

Figure 10: CAMEO template (Source: Manual CAMEO software (EPA, 2010)).

The cognitive factor is related to the employee's capacity to interpretation, analyze, assess and take a decision that enables them to carry out tasks without any health damage.

The psychological factor is related to the employee's capacity to deal with emotions related to workplace and tasks that enable them to carry out tasks without any health damage.

The sociological factor is related to the employee's social context in and out of the workplace that enables them to carry out tasks without any health damage.

Whenever it is possible, ergonomic analysis must begin with concept and project phase in order to avoid employee's health damage in the operational phase. Indeed, such application in design phase regards more physical factors than other ergonomic factor, because during the concepts and project phase, we do not know the psychological and sociological employee's features. Whenever there is a lack of physic employee's population's characteristics we can study similar population to project ergonomic workstations.

The ideal situation is that such ergonomic factors are balanced well enough to avoid bad influence on the employee's performance and health damage. Unfortunately, that's not happening in many workstations projects cases. In fact, to achieve the perfect level of workload that enables employee's to carry out their activities in high performance without healthy damage is very hard for a long period of time. Nevertheless, the real problem is when such ergonomic factor is not balanced for a long period of time and consequently those employee's start to have health problems.

Once the workplace conditions and the employee's capacity are understood, it is theoretically feasible to define the best workload level for each employee. Actually, such definition is not so easy because firstly, the employee's capacity vary a long time depends on many factors like for example personal problems, health problems, age, social context, happiness in personal life, relation with a group in a work place, relation with boss in the workplace. Secondly, the workload varies a long time as well and the quantity or quality of work that may represent different influence in employee's ergonomic factors workload.

Depends on situation, different activities have different level of ergonomic factors. For example, employee's who work in construction have tasks which are more intensive in physical and psych factors than cognitive. In this case, depends on the work environment, such employee's performance may be more affected by higher levels of stress than employee's bad fitness condition.

By the other hand, other carriers are more intensive in other ergonomic factors like for example, engineers who work in an office more than 8 hr per day in a project but they know very well all technical issues. When compared with other new project that is much more complex, it will be required much more from engineer's cognitive capacity and they delay more to carry out analysis and take decision. Such situation may also increase the psychological workload because they know that in this new project will be harder to finish on required time.

Such examples give a short idea about how complex is to achieve a balance between employee's capacity and workload and consequently to achieve high performance without any health damage. Furthermore, the ergonomic analysis must be constantly reviewed specially when is necessary to modify activities or new issues take place.

Regarding health damages, the main problem is to be able to detect all time situation or activities that require from employee's over capacity and consequently affect their health. In order to solve such problem is necessary to review constantly the ergonomic analysis, but in fact, many ergonomic analysis is reviewed only when it is detected some employee's health damage. The problem with such reactive approach is that some health damages, delay years to come out and affect employee's performance over time.

In addition, to define employee's performance it is necessary to understand employee's work capacity in order to plan tasks workload correctly.

Basically, performance can be understood like productivity, which means the relation between how much was done per how many were planned to be done. Such concept was very well applied to production line from the beginning of a modern production system. After the quality concept come out in 80th, whenever some tasks are carried out the product or service result of such task is associated with some level of expected quality that in many cases it is not totally implicit. In the 90th ages such quality concepts was incremented by environment impact concept, which means task must be carried out without cause, any bad environmental impact, in other words, bad modification in environment characteristic. At the end of 90th ages, the safety concept finally gets more importance and from this stage and on was required to carry out activates and task safely in addition to quality and environment requirements. Recently is being discussed the social effect of such companies' activities and requires a good social impact internally and externally.

Such requirements are a big challenge for organization nowadays and became an additional competitive factor to be taken into account.

Furthermore, we cannot forget that the social context and the company's culture define what it is a good performance as well.

In many companies all over the world, it is understood that a good employee performance are those employee's who works over than eight or ten hours per day. In many cases, such employee's are not really efficient because they take more than eight or ten hours to carry on tasks that would be done in less than eight if they had more focus in their tasks or if there were a better planning for their activities.

Another good example is when a good employee's performance is understood as those are able to be involved in many tasks, problems, workshop groups and projects even when such issues are not related to their department. In such cases, most of the times such employee's are not able to give a good answer for any of those problems that they are involved or give not a complete answer or not fulfill tasks efficiently.

Another example is those companies that worry to control the time that employee's spend in the office. In such cases, companies define, for example, at minimum 8 hr per day with 1 hr for lunch. The point is that many companies focus in such control and in many cases do not pay too much attention that some employee's are very efficient and are able to carry on their task in 6 hr and whenever they take over than one hour in lunch time it will not affect their performance. By the other way round, other employee's take exact time defined by the company, but are not efficient to carry on their tasks.

The more accurate concept with efficiency is achieved the task target at the time required, with the quality required using the resources defined. Efficiency depends on the quality of resources available to fulfill a task as well as the ability to carry out such task. Whenever such tasks are carried out with higher quality than expected or lower time that expected or with lower resource or a combination of such factors the efficiency is higher than expected. In fact it is necessary to have a balance about such factor to achieve the efficacy target and it must to be paid attention that hardly we complete tasks in a shorter period of time and high quality with the same resources.

The main point in this discussion is that to achieve such high performance no matter how companies define efficiency, it is necessary to effort of employee workload and in many cases, work over than 8 hr per day, or have lower employee's in a workstations cause employee health damage that will show up in a short or long period of time.

Such assumption seems to be obvious, but the actual culture in society as a general is to produce more and more in the workplace and that result in a "Time disease" that is a sensation that everybody in the workplace have never time enough to carry out their task and they are always overloaded with tasks. That is a big challenge to modern society to change such concepts of efficiency and productivity in the workplace.

In addition, basic issues like planning activities are forgotten for many organizations, but in reality the real reason to not dedicate to much effort to plan activities is that once the idea is to produce more and more how it is possible to assume that is not possible to produce more, or not possible for employee do not accept one specific task from his boss.

Such issues discussed are a big challenge for ergonomic science and all of them affect employee's health as well as their performance. Therefore, each ergonomic factor, workplace issues and employee's physical and physic, must be clearly understood. The next item we will discuss each one of them such ergonomic factor on in order to understand the importance and impact of each one in employee's health and performance at work.

1.5.1. Physical Factors

The Physical factor was the first one to be discussed in ergonomic and it seems to be one of the easiest to be balanced because depends only to measure employee's physical physiology, anthropometrics features as well as regards biodynamic limits. Furthermore, is necessary for analysis workstation features fit it under the employee's physical characteristics.

The preventive ergonomics approach emphasizes on early identification of physical factor related musculoskeletal disorders to mitigate risk factors. Therefore, during the concept and design phase all characteristics of work place such as tools, activities, workstation layouts and materials must be taken into account in ergonomic studies.

A proactive approach to Ergonomics will ensure that:

- The ergonomics factors will be taken into account to workstation and facility layout.
- The facilities and workstation projects are flexible to be adapted to the employee population physical and psych characteristics.
- The work demands will fit on worker´s physical and psych capacity.

Whenever we are planning and design a workplace station or even activities we must take into account anthropometry and biomechanical factors.

Anthropometry can be defined as the measure of physical human traits which is considered in workstation projects.

Biomechanics can be defined as the study of the human body considering it capacity limits.

Both aspects are very important when design a product that will be used for employee's like tools, vehicles, and workstations. The main problem when such design is being carried out is to not consider the most of population anthropometry that is some case means more than 90 percent. In many cases, such design is based on average and do not apply to many people outside of such average limits. Consequently, many employee's may have health problems a long term time. In fact an anthropometry research is not easy to be carried out, but that is a key factor in tolls and workstation design.

The human biodynamic can be described in operational terms like:

- Positioning movements are those in which the body moves from one specific position to another.

- Continuous movements are those that require muscular control during the movement.
- Manipulative are fingers or hand movements which involve the handling of objects.
- Repetitive movements are those in which the body (or part of the body) repeats the same movement.
- Sequential movements are sequences of independent body (or part of the body) movements.

The Fig. **11** shows the hands biodynamic possible movements.

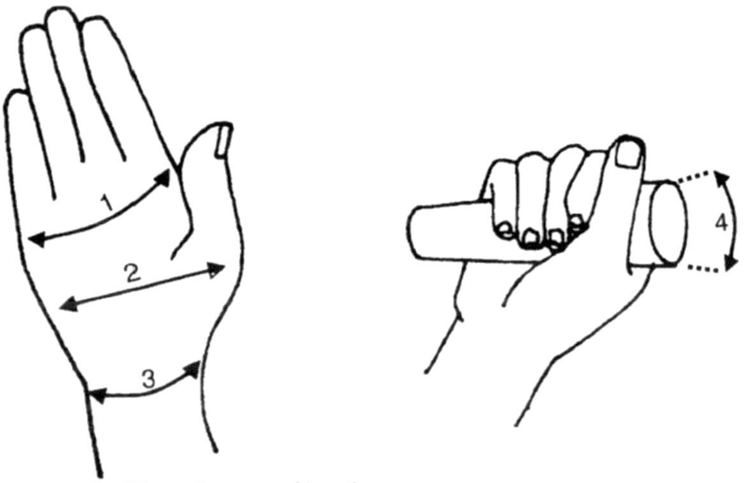

Figure 11: Important hands dimensions (Source: Cherie, 2009).

In order to avoid health damages it is necessary some cares about posture whenever some equipment or product are being lifted or carrying on like shows Fig. **12**.

Whenever employee's are in stand up position dealing with some material is advisable that all body joints must be in a neutral position. In such position the muscle and ligament, which span the joint, are stretched to the least possible extent (Fig. **12** A).

Once the material is on the table and will be carried to some place it is necessary to keep the material close to the body otherwise the arm will be outstretched and the trunk bend over forwards (Fig. **12** B).

In situations where is necessary to take some material in some box is advisable to avoid bending forward. The upper part of adult body weight around 45 kg, consequently whenever the trunk is bed forward harder is the back muscle and ligament to maintain the upper body in balance (Fig. **12** C).

In some tasks, it is necessary to move some material from one specific place to another spinning around the trunk. The twisted posture of the trunk causes undesirable stress to the spine (Fig. **12** D).

In all situations it is necessary to avoid excessive weight when reaches a material. It is necessary to limit the extend of forward and sideways reaches to avoid having to bend over or twist the trunk (Fig. **12** E).

In tasks that it is necessary to carry some material, it is advisable to avoid carrying such material above the shoulders. The hands and elbows must be below the shoulder (Fig. **12** F).

In situations when it is necessary to lift some material it is advisable to observe the material weight (Fig. **12** G). It is also important to be careful when carrying some material and it advisable to avoid doing that with only one hand because in this case the body is subjected to mechanical stress (Fig. **12** H).

In general terms, whenever is hard to carry some material a possible solution is to use transport accessories. That avoids health damages and makes those activities more efficient in many cases (Fig. **12** I). It must be careful because transport accessories come out another occupational physical risk as discussed previously.

Finally, in case of static position it is necessary that such position don't maintain for a long period of time no matter what if the workstation is well design (Fig. **12** J).

Indeed, there is a mathematic approach to calculate the maximum lift weight that one employee can lift regarding the ideal working conditions that is eight hour workday, temperature between 19 $^{\circ}$C and 26 $^{\circ}$C, humidity between 35% and 50%. The NIOSH proposes an equation that regards different factors like the Horizontal Multiplier factor, the Vertical Multiplier factor, the Distance Multiplier factor, the Frequency Multiplier factor, Distance Multiplier factor, Frequency Multiplier factor, Asymmetric Multiplier factor, Coupling Multiplier factor and Recommended Weight Limit.

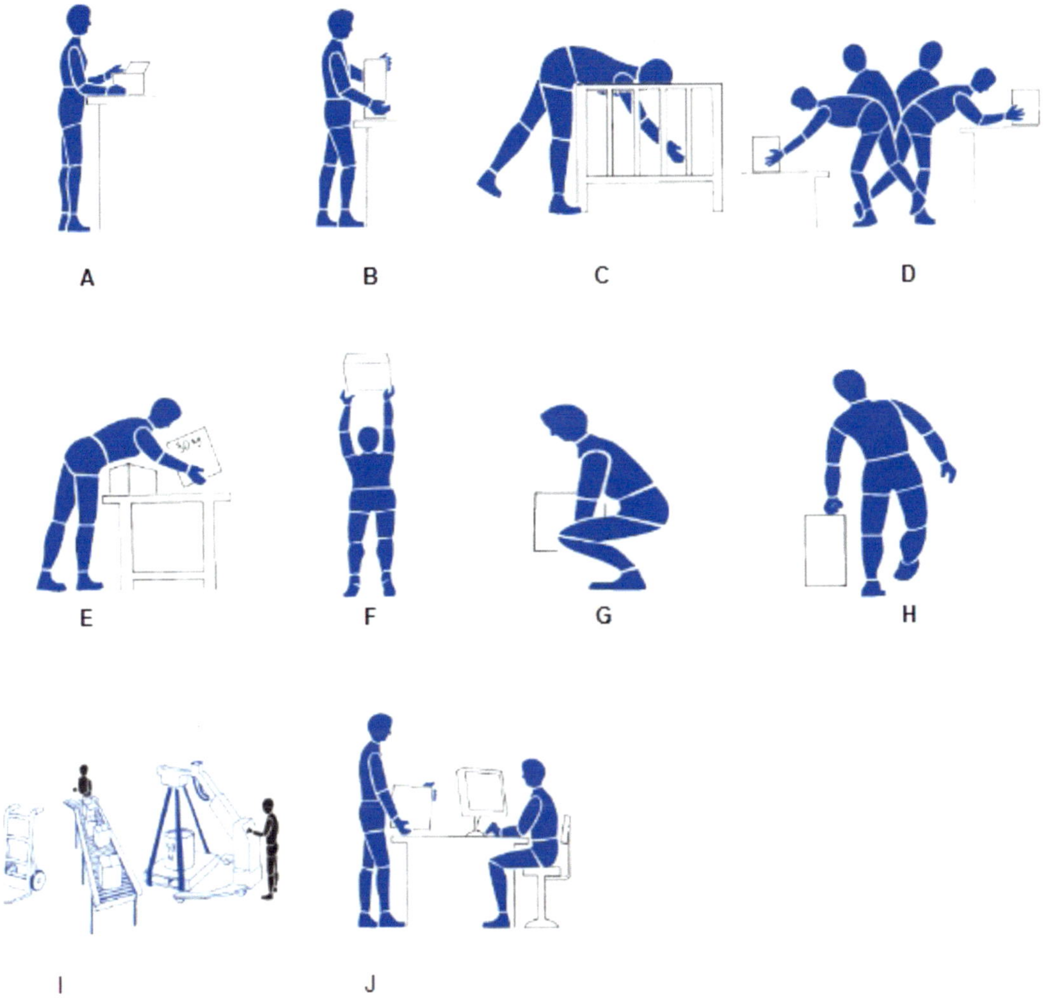

Figure 12: Ergonomic care in posture (Source: Cherie, 2009).

The lift NIOSH equation is:

$$RWL = LC \times HM \times VM \times DM \times AM \times FM \times CM$$

where:

RWL = The Recommended Weight Limit.

LC = OBJECT Lift Weight.

HM = Horizontal Multiplier factor.

VM = Vertical Multiplier factor.

DM = Distance Multiplier factor.

AM = The Asymmetric Multiplier factor.

FM = Frequency Multiplier factor.

CM = The Coupling Multiplier factor.

For most of such factors, there is a specific equation like:

HM=25/H. The "H" parameter is the horizontal distance (in cm) from the midpoint between the ankles with the hands while holding the object. Whether the distance measure is "in (the equation will be "HM=10/H".

VM=1-0,003 x [V-75]. The "V" parameter is the vertical distance (in cm) of the hands from the ground at the start of the lift. Whether distance measure is "in" the equation will be "VM=1-0,0075 x [V-30]".

DM=0,82 +4,5/D. The "D" parameter is the vertical distance (in cm) that the load travels. Whether the distance measure is "in" the equation will be "DM=0,82 +1,8/D ".

AM=1-0,0032 A. The "A" parameter is the twisting angle of the body while lifting, measured in degrees).

The Fig. **13** shows the horizontal and vertical distance measurement.

Other factors like FM and CM are qualitative and based on tables values. In case of FM factor, the frequency (F) of lifts and the duration of lifting (in minutes or seconds) over a work shift can be defined based on Table **5**.

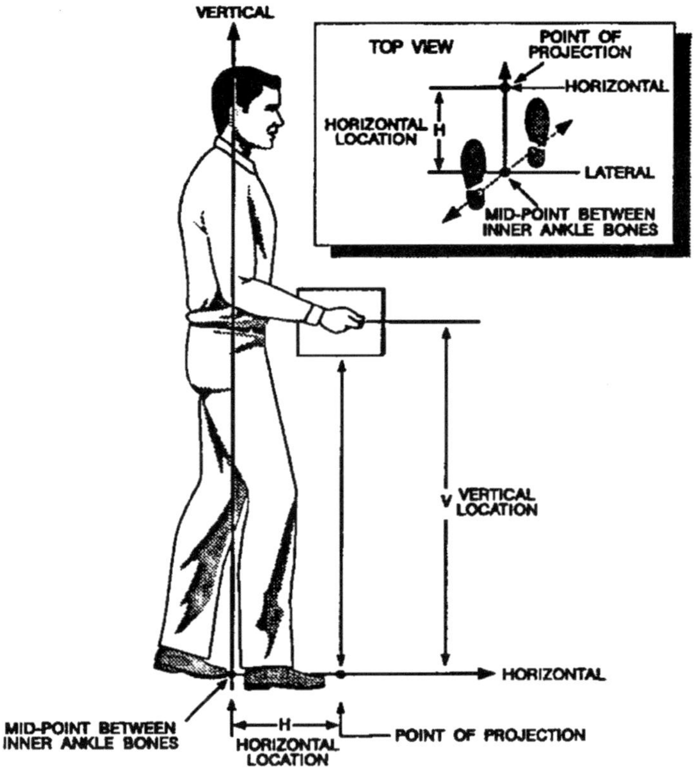

Figure 13: Graphic Representation of hand location (Source: Water Thomas, 1994).

Table 5: FM factor (Source: Water Thomas, 1994)

F = Time Between Lifts	FM Factor			
	Lifting While Standing:		OR Lifting While Stooping:	
	One Hour or Less	Over One Hour	One Hour or Less	Over One Hour
5 min	1.0	0.85	1.0	0.85
1 min	0.94	0.75	0.94	0.75
30 sec	0.91	0.65	0.91	0.65
15 sec	0.84	0.45	0.84	0.45
10 sec	0.75	0.27	0.75	0.27
6 sec	0.45	0.13	0.45	-
5 sec	0.37	-	0.37	-

In case of the coupling factor (CM), that reflect the feasibility to catch up the material the categories are shown in Table **6**. The CM is classified as good, fair or poor.

Table 6: FM factor (Source: Water Thomas, 1994)

C=**Crasp**	**CM factor**	
	Standing V<75cm (or 30 in)	**Stooping** V≥75cm (or 30 in)
Good	1.0	1.0
Fair	1.0	0.95
Poor	0.9	0.9

The category good means handles or hand cuts out of the optimum design, comfortable grip in which the hand can be easily wrapped around the object.

The category fair means handles or hand cut out of less than optimum design, comfortable grip in which the hand can be easily wrapped around the object.

The category poor means handles or hand cut out of less than optimum design or loose part of an object that is hard to handle the object.

Indeed, we need to aware about such approach limitations. Thereby, in activities that more than 10% of total activity is not lift it is necessary to assess the other activities in terms of energy demanded in order to analyze the effect on health. In general terms the limitations of such approach are:

- Lifting by only one hand.
- Lifting activity by over than eight hours.
- Lifting while carrying, pushing or pulling.
- Lifting with high speed motion.
- Lifting in unstable floor.

An example of the NIOSH lifting equation can be applied to an employee who lifts a 23 kg box poor handle, with 25 cm horizontal distance, 30 cm vertical distances, 40 cm vertical load travel and zero twist angle. In average the employed make one lift for each 5 minutes in total duration less than one hour.

RWL = LC x HM x VM x DM x AM x FM x CM

RWL = 23 x 1 x 0.87 x 0.93 x 1 x 1 x 0.9=16.75

In addition, to biodynamic and workload that characterize the dynamic workload there is as well a static workload that is also an ergonomic study objective. The most common situation that static workload is assessed is in office workstation.

All over the words, for many years the office ergonomic issues and in particular static workload wit health damage is caused by bad posture has been a big challenge due several factors like:

- Safety Culture.
- Ergonomic awareness.
- Cost *versus* benefit investment.

The first aspect affects many issues in Occupational management as well as risk management in many countries. Culture can be understood as the employee's values reflected by their attitude and behavior. The safety culture is nothingness than preventive culture. That means all effort will be done to avoid unsafe conditions. Such preventive action depends on investment as well as leadership to support them as priority actions to be taking place. In many cases, the leadership does not support enough the preventive action, and in other cases the safety culture does not achieve the whole company's level.

The second point is lack of Ergonomic awareness. Actually, many companies with high level of safety culture have lack of ergonomic awareness because they focus on avoiding accident and do not pay attention to occupational issues that affect employee's health a long time like ergonomic.

The third point is the cost *versus* an economic benefit that mostly evaluates safety issues regarding only the direct cost of the accident. Many leaders do not realize the value in ergonomic investment because in many cases, they are not able to see the consequences of bad ergonomic design in workstations and workplaces.

Despite such barriers, like other activities or workplace, office workstations must take into account the type of activity as well as the employee's anthropometry. The Fig. **14** shows the correct body position in office workstation.

Whenever we are discussing about workstation like offices theoretically is easier to achieve the ergonomic balance because such workplace has not modifications a long time as well as employee's physical features. Even though, it's advisable to project a workstation flexible enough for a range of physical height. The main reason to have a flexible workstation is that from time to time employee's change their workstation and it is necessary that such work station be flexible enough to be adapted to new employee's.

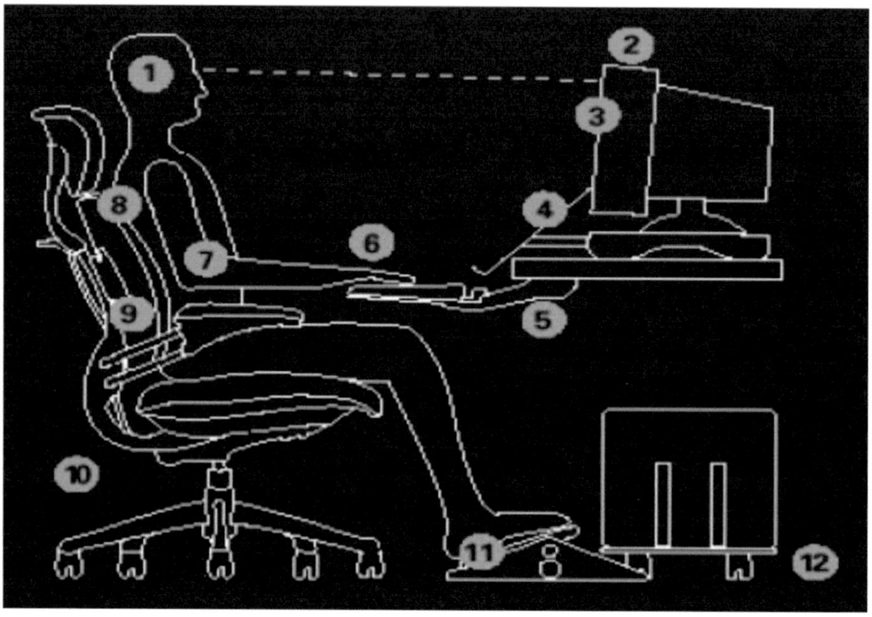

Figure 14: Correct position in offices workstation.(Source: www.safety.rochesrter.edu).
Legend:
1 – Top of monitor at or below eye level
2 – Monitor and Keyboard centered in front of you
3 – No Glare on screen
4 – Documents in Line with keyboard and Monitor
5 – Negative tilt keyboard support
6 – Wrist Flat and Straight
7 – Arms and elbows close to the body
8 – Change posture often
9 – Work in a reclined position
10 – Take frequent short breaks
11 – Feet flat on Floor or footrest
12 – CPU off desk

1.5.2. Cognitive Factors

The cognitive ergonomics objective is to understand how the human mind percepts, process and use information to take decision on work activities The other elements of cognitive ergonomics are the human mind sensation and memory.

The cognitive ergonomics challenge is not only understanding the decision process, but also understand the human mind into some context, such as human-computer interaction; human-machine interaction; mental workload; decision making and information usage.

The significant importance to analyze the cognitive process is to understand the decision based on cognitive process that in some cases lead to human error that in an industry may cause accidents, environmental impacts and loss of production.

In order to conduct a cognitive task analysis, it is necessary to have a set of factors that influence on the cognitive decision process and describe them (Wayne W. Zachary, Joan M. Ryder, and James H. Hicinbothom). Those factors are job task environment, person machine interface and knowledge. There are four types of knowledge that are perceptual knowledge, declarative knowledge, procedural knowledge, knowledge and action.

- Perceptual Knowledge is based on perceptual experience in the following sense refers to how to code specific problem based on information context from environment that come out from others team member or from machine interface.

- Declarative knowledge refers to the person's internal representation of some problem or situation.

- Procedural knowledge refers to prior actions and task as well as defined specific goals.

- Action knowledge refers to how perform a specific task based on previous background to solve specific situation.

The Fig. **15** shows the cognitive net framework decision based on environment perception and knowledge applied to turn into such information in the final decision. The environment perception is information related to teammates or machines. The knowledge is the four types of knowledge explained above applies to use input information, previous background and procedural information to decide steps, target and take decisions in order to perform a task.

The cognitive factor can be assessed in different activities from the simplest to more complex one. Regarding the human capacity do collect information, process such information and take decisions we can say that mostly the activities which involves human-machine interaction are more complex to human and required more from their capacity to assess information and take decision. The usual example of human-machine interaction is the task with a computer. Many types of human machine interactions are applied daily in our lives like information system like temperature sensor, velocity indicator, pressure indicator and others operational conditions applied in car, buses, boats, planes as well as in industrial process that are designed to safe and reliable operation and human interface condition. Despite that, it is always a challenge to design such systems in order to avoid cognitive workload that result in human error.

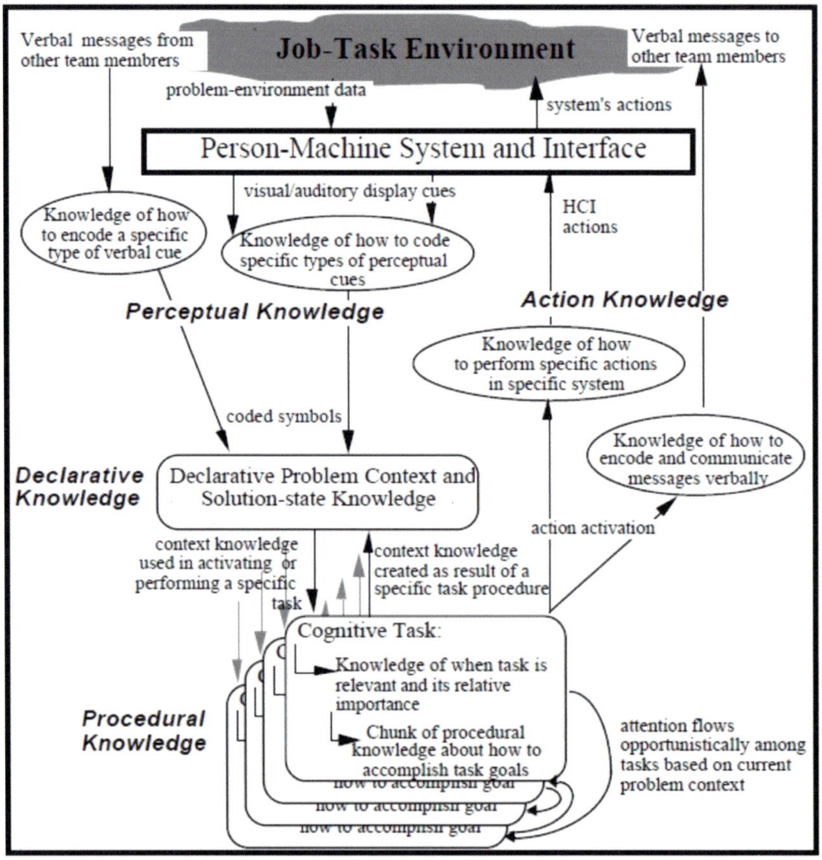

Figure 15: Cognet knowledge framework (Source: W Zachary, JM Ryder, JH Hicinbothom, 1998).

Actually in some cases, the excess of system interaction cause cognitive workload and consequently human error. That is usual in modern system where a lot of controls take place to give information to support human decisions. The main point here is that human are not able to get all information, process them and take decision. An example of such situation is a Plant shutdown where a lot of components are highlighted in control panel that make operator decision harder. In such situation, in addition to the cognitive workload there is psych factor that contributes to human error. Indeed, complex system design more and more faces the paradigm to give more control to human or take decision for them.

In the first case, in many situations it is not necessary to more controls to take decisions like for example a tank operation control that essentially is necessary to know the level, temperature, and flow. In many cases, such controls are implemented in the control room and in local equipment and further redundant controls are implemented as well not for design decision but in some cases for client requirement.

In fact, based on the experience of many years in operational plants, many experts and senior operators advise that it is necessary to analysis case by case. Therefore, whenever plants are in the design phase, the hazard operability analyses discuss such issues in order to avoid unsafe process condition. After many years, many devices like safety instrumented functions, relief valves and sensors were developed and applied as recommended. Nowadays, the actual plants face a curious fact that many operators learn all such automatic systems, but have no experience in local control. That means, in case of such automatic device are not available, such new operators are not able to control plant as well as the older operators that were involved in all technology process development along the last 30 years.

Nevertheless, the normal tendency is to implement more and more control in the process, but in many cases nowadays during hazard and operability analysis many experience operators recommend to take out such excess redundant controls due to cognitive workload or even that is not necessary and force operators to check the equipment condition in local.

By the other way round, some systems are designed nowadays to have more and more self-controls in order to avoid human error caused for bad decisions. Depends on industry and application, different level of system has auto-control or human control is designed. In railway industry, mostly the train is under driver control even when sophisticated systems like Total Control Management System (TCMS) are designed to give them information about the system. The last catastrophic Train accident that happens in Spain (Fig. **16**) shows that in some cases systems must assume control whenever unsafe condition are detected.

Figure 16: Train Derailment in Santiago de Compostela (Spain 2013) (Source: www.bbc.uk.com) (http://www.bbc.co.uk/news/world-europe-23449336) - 25 July 2013 Last updated at 14:35.

In this particular case, all attention of accident analysis was given to driver human error that allows the train run in high velocity, but the real point to discuss here is why there is not an automatic control to reduce train velocity once is detected a dangerous curve in the track. No matter how avoidable was the human error, guilty the driver will not solve the problem to avoid that thousands of high velocity train around the word to derail due high velocity in specific track.

Regarding system decision control, some system is designed to have own decision in order to avoid human error. Such philosophy is adopted by Nuclear Power Plants in order to avoid human error in case of high cognitive workload in emergency situations. In this case, The Fault Tree Analysis is carried out to define all events which lead to accident and such logic is implemented on process in order to detect automatically the unsafe conditions as shows Fig. **17**.

Based on all discussion above it is clear that systems and product design are at the heart of the profession of ergonomists (Dowell and Long, 1998). The design of a new system is the process that happens from the conceptualization phase only when it is used by the people for whom it is intended. From the point of view of cognitive ergonomics, there are two aspects of interest in system design (Carroll, 1991).

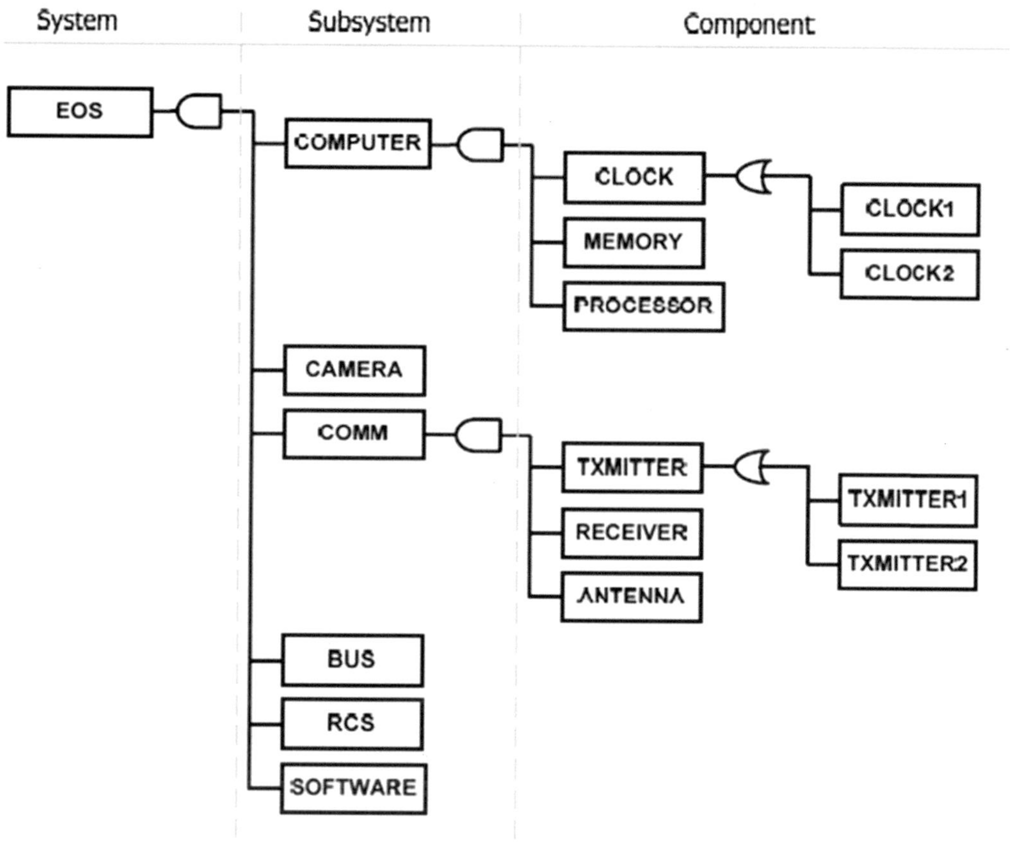

Figure 17: Nuclear Plant Fault tree (Source: Hamed, 2007).

On the one hand, they are interested in "the process" of design itself. That is, cognitive ergonomics want to understand how people devise a new system, and what are the individual and group factors involved in making decisions that lead to certain solutions defining the system. Furthermore, cognitive ergonomics would like to know whether the solutions adopted to suit the needs and characteristics of users. Their main role in this sense is to describe the human being at all levels of functional organization appropriate for the system being designed (Velichkovsky, 2005; Wickens and Hollands 2000).

Actually, such concepts are based on the fact that errors are not caused by irresponsible behavior or defective mental functioning. They may be rather the consequence of not having taken into account how a person perceives, attends, remember, makes decisions, communicates and acts in a particularly designed work system. This standpoint suggests investigating the causes of human errors by analyzing the characteristics of human information processing. Here, the first step has been the classification of errors according to the level of processing involved in the behavior that led to the error (Velichkovsky, 2013).

Although there are more elaborated classifications today, it is possible to make a synthesis based on the classical scheme proposed by Jens Rasmussen (1983). He distinguishes three types of errors depending on the level and degree of cognitive control involved in the erroneous behavior. The three types of errors can be largely attributed to the familiarity that the person has with the systems are:

- Error based on skill, which means activities are carried out based on experience.
- Error based on rules, which means activities are carried out based on procedures and established steps.
- Error based on knowledge, which means activities are carried out based on capacity to understand the situation and take decisions based on knowledge.

In order to predict human error based on such categories is also important to understand the performance shape factors, which means, factor that take influence on human error. In fact, different Human reliability analysis methods are applied to analyze and define the human error probability. Thereby, such approach is being carried out for a very long time, but it is always a challenge to comprise all factors which take influence on human error in Human reliability analysis.

Indeed, modern approaches comprise the person and her working environment as a highly interactive joint cognitive system (Hollnagel and Woods, 2007). The interaction between the two components is of a crucial importance for any ergonomic analysis. Based on these assumptions, several authors have proposed methodology for estimating the probability of human errors depending on the specific situation in which human-machine interaction occurs. The methodology presupposes two steps of analysis:

- First, to identify the types of errors those are possible for a specific task in a given scenario of event development;
- Second, to classify these types of errors of their ranges of probability to identify which are the most probable and which are the least

probable within the given joint cognitive system (Hollnagel, 1998; Cacciabue, 2004).

The chapter five demonstrates different human reliability methods with examples. The next item will discuss the social context and its influence on ergonomic.

1.5.3. Social and Organizational Factor

The social and organization factor objective is to understand and improve on social-technical work systems, including their structures, policies and organizational processes. Thus, ergonomics must be involved in the social design of communication systems, interaction routines within the working groups, times and shift schedules in a company, and other related issues. In order to have a clear understanding about such complex social-technical work system is necessary to understand the context, goals, knowledge, communication and other process which takes influence on social-technical performance.

Indeed, such analysis is in most of cases complicated because many of such aspect are not formally defined and followed by members in a group. In addition, it is import to understand how leadership influence in social-technical systems and team performance. The social-technical system can be defined by three main factors that are task network, knowledge network and social network.

The Task network is related to all tasks carried out by an individual in a social-technical system. Such tasks are dependent of hierarch, function, skill, experience of each member and have the main objective to achieve the social-technical system goals. Indeed, we can also see in some cases that team members carry out tasks that are not related to their team goals, but with another team or individual particular interest of someone in or outside the group.

The knowledge network is that enable member to carry out their tasks and in many cases such knowledge must be spread out for different group members.

The social network, enables member communicates and spread out the knowledge in order to achieve team goals. Those three factors are represented in Fig. **18**.

Whenever the task analysis takes place some elements such as communication, coordination activities, holes, and team goals must be taken into account.

Once the tasks are well represented and understood, it is necessary to understand how members will achieve the goal. Therefore, the key factor in the social -

technical system is to understand the team performance and by this way it is essential to first understand the individual and team competence, as well as knowledge is disseminated.

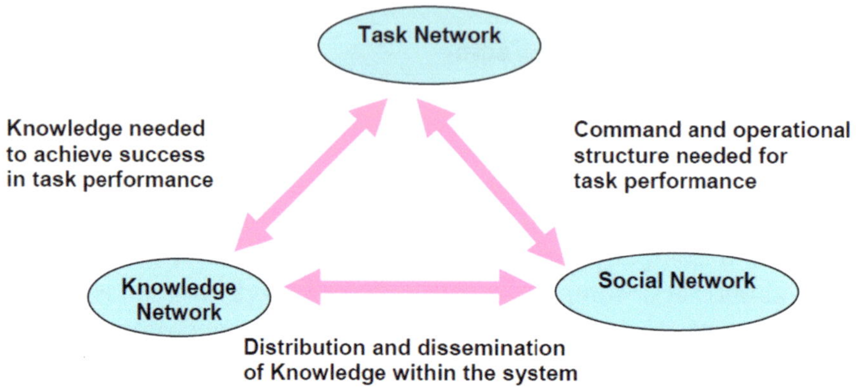

Figure 18: WESTT Triangle (Source: Houghton, 2005).

In real life that's not meant that individual will share knowledge in benefit of another individual mainly when the team goals are not clear or if each individual has different goals to achieve in a team. Indeed, many teams have individual member goals and not team goals. Many organizations believe that the team goal is the sum of individual goals, but in this case if member do not understand the influence of his own goal in other member or team goal the knowledge will be concentrated in each individual to allow them to achieve perform tasks in order to achieve their goals. In addition, in contrast of social network ideal, many organizations stimulate members to compete with one each other because only members who achieve their individual's goal will be awarded.

Actually, the team member's relationship is complex to understand because there are internal conflicts that are not declared and take influence on the team's performance. The leadership team has an important hole in reducing the member teams' conflict in order to allow a team achieves their goals. Despite such important hole, in some cases the leaders drive the member team conflict when they stimulate the team member competition to each one achieve high performance by itself.

Thereby, all issues related to sociotechnical systems must be taken into account when we try to understand the social network, knowledge network and task network.

In order to start such understanding, a diagram regarding all team member roles, knowledge and sources can be applied. The Fig. **19** represents a social network which describes a project to support client requirements. Such a project requires a safety engineer and reliability engineer analysis to support client decision. The different social network agent's holes are represented by different colors and stripped nodes represent knowledge shared among members.

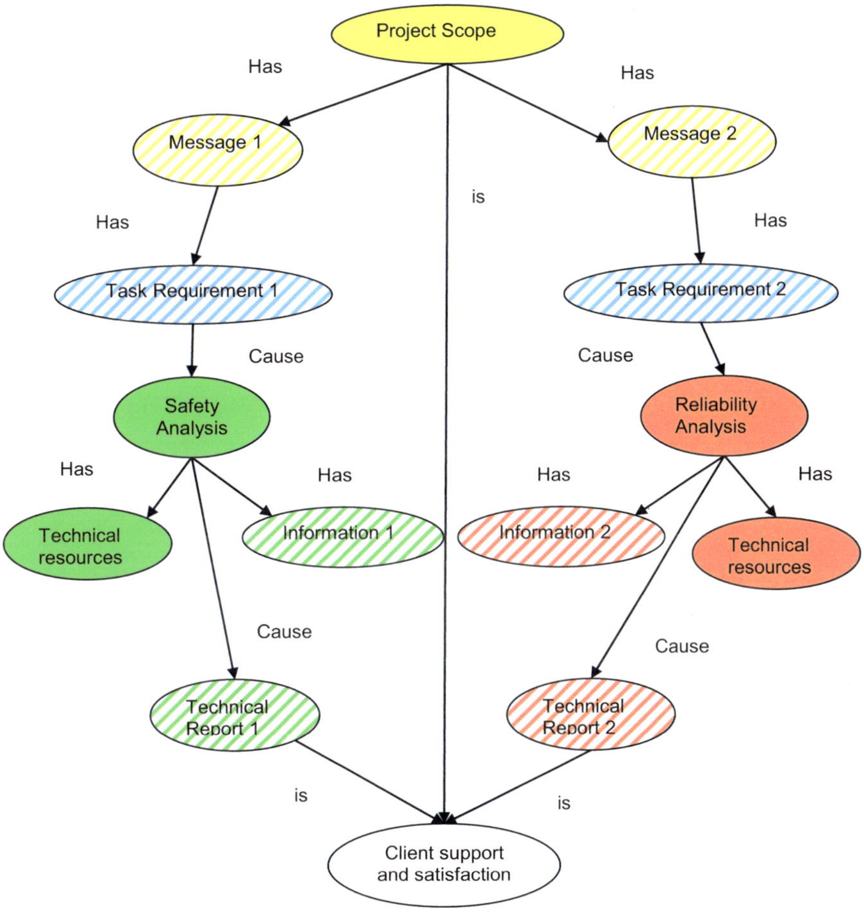

Figure 19: Social Network.

In this example the project manager receives the requirement from client to define risk of old pipeline in order to take place it. The project manager contacts the Safety and Reliability team leaders in order to support client decision to define the risk of accident. The safety team leader selects one safety engineer to carry out the quantitative risk analysis. The reliability team leader selects one specific reliability engineer to carry out the pipeline reliability analysis. The safety engineer needs resources like software to model the consequences of product spill as well as information to fulfil the quantitative risk analysis. The reliability engineer needs resource like software to carry out reliability analysis as well as information about historical failures on the pipeline. The result of each engineer analysis will be a technical report that support client decision. Some information is shared among agents like:

- Message 1 (between project manager and safety team leader).
- Message 2 (between project manager and reliability team leader).
- Task requirement 1 (between safety team leader and safety engineer).
- Task requirement 2 (between reliability team leader and reliability engineer).
- Information 1 (between safety team leader and safety engineer).
- Information 2 (between reliability team leader and reliability engineer and safety engineer).
- Technical report 1 (between safety team leader, safety engineer, project manager and client).
- Technical report 2 (between reliability team leader, reliability engineer, project manager and client).

Indeed, there are different possible social network configuration depends on sociotechnical system that is being represented. Such diagram is only the first step to help understanding the complex social and organizational network, which influence on organizational performance and must be taken into account by an ergonomist.

1.6. OCCUPATIONAL RISK ANALYSIS

As we have discussed in several items above, there are different occupational risk which takes influence on employee health and depends on workplace some of them are predominant. The main objective of this item is to carry out a simple instances of occupational risk analysis take into account all types of risk in which some employee is exposure on their workplace as an example of holistic occupational analysis. In order to analyses occupational risk it is necessary to identify hazards and classify the risk based on qualitative severity and frequency criterion. The risk is a combination of frequency and severity. On Table **7**, three severity categories are defined as an example.

Table 7: Severity Classification

Severity Classification		
Low	The severity impact results in superficial health damages. There is not a loss of more than one workday to recovery health. The health damage are reversable and take no influence on phisical and mental work capacity.	1
Medium	The severity impact results in significant health damages. There is a loss of more than one workday to recovery health. The health damage are reversable with health care to remove the temporary influence on phisical and mental work capacity	2
High	The severity impact results in high health damages. There is a loss of more than one workday or permanently loss of work capacity . The health damage are not reversable and take influence on phisical and mental work capacity forever. In Some cases may cause	3

The frequency category is defined on Table **8** regarding the exposure to occupational hazard.

Table 8: Severity Classification

Frequency Classification		
Low	Less than 40 hours exposure per month. Usually the employee is not exposure to hazard or they are exposure to very low toxic concetration.	1
Medium	Between 40 and 160 hours of hazard exposure per month. Usually the employee is not exposure to hazard or they are exposure to medium concentration.	2
High	Higher than 160 hours per month of hazard exposure. The employee is currently exposure to hazard or they are exposure to high toxic concentration.	3

The Table **7** and **8** shows the severity and frequency classification. The occupational risk can be clarified as severity and frequency combination as show Table **9**. The occupational risk has three risk categories that are unacceptable, moderate and not relevant. The risk category levels are represented in the risk matrix by different colors. Indeed, when risk is intolerable it is necessary to mitigate such risk at least to moderate level.

In order to exemplify the occupational risk analysis the palletizing process (palletizing disc operation) is described in Table **10** regarding four types of hazard exposure and take into account the frequency and severity for each of them. In addition, the occupational risk is analyzed as well as a control measure and preventive actions.

Table 9: Occupational Risk Matrix

		Severity		
		1	2	3
Frequency	3	3	6	9
	2	2	4	6
	1	1	2	3

Table 10: Pelletizing Process Occupational Risk

		Occupational Risk Analysis									Local Pellet Plant	Area Operational
Process: Pelletizing					Acessed by: Eduardo Calixto		Aproved by:					Review
Identification												
PROCESS	Activity	Task	Safety and Health	Rotine	Not Rotine	Consequence	Frequency	Probability	Severity	Risk	Control measure	Preventive Action
1-Pelletizing	Pelletizing disc operation	Monitor, control and operate disc	Arm or hand trapped			Death or loss of body part	1		3	3	Check the disk operation condition before any intervention take place	Stop the disc and certify that is locked in safe position whenever physical contact will be necessary
			Noise exposure			Loss or disturbance or audition capacity	2		2	4	Monitore noise as well as the use of ear protection.	Use ear protection in pelletizing operational area
			Dust exposure			Damage to respiratory capacity	2		2	4	Monitoring the use of mask as well as the pelletizing disc particules emission	Use mask protection in pelletizing operational area
			Phisic Ergonomic factor			Muskolosesquel etal disorder	2		2	4	Check ergonomic conditions on operator room regarding operator phisical and	Ajust ergonomic conditions to operator phisical and biodinamic conditions

Another work process that is common all over the word is cleaning activities. Mostly, such tasks related to such process face different occupational risk like physics, chemic and ergonomic. An example of cleaning process occupational risk can be seen in Table **11**.

Table 11: Occupational risk in cleaning process (C- chemical, B- biological, E- Ergonomic)

		Occupational Risk Analysis							**Local Hospital**			**Area Patient room**
Process: Cleanning				**Accessed by: Eduardo Calixto**				**Approved by:**				**Review**
			Identification									
PROCESS	**Activity**	**Task**	**Safety and Health**	**Rotine**	**Not Rotine**	**Consequence**	**Occupational**	**Frequency**	**Severity**	**Risk**	**Control measure**	**Preventive Action**
1-Cleaning	Prepare cleaning solution	Mix chemical products	Exposure to toxic product			Irritative to skin, respiratory system; Neurotoxic or reproductive toxic agents	C	2	2	4	Control and define product applied to cleanning activities	Use gloves to protect hands
	Clean bathroom	Use soap and detergent	Exposure to toxic product			Skin, eye and mucous membrane irritation	C	3	1	3	Control use of detergent and proceed mixture with water	Use gloves to protect hands whenever it is possible
	Clean room	clean humid room	Mould exposure			Dermal allergies, asthma, Sick-Building Syndrom	B	2	2	4	Carry out inspection in room to avoid high mould concentration	Use mask and avoid high mould concentration in rooms
	Clean room	clean contamined room	bacteria exposure			Allergic dermatitis; purulent infections; inflammatory conditions of respiratory and other organs	B	2	2	4	Access contamination level and isolate room whenever contamination level is high.	Use mask and gloves
	Clean floor	Mopping	forced posture			High static load on the upper arm and back muscles	E	3	2	6	Check activity workload and define intervals to rest	Avoid repetitive task for a long period of time
	Clean floor	Mopping	biodynamic movement			Movement controlled by wrist, requires high forces. This combination of repetitive movement and high forces can lead to damage in hand orwrist area.	E	3	2	6	Check activity workload and define intervals to rest	Avoid repetitive task for a long period of time

The occupational risk assessment can be carried out by bow tie analysis that has a good graphic representation as shown in Chapter 1. Despite such representation, the occupational risk must be documented in order to enable consultant, easy update and internal audit.

REFERENCES

Abrams, H.K. (2012). A Short History of Occupational Health. *Journal of Public Health Policy, 22*, 34–80.

An Account of the Founding of H.M. (2013). Inspectorate of Mines and the Work of the First Inspector Hugh Seymour Tremenheere. http:/ /www.Ncbi.nlm.nih.gov.

Annett, J.; Cunningham, D. & Mathias-Jones, P. (2000). *A Method For Measuring Team Skills. Ergonomics, 43*, 1076-1094.

Carroll, J.M. (1991). Designing interactions. New York: Cambridge University Press.

Construction Safety and Health (2012). Workplace Safety & Health Topics. National Institute of Occupational Safety and Health. Retrieved from: 3rd August, 2012, from: http://www.cdc.gov/niosh/topics/construction/.

Cooke, N. J.(2005) Measuring Team Knowledge. In: Stanton, N. A., Hedge, A., Salas, E., Hendrick, H. & Brookhaus, K. (Eds) *Handbook Of Human Factors and Ergonomics Methods*. London: Taylor & Francis, 49-1-49-6

Dowell, J., and Long, J.B. (1998). *Conception of the cognitive engineering design problem. Ergonomics, 41,* 126-139.

Dźwiarek, M. Jankowski, J. (2012). The methods of virtual modelling of dangerous zones and safety devices to Support risk assessment in machinery design. Sopot: Working on Safety.

Employers Safe Working Practices, Health & Safety Policy. (2013). Retrieved from: 15th February, 2013, from: http:/ /www.Citation.co.uk.

Health and safety at work statistics.(2012). eurostat. European Commission. Retrieved from: 3rd August, 2012, from: http://epp.eurostat.ec.europa.eu.

Health and safety in the Construction Industry. (2013). Veritas Consulting. Retrieved from: 20th March, 2013.

Injuries, Illnesses, and Fatalities. (2012). Bureau of Labor Statistics. Retrieved from: 9 August, 2012. http://epp.eurostat.ec.europa.eu.

International Hazard Datasheets on Occupations (HDO). International Labour Organization. Retrieved December 26, 2012. http://epp.eurostat.ec.europa.eu.

NIOSH Workplace Safety & Health Topic: Agricultural Injuries. (2013). Retrieved from: 15th February, 2013, from: http://www.cdc.gov/niosh/topics/agriculture.

NIOSH Pesticide Poisoning Monitoring Program Protects Farmworkers.(2013). Retrieved from: 15th February, 2013, from: http://www.cdc.gov/niosh/docs/2012-108.

NIOSH Alert: Preventing Deaths, Injuries, and Illnesses of Young Workers (2013). Retrieved from: 15th February, 2013, from: http://www.cdc.gov/niosh.

Oak Ridge National Lab Safety Document(2014). http://www.ornl.gov/.

Salas, E. (2005) Team Methods. In: Stanton, N. A., Hedge, A., Salas, E., Hendrick, H. & Brookhaus, K. (Eds) *Handbook Of Human Factors And Ergonomics Methods*. London: Taylor & Francis, 43-1 - 43-4.

Missala,T. (2012). Paradigms and safety requirements for new generation of workplace equipment. Sopot: Working on safety 2012.

W Zachary, JM Ryder, JH Hicinbothom. (1998). Cognitive task analysis and modeling of decision making in complex environments. Published In J. Cannon-Bowers & E. Salas (Eds.), *Decision making under stress: Implications for training and simulation*. Washington, DC: American Psychological Association.

Wenger, E. (1987). Transactive Memory: A Contemporary Analysis of The Group Mind, In B. Wellman and S. Goethals (Eds.) *Theories Of Group Behavior*. NewYork: Springer, 185-208

Velichkovsky, B.M. (2005). Modularity of cognitive organization: Why it is so appealing and why it is wrong. In W. Callebaut & D. Rasskin-Gutman (Eds.), *Modularity: Understanding the development and evolution of natural complex systems* (pp. 335-356). Cambridge, MA: MIT Press.

http://www.usatoday.com. Accessed on October 21st, 2014.

www.bbc.co.uk/news/world-africa. Accessed on 30th December 2014.

CHAPTER 2

Qualitative Risk Analysis: Concepts and Methods

Abstract: The risk analysis methods encompass different types of qualitative and quantitative risk analysis, which is applied during risk assessment in order to support decisions to mitigate the risk whenever it is necessary. Depending on the type of industry the risk requires a systematic risk management, which is supported by risk analysis methods. During recent decades, many procedures, laws and methods have been applying such risk methods. However, substantial effort is still necessary in order to mitigate the industrial risk in some industries such as Transportation, Aerospace, Chemical, Oil & Gas and Nuclear because those are responsible for the worst accidents in terms of consequences for employee's health, environment and society. Therefore, the first step to avoid or mitigate these risks are to apply a systematic risk analysis to enable identify, analyze, evaluate and mitigate such risks. This chapter describes the best risk analysis methods to support the risk management process.

Keywords: Risk, risk matrix, risk mitigation, risk management, preliminary hazard analysis, failure mode analysis, hazard and operability analysis.

2.1. INTRODUCTION

Risk Analysis and Management started around middle of twenty centuries in different industries with different approaches like:

- In 1960's - Aerospace Industry with Quantitative Risk Assessment methods, Nuclear Industry with Probabilistic Risk Assessment approach,
- In 1970's - Chemic Industry with Quantitative Risk Assessment and Seveso directive,
- In 1980's - Oil and Gas Industry with Quantitative Risk Assessment and Safety Case,

The other industries followed the established approaches, methods and standards at long last in the final decades of twenty centuries and adapted the risk analysis methods and management to their characteristics and requirements.

The remarkable point is that in most industries such Risk Analysis and Management methodology came out after catastrophic accidents. That contradicts the Safety preventive sense, but that is the way that mankind has learned to avoid loss of life in industries with the last few decades.

Eduardo Calixto

Nowadays, most of industry with catastrophic accident consequences like: Oil and Gas, Aerospace, Aeronautical, Chemical and Nuclear have a high level of Risk Management and Risk Methods applications. Even though, Risk management is always a challenge and even nowadays with high level of Risk Management, major accidents continue to happen.

In fact, that is a very interesting by the methodology point of view and that also explains the probabilistic nature of risk analysis and management that is confirmed by those eventual major accident occurrences despite all risk analysis and management implemented.

The Risk Management main objectives are to keep risk under acceptable level and to do so, a sequence of activities like plan risk analysis, identify hazards and assess risk, control and mitigate risk, communicate risk are carried out. In addition, risk documentation and periodical risk revision are required to certify that the risk is below acceptable level limits.

In general terms, occupational risk management comprises process risk which employees face in the workplace. In fact, most of risk analysis methods and even risk management are driven by safety process. Occupational Risk mostly has one specific program and do not apply all risk analysis methods. Safety process risk is related to major accidents, that means rare accidents with catastrophic consequence.

The safety process management is applied to maintain process risk under acceptable levels and avoid whenever it's possible incidents and accidents. In order to mitigate process risk, different layer of protection with significant reliability are applied to assure that in case of unsafe process condition the system will be back to a safe condition. Therefore, the acceptable risk level in industrial process relies on multiple layers of protection.

There are different types of layer of protections such as design features, control system, Safety protection functions and emergency response plan. The best approach to maintain an acceptable risk level is to start risk management in the design phase. Moreover, it's also necessary to implement risk analysis' recommendation throughout the asset phases and update risk analysis whenever some process modification takes place.

In fact, during the project phase, there are more flexibilities for modifications and incorporate new ideas to improve asset safety rather than operational phase. The major approach to inherently safer process designs is divided into the following categories (Crow and Louvar, 2002):

- Intensification;
- Substitution;
- Attenuation;
- Simplification;

Intensification means minimize risk whenever possible based on the less hazardous equipment and products. Substitutions means take place whenever it's possible with more safety equipment and products. Attenuation means process designed to have accident effect mitigated. Simplification means establish process control which enables to control process easily in case of incident.

Such decisions are best conducted when a Risk management process is established and in general terms such risk management process is well described as shows Fig. **1**.

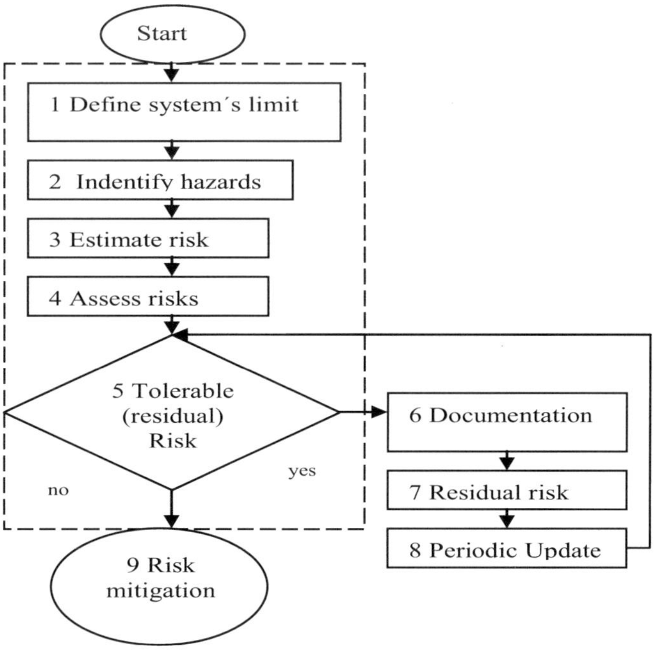

Figure 1: Risk management Steps (Source: Moergeli, 2006).

Item 2.3 will be described in more detail regarding when risk analysis methods must be applied to enterprises phases and which are the different ways to manage risk.

As mentioned before, in order to assess risk, it requires that risk analysis methods which are established approach to assess hazards and quantify qualitative or quantitative their risk. Such methods have qualitative and quantitative concept. In the first case, a group of specialists identifies hazards and qualify risk based on their opinion. In the second case, despite hazards being identified qualitatively based on specialist opinion, risk is calculated by mathematical methods. Independent to be qualitative or quantitative, risk analysis is deductive or inductive. On a Deductive risk analysis case, the first step is to identify hazards or incidents and then their causes, consequences and when necessary propose recommendation. In an Inductive Risk analysis, the first step is to define process deviation, equipment failures or events that turn into consequences like an accident. The most usual qualitative Risk analysis methods are:

- PHA (Preliminary Hazard Analysis) is a qualitative inductive method which identifies hazards, causes, consequences, detections and propose recommendations. In some cases, PHA has a risk assessment based on risk matrix, where probability (or frequency) is related to the causes and severity that is related to consequences. PHA is one of the most applied risk analysis methods and can be applied to whole asset life cycle to risk assessment in complex process or simple tasks.

- HAZOP (Hazard Operability) is a qualitative inductive method which identifies hazards, causes and finally Hazard consequences. In HAZOP case, hazard means process deviation such as high temperature, high pressure, high level, and high flow. In fact, the HAZOP method defines guide words such as high and low and combine with process parameters like temperature, level, pressure, flow. The first HAZOP step is to define the system, subsystem and boundaries and the second step is to split subsystem into nodes, which means, part of the subsystem which comprises a group of equipment that will be assessed. For each node, all process deviations are assessed as well as causes, consequences, safeguards and layers of protection which are devices that help to bring the system to safe condition. The HAZOP method is very well applied on project phase (Basic project) that came out of recommendations to have a more safe process but can also be applied to the operational phase.

- FMEA (Failure Mode Effect and Analysis) is a qualitative inductive method which identifies equipment failure modes, causes, detection and finally consequences. In FMEA case, when focus is safety, will be regarded unsafe failures, that means failure that cause an unsafe condition of equipment that can trigger an accident. The first FMEA step is to define a system, subsystem and boundaries and the second step is to define subsystem's equipment and components. Thus, for each specific equipment component, the failure modes are assessed as well as causes, consequences and when necessary recommendations are proposed. When FMEA is applied in products project phase is called DFMEA (Design Failure Mode Effect and Analysis). In addition, PFMEA (Process FMEA) is applied to detect failures caused by manufacturing. The System FMEA is also applied to consider the failures during the operational phase, which influence on systems safety and performance.

Another variation of FMEA is FMECA (Failure Mode Effect and Criticality Analysis). The FMECA applies a criticality assessment that comprises probability (frequency), severity and detection. Thus, in addition to consider risk, is also considered the detectability of failure and in this case, as the easier as failure can be detected the lower is the detectability value. The criticality results of multiplication of probability (frequency), severity and criticality. That enables to come out a criticality rank about equipment failure modes in order to prioritize recommendations.

2.2. RISK CONCEPTS

By definition, the risk is the combination of an event of hazard and its consequence. In order to analyze and evaluate the risk, the qualitative and quantitative approach can be performed. In fact, when risk is assessed and evaluated based on qualitative methods, such assessment is performed qualitatively based on specialist opinion regarding a risk matrix with the frequency and consequence criterion established.

There are different configurations of risk matrix and such configuration must reflect the law and companies risk policy. In fact, before make up an appropriated risk matrix is necessary to define clearly the frequency or probability category as well as severity. There is a very hard discussion between risk specialists about which is best applied in risk matrix: frequency or probability, but in reality, most of the time it is easier for the specialist who take part in risk analysis to predict frequency than probability about the causes of accidents. The Fig. **2** shows an example of risk matrix with four severity categories and six frequency categories. Again, the risk matrix configuration must fit well to companies and their process and even in some cases is necessary to use different risk matrix for different process in the same company.

		Frequency Category					
		A (Extremely Remote)	B (Remote)	C (Little Frequent)	D (Frequent)	E (Very frequent)	F (Extremely frequent)
		At least 1 between from 1000 to 100.000 years	At least 1 between from 50 to 1000 years	At least 1 between from 30 to 50 years	At least 1 between from 5 to 30 years	At least 1 between from 1 to 5 years	At least 1 in 1 years
Severity Category	IV	M	NT	NT	NT	NT	NT
	III	M	M	NT	NT	NT	NT
	II	T	T	M	M	M	M
	I	T	T	T	M	M	M

Figure 2: Risk Matrix (Source: Calixto, 2011).

In addition, severity classification must describe all parties affected in the case of an accident like employees, community and environment as well as company installations cost. The Table **1** shows an example of severity category.

The Table **1** shows four severity categories regarding aspects like personal safety, installation, environment and image as well as social impact that is measured by impact in economic activity.

The other important concept in risk management is risk perception that means how much employees and other affected parties like community are aware about risk to which they are exposed. The risk perception is related to risk communication that is a very important task of risk management. In fact, risk communication is hard to be applied because it requires different ways to be applied for different groups of employees or even society. A powerful tool to communicate risk is "Safety Dialog" which is a discussion about one safety related issue that is carried out for some employee. The main objective is to make group awareness about such issues and enable discussion about that. In order to communicate process risk to operators a safety dialog is appropriate because enable a discussion about the risk rather than a electronic message. On same way, whenever a meeting is carried out with the community, a communication about risk can be done and most of the cases such communication is about emergency procedures that are very important to the community in case of an accident. Despite a technical feature of risk communication whenever is it possible the Communication Management must be involved in such process because they know the best ways to communicate and to deal with the information within the company.

Table 1: Severity Category (Source: Calixto, 2011)

			Severity Description			
			Personal Safety	Installation	Environment and Image	Social
Severity Category	IV	Catastrophic	Catastrophic injures with death, its possible to affect people outside	High equipment damage and high loss of production	High environment impact which affect company reputation	High economic effect in local community business, tourism and loss of life quality (Between $101.000.000,00 to $336.000.000,00)
	III	Critical	critical injuries, employee stay a period of time out of workplace	Equipment serious damage with high repair cost	Critical effect on environment hard to recovery and bad compay reputation	Economic effect in local community business, tourism and loss of life quality (Between $2.500.000,00 to $101.000.000,00)
	II	Marginal	Moderate injuries with first aid assistence required	Equipment small damage with small repair cost	Not serious effect on environment that can be recovery under human action bu cause impact on compay reputation	Low economic effect in local community business, tourism and loss of life quality (Between zero to $ 2.500.000,00)
	I	No effect	Minor injuries with minimum first aid assistence required	Equipment very small damage with very small repair cost	Minor effect on environment that can be recovery under human action and cause no impact on compay reputation	No economic effect in local community business, tourism and loss of life quality

The risk communication has a high influence on risk perception but do not guaranty that risk perception will trigger preventive behavior on workplace because that's depends on safety culture as well. In addition, even though employees and society realize the risk they are exposed, there is a third factor that influences their behavior defined by their risk profile. In general terms, risk profile can be aversive, neutral or seeking and that varies from people to people and depends on the situation; a single individual can have a different attitude. Such risk profile is very important to understand the leader's attitude to prevent and mitigate the risk.

Finally, risk analysis result has to be considered as input in Emergency Plan and such plan must be part of risk management. An emergency plan is a set of activities defined to be carried out in case of an accident as well as resources and responsible for each task. A well-defined Emergency Plan is essential to have a good emergency response in case of accident, but in addition is necessary to carry on periodically emerges simulating that is a practical exercise of emergency plan application regarding an accident scenario. Thus, it is very important to take into account the risk analysis results in emergency plan otherwise the emergency response team will not be prepared to effectively respond to a predicted accident scenario. Actually risk analysis does not cover all accident scenarios as well as emergency plan, but the challenge is to be prepared for all possible events, even natural catastrophes and terrorist attacks that must be covered by risk analysis and included on emergencies plans.

2.3. RISK ANALYSIS METHODS

In order to assess hazards and quantify their risk it is required to use appropriate methods which support the risk management process efficiently. Such methods are known as Risk Analysis and can be qualitative and quantitative. The Qualitative Risk analysis methods are supported by specialist opinion to define hazard, cause, consequences, detections, protections and propose recommendation. Mostly, to assess risk qualitatively using a risk matrix as mentioned before. The Quantitative Risk analysis methods may also define hazard, initiate event or top events by specialist opinion, but calculate mathematically their probability or frequency as well as consequences.

In fact, it depends upon the objective of the risk analysis that different methods can be applied. From simple tasks or usual activity, for example, where the most important thing is to identify hazards and implement the proposed recommendation to avoid accidents, the qualitative approach can be applied. Contrarily, when it is assessed that the process or activity with a probable accident with high severe consequence more complex methods are applied.

The Risk Analysis specialist can support and define which methods are more appropriate, but even specialists prefer one technique more than another and can influence such decision. In general terms, the risk analysis methods can be applied in different asset phases with different objectives as shown in Table **2**.

In order to describe each risk analysis method features the follow section will discuss each of them with examples.

Table 2: Risk Analysis Methods Application

Situations	**Enterprise phase**				
	Pre-Design	Design	Basic project	Operational	Deactivation
1 – Identify hazards in usual activity to avoid accident				PHA FTA* ETA* Bow Tie*	PHA
2 – Identify hazards in the process or equipment to mitigate risk	DFMEA	DFMEA	HAZOP FMEA/FMECA PHA LOPA SIL ETA FTA	PHA FMEA/FMECA PHA LOPA SIL ETA FTA	PHA
3 – Define Vulnerable Area, ISO-Risk and Societal risk			Consequence and effect analysis	Consequence and effect analysis	
4 – Decide which technology has lower risk	Consequence and effect analysis* FMECA*	Consequence and effect analysis* FMECA*			

FTA* (Use only the FTA logic diagram without quantifying probability)
ETA* (Use only the ETA logic diagram without quantifying probability)
Bow Tie* (Use only the Bow Tie logic diagram without quantifying probability)
Consequence and effect analysis* (Use analysis of similar plants)
FMECA* (Use analysis of similar equipment)

2.4. PRELIMINARY HAZARD ANALYSIS (PHA)

The PHA analysis come from the military industry application as a reveal technique applied to check missile system launch. In that case, 4 of 72 missiles intercontinental atlas were destroyed with high cost.

Nowadays, the PHA is applied in many industries in operational activities or project conceptions. Thus, no matter of the application, the main objective is to support decision in order to avoid accident and eliminate unsafe conditions.

The PHA analysis can also have a specific focus on environmental or safety. Such method is a good tool application in terms of risk analysis in operational areas, but in case of a project that complexity increases because the complexities of process with different hazard and in addition must be taken into account in such analysis safety, environment, company image and accident cost.

In general terms the PHA steps are:

- 1 (Step – Collect information about the system).
- 2 (Step – Define risk analysis team).
- 3 (Step – Define risk matrix when is applicable).
- 4 (Step – Define scope (system, subsystem and process)).
- 5 (Step – Proceed PHA).

In the first step is necessary to collect all information about System that will be assessed that means technological as well as accident and incident historical information.

On the second step, once there's a great idea about system, it is necessary to define a PHA team to take into account all specialist knowledge to carry out a PHA. Mostly such specialists are professional from safety, operational, maintenance, instrumentation, environment as well as the engineer responsible for equipment and process Plant (or project).

The third step, before start up of a PHA, it is necessary to define which risk matrix will be applied to a PHA. That is a very important task and if attention is not paid in such detail the PHA can be rejected by companies or even by authorities when for example the PHA for an environmental license. When a PHA is carried out to get an environmental license it is very important to compare the risk matrix that is intended to be used with that is defined by Environment Agencies.

The fourth step is necessary to define the analysis scope and in some cases is clear which system and subsystem will be assessed in other cases not. For example, when a modification is carried out and there are new hazards or processes, probably such hazard effect can be local or not. Thus, it is very important to assess the impact on modification of the whole process in order to define the scope of analysis regarding all system and subsystem affected.

The fifth step in the PHA is proceeding and is necessary to pay attention to frameworks issues like the room and electronic devices necessary in the PHA as well as to plan the required time to carry on analysis regarding all professionals' time availability. In many cases it is hard to maintain professional from operations and maintenance for a long period of time (over two weeks). In fact, all effort must be done because change professional on a long PHA, no matter how qualified they are, is hard because different point of view and initial assumption must be discussed again and again.

The PHA can be applied to software or usual office tool (word or excel) and the file configuration must have at minimum: Hazard, cause, consequence, probability (frequency), severity, risk, the detection and recommendation.

Hazard means unsafe condition, product, process or equipment condition which can cause any health damage to an employee. In many cases a hard discussion about what is a hazard happens among safety professional. Actually, there is always a sequence of cause and consequence, but to be clear the last consequence level must be described and be clear to classify severity. By this way, the hazard will be the event before consequence and all causes are events between hazards and consequences like external events, equipment and human failure or process deviation. For instance, in the chemical and Oil and gas industry toxic product leakage can lead to different consequences like jet fire, toxic cloud, cloud explosion and fireball. Each one of such consequence has different severity.

The other important point that creates hard discussion are safeguards and layers of protection. Layer of protection are devices that can avoid accidents or even mitigate their effect independently of human intervention. In many cases, human intervention is also considered a layer of protection, but not in terms of alarms or instruments.

In some cases after recommendation risk is assessed again in order to show how much is expected to be mitigated by recommendation implementation. An example of Platform PHA is shown in Table **3** below.

Table 3: Preliminary Risk Analysis (Platform) (Source: Calixto, 2007)

<table>
<tr><th colspan="15">Preliminary Hazard Analysis</th></tr>
<tr><td colspan="2">Platform: P-90</td><td colspan="11">System : Load</td><td colspan="2">DATA:26-07-2009</td></tr>
<tr><td colspan="2">Subsystem: 1.1 -Load</td><td colspan="11">Description: From well SDV until of production mainfold</td><td colspan="2">Draw :xx.xxx.xxxx</td></tr>
<tr><td rowspan="2">Hazard</td><td rowspan="2">Cause</td><td rowspan="2">Consequence</td><td rowspan="2">Detection/ safeguard</td><td rowspan="2">Freq</td><td>Personal</td><td></td><td colspan="2">Install</td><td colspan="2">Environ</td><td colspan="2">Social</td><td rowspan="2">Recommendation</td><td rowspan="2">AH</td></tr>
<tr><td>S</td><td>R</td><td>S</td><td>R</td><td>S</td><td>R</td><td>S</td><td>R</td></tr>
<tr><td rowspan="7">Huge Oil and Gas spill</td><td>Pipeline rupture</td><td rowspan="3">Puddle Formation, atmospheric explosive formation, oil spill on the sea, health damages</td><td rowspan="3">Visual, Gas detector, pressure detector and other detectors</td><td>C</td><td>IV</td><td>NT</td><td>III</td><td>M</td><td>II</td><td>M</td><td>II</td><td>M</td><td rowspan="3">R01) Follow procedure for PIG operation. Action by: Operation Group</td><td>1</td></tr>
<tr><td>Connections rupture</td><td>C</td><td>IV</td><td>NT</td><td>III</td><td>M</td><td>II</td><td>M</td><td>II</td><td>M</td><td>2</td></tr>
<tr><td>PIG Operation failure</td><td>C</td><td>IV</td><td>NT</td><td>III</td><td>M</td><td>II</td><td>M</td><td>II</td><td>M</td><td>3</td></tr>
<tr><td>Submarine connection rupture</td><td rowspan="4">Atmospheric explosive formation, oil spill on the sea, health damages</td><td rowspan="4">Visual and pressure detector</td><td>B</td><td>III</td><td>M</td><td>III</td><td>M</td><td>IV</td><td>M</td><td>IV</td><td>M</td><td rowspan="2">R02) Follow procedure to Load platform. Action by: Operation Group</td><td>4</td></tr>
<tr><td>Riser component failure</td><td>B</td><td>III</td><td>M</td><td>III</td><td>M</td><td>IV</td><td>M</td><td>IV</td><td>M</td><td>5</td></tr>
<tr><td>Ship collision</td><td>B</td><td>III</td><td>M</td><td>III</td><td>M</td><td>IV</td><td>M</td><td>IV</td><td>M</td><td rowspan="2">R03) Riser robust design to avoid fatigue. Action by: Project Group</td><td>6</td></tr>
<tr><td>Subsea material fatigue</td><td>B</td><td>III</td><td>M</td><td>III</td><td>M</td><td>IV</td><td>M</td><td>IV</td><td>M</td><td>7</td></tr>
</table>

The partial PHA above highlights possible hazards on a Platform Load Subsystem as well as all causes, consequences, detection and safeguards, risks and recommendations. In this case, risk was assessed based on Fig. **2** and for each consequence, there is a severity category which results in different risk classification. It's also possible to have one frequency category for each specific cause and in this case there will be several risks assessed. The last column shows the number of accident hypotheses or in other words accident scenarios. Such accident hypothesis is used to simulate consequence and define a vulnerable area.

The additional example of preliminary hazard analysis is applied on metallurgy Industry and assess hazards in a palletizing process as described in Table **4**. In this example, the PHA focuses on environmental aspects that means a process or product that interacts with the environment and can modify it characteristics. In this case, an emission gas from pelletizing process is the significant environmental aspect in such process. Thus the preliminary hazard analysis template comprises hazard, cause, effect, frequency, severity, and recommendation. The risk classification is based on the risk matrix on Fig. **2**. In this case, the hazard is the environmental aspect “huge pellets particulate emission” and different from previous case has a delayed effect on employees and population health. In order to mitigate such risk (catastrophic) three actions are proposed that the first and third are preventive and the second one is reactive. In many cases this preliminary Risk analysis is also called Aspect and Impact analysis when environmental hazards are being assessed. When safety is being assessed, such analysis can also called hazard and damage analysis and both activities and tasks in operational phase and is an excellent tool to support Occupational Risk Assessment.

Table 4: Preliminary Risk Analysis (pelletizing process)

<table>
<tr><th colspan="7">Preliminary Risk Analysis</th></tr>
<tr><td colspan="2">System: Pelletizing</td><td colspan="4"></td><td>Draw: D21.11.001.3</td></tr>
<tr><td>Hazard</td><td>Cause</td><td>Effect</td><td>F</td><td>S</td><td>R</td><td>Recommendation</td></tr>
<tr><td rowspan="2">Huge pellets particulate emission</td><td>Total failure on electrostatic precipitator</td><td rowspan="2">Bad air quality with effect on employees and population health in medium and long time</td><td>E</td><td>III</td><td>NT</td><td rowspan="2">1 - Install automatic system to shut down the pelletizing process whenever Electrostatic precipitator shut down. (Action: Engineering Management)
2 - Establish procedures to control emission in case of electrostatic precipitator failure.
(Action: Operational Management)
3 - Establish preventive maintenance, predictive maintenance and inspection plan to maintain electrostatic precipitator with high availability.
(Action: Maintenance Management)</td></tr>
<tr><td>Partial failure on electrostatic precipitator</td><td>C</td><td>III</td><td>NT</td></tr>
</table>

Mostly, when assessing processes in a project or operational phase, it’s necessary to qualify hazard risk in order to have priority to implement recommendations. In such cases, whenever such recommendations are implemented it is possible to assess risk again in order to know how much risk was mitigated. Depending upon the nature of activity or process when a PHA is carried out risk assessment based on risk matrix is not necessary because all recommendations must be followed up to have a safety activity. That happens in the Construction industry because the hazard in each activity must be controlled that means risk minimized and whenever is possible eliminated. An example of Preliminary Hazard Analysis is a roof repair which comprises several activities as shown in Table **5**.

Table 5: Preliminar Hazard Analysis (roof repair)

Preliminary Hazard Analysis				
Activity: Roof Repair				Draw: C01.111.002.5
Task	Hazard	Cause	Effect	Recommendation
1 – Remove roof	Material falls down	The material moved not correctly from the roof	Death or serious employee health damage	1 – Isolate the area where material is being taken out. (Action: Operator) 2 – Use Individual equipment protection (helmet, boots and gloves) (Action: Operator)
2 – Install temporary roof	Material falls down	Not appropriated removed from the roof	Death or serious employee health damage	3 – Follow procedure to remove material from the roof. (Action: Operator). 4 – Use Individual equipment protection (helmet, boots, gloves and belt to hang on scaffolding)
	Employees fall from the roof	Not using safety protections on Scaffolding. Scaffolding not built up correctly		
3 – Build up new roof	Material fall down	Not appropriated removed	Death or serious employee health damage	1 – Isolate the area where the material is being taken out (Action: Operator)
	Scaffolding fall down	Scaffolding not builds up correctly	Death or serious employee health damage	1 – Isolate the area where the material is being taken out. (Action: Operator)
	Waste material of construction	Not following procedure to remove waste from the construction area	Construction waste, environmental impact on the local area	5 – Follow procedure to remove construction waste to adequate place. (Action: Operator)
	Burning when performing welding work	Not use appropriated equipment protection	Medium or serious damage to employee's health	4 – Use Individual equipment protection (helmet, boots, gloves and belt to hang on scaffolding)

These three examples show the main PHA features that are simple and applicable in all enterprise phases. The Preliminary Hazard Analysis has advantages:

- Easy to be applied in different industries.
- Can be applied in different enterprise phases. That means once applied in project phase such PHA's can be updated on long enterprise phases.
- Clear to understand and define a rank, of which are the most critical process, activities or task in term of risk which support to priories the recommendation.

The Preliminary Disadvantages are:

- As qualitative analysis the risk can be underestimated and recommendation or mitigation actions may not be implemented.
- Depends on specialist experience and historical data that means if specialists do not realize one specific hazard such hazard will not be assessed.
- Does not explain in details the equipment and process failures and because of that do not define specific action in such situation.

The advantages and disadvantages must take into account when such PHA is carried on and in some cases, additional qualitative analysis must be carried on with the equipment or process focus.

2.5. HAZARD AND OPERABILITY ANALYSIS (HAZOP)

The HAZOP means hazard and operability and it's a famous risk analysis in Chemical and Oil & Gas industry. This technique was introduced for ICI Chemical company engineers in 1970 in order to prevent process deviation. It consist in very structural methodology with specific world which provide a guideline to assess process deviation, causes, consequences and whenever it's possible to propose recommendation to mitigate risk.

The HAZOP analysis mostly is carried out on Basic Project phase in order to have time enough to implement the recommendation in time during the project. The HAZOP can also be carried out for plants in the operational phase, but that is not usual because is not expected to implement or do a lot of modification in the plant on operation phase. By the other way round is very important to carry on HAZOP analysis on operational phase whenever on specific modification in process is implemented. In this case, is also necessary to take into account the impact of such modification in another subsystem and consider them in HAZOP. Therefore, is also important after HAZOP, communicate the new process risk for all specialists involved in such plant operation and maintenance.

The HAZOP methodology consists in defining which are the consequence of process deviation and try to mitigate the risk implement recommendations as layer protection. That is a good structural analysis, which specific process deviation, guide world and concepts.

The first step in HAZOP analysis defines the system, subsystems and in each subsystem are necessary to define nodes. Those nodes will limit the assessment of consequence process deviations and include a group of equipment, alarms, and valves and so on. Depends on coordinator it's been considered the causes and safeguard into the nodes, out of node or both of them. In fact, if it's been considered the process deviation consequence into the nodes and causes and safeguards in anywhere, the focus is on a node without forgetting any important issues out of that.

The second step is to ask for group about process deviation as pressure, level, temperature, flow and contamination, but to do that some good words are necessary. The mean guide words, terminology is shown in Table **6**.

Table 6: HAZOP Guide Words

WORD GUIDE	MEAN
None	There´s no parameter
Less	Quantitative reduction
More	Quantitative increase
Part of	Qualitative reduction
Either	Qualitative increase
Reverse	Opposite flow than usual
Other	Complete substitute

Based on those words, is asked to a HAZOP Analysis group about the effect of process parameter deviation like low and high pressure, low and high temperature, low and high level, low (no) and high flow. After that, the causes, consequence and safeguards are assessed and is assessed the necessity of additional safety function to mitigate the risk. The safeguard is considered the equipment which functions acts independently to bring the process to safe condition or catch up operator attention for unsafe conditions. The difference between safeguarding and layer of protection is that "Layer of Protection" operates independently without human action to bring the process to safe condition. Thus, all layers of protection are safeguards but not all safeguards are layers of protection like alarm for example. The HAZOP steps can be summarized in Fig. **3**.

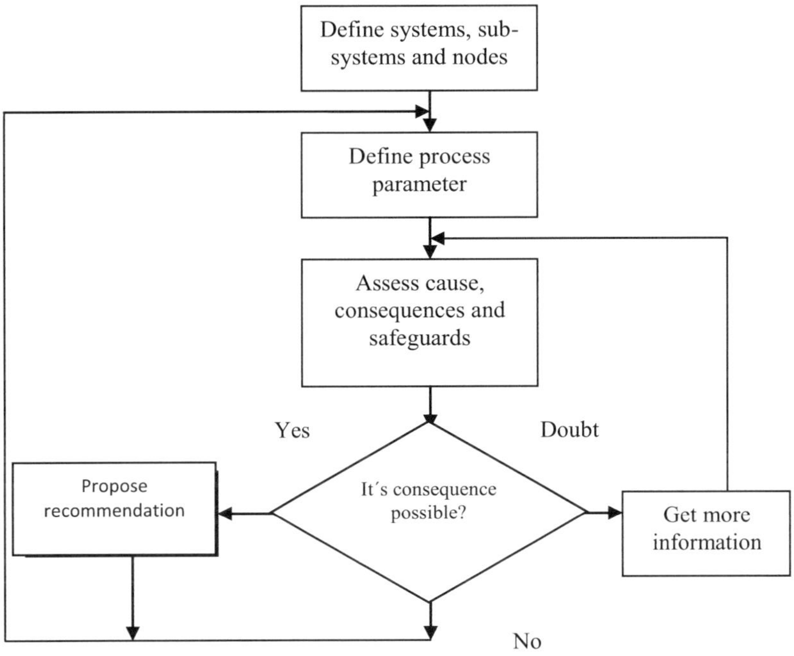

Figure 3: HAZOP steps (Source: Calixto, 2007).

Is very important to have a successful HAZOP, to have a multidisciplinary group, that means different specialist from operation, maintenance, instrumentation, process, project and safety. In addition is also necessary to have a HAZOP leader that coordinates HAZOP Analysis. Mostly, when HAZOP analysis is carried out for a whole process it take one week or more and that is a big challenge to the HAZOP leader to keep specialist OS HAZOP analysis along this time. What happens in many cases is that the specialization has taken place a long HAZOP that makes harder the analysis because whenever a new member come to the group is necessary to clear up all assumption. The remarkable issue on HAZOP is logistic and framework like media and electronic devices as well as adequate room. As all risk analysis, documentation is also very important and that means P&D draw, process description, equipment data as well as accident historical data. It's so important to have such document updated and make such references to Hazop files. In many cases, in order to not delay HAZOP analysis the document is not updated and HAZOP leader must pay attention to it and make reference on HAZOP only about the document that they have. That means if there is not some devices in P&D draw, even though specialist confirm such equipment in next P&D draw version, the equipment must be considered in HAZOP as a recommendation. That usually happens for example, in case of missed alarm or Safety Instrumented Function. The Fig. **4** below, shows an example of HAZOP file that is assessed a low level in a vessel on hydrogen generation plant.

XXXXX		**HAZOP (Hazard Operability)**		**XXXXXXXX**
Unit	**UGH**	**System**	**Feed treatment**	**Date: 06-04-2006**
Subsystem: Pressure control feed treatment Node: From D-03until out of PSA, passing by D-03, C-04, C-05, C-06, F-05				Draw n°:xx.xxx.xxxx
Hazard	Cause	Effect	Safeguard	Recommendation
Low Level	- Control failure in LIC-02 - Open failure in LV-012	-Send H2 to E-01	-LAl-01 -LSHH-02	- Install logic control on FIC - 02. (Action: Project Group)

Figure 4: HAZOP Example (Source: Calixto, 2007).

In that example a low level deviation in a vessel due to level control failure or open failure valve LV-012 cause H2 liberation. Despite there are two layer protection, one low level alarm and one low level shutdown control, Hazop analyze group thought necessary more one protection and recommended one logic control (FFIC-02) implementation. There is not a rule to limit the recommendations on HAZOP but the question is how much each recommendation mitigates risk and how much is cost beneficial.

In order to find out how much risk is mitigated with the recommendation, additional quantitative risk analysis like LOPA, SIL analysis as well as a consequence and effect analysis must be carried on.

Some specialist tries to use the risk matrix on HAZOP in order to assess risk and prioritize recommendation. That is not an effective best practice because mostly in HAZOP there will not be catastrophic accident due to process control establish by process engineer as well as a layer of protection.

By this way, most of process deviation risk will be classified as moderate and in this situation is not necessary to implement additional actions to mitigate more the risk. Risk is under control in acceptable risk.

A best practice is carried out SIL analysis regarding all layers of protections in order to classify correctly the SIF (Safety Instrumented Function).

Mostly, catastrophic accident is detected in PHA and a risk matrix is used to assess risk in order to identify which accident scenario must be assessed by consequence and effect analysis in order to calculate individual and societal risk.

When there is not a PHA an alternative is carried on at HAZID analysis that means hazard Identification as show Fig. **5**. Basically, that a same conception of PHA the difference is that all System, subsystem and node files used in HAZOP are also used in HAZID analysis. In the end, is possible to compare HAZOP and HAZID recommendation and will be clear which are the catastrophic event on plant. The HAZID can be carried out by another group in same time that HAZOP is being carried out, but the best approach is carried out HAZID before HAZOP by same specialist.

<table>
<tr><td colspan="2">XXXXX</td><td colspan="5">HAZID (Hazard Identification)</td><td>XXXXXXXX</td></tr>
<tr><td>Unit</td><td>UGH</td><td>System</td><td colspan="4">Feed treatment</td><td>Date: 07-05-2006</td></tr>
<tr><td colspan="7">Subsystem: Pressure control feed treatment
Node: From D-03 until out of PSA, passing by D-03, C-04, C-05, C-06, F-05</td><td>Draw n°:xx.xxx.xxxx</td></tr>
<tr><td>Hazard</td><td>Cause</td><td>Effect</td><td>Safeguard</td><td>F</td><td>S</td><td>R</td><td>Recommendation</td></tr>
<tr><td rowspan="2">Huge H2 leakage</td><td>- Corrosion on vessels</td><td rowspan="2">-H2 cloud,
-Jet fire,
-Explosion,
-Fire ball</td><td rowspan="2">-Gas detector
-LSHH</td><td>A</td><td>IV</td><td>M</td><td rowspan="2">- Carry on inspection periodically on vessels and pipelines. (Action: Maintenance)
- Performance test on LSHH periodically. (Action: Maintenance)</td></tr>
<tr><td>- Corrosion on pipelines</td><td>A</td><td>IV</td><td>M</td></tr>
</table>

Figure 5: HAZID Example

In general terms, we can say that HAZOP drawback is:

- As qualitative analysis the process hazard can be sub estimated and a recommendation or mitigate action may not be implemented.

- In some cases a lot of recommendation cannot be implemented because of cost and there is not a clear idea how much impact on process risk will be when such recommendations are not implemented.

- Do not explain in details the equipment failures and because of that do not define specific action for equipment.

- Do not give a priority in terms of recommendation implementation.

- The HAZOP advantages are:

- HAZOP is a very well defined Risk analysis and covers all process, system and subsystem in P&D draw ;

- Is possible to assess the impact of one process deviation in others subsystems;

2.6. FAILURE MODES AND EFFECT ANALYSIS (FMEA)

The FMEA Analysis was carried out firstly by USA army. In 50th ages was developed the procedure MIL-P-1629 and latter on other industries such as aerospace, railways, Oil and Gas and Nuclear has also applied the FMEA in their asset management process.

In fact, FMEA is applied in different asset phases in order to avoid failure caused by wrong design and human error during manufacturing, installation and operation phases. In the design phase, FMEA is known as DFMEA that means Design Failure Mode and Effect Analysis and it will be described in more detail in the next section.

In addition, during the manufacturing phase, the FMEA is known as PFMEA (Process Failure Mode and Effect Analysis). Thereby, it's also included failure modes triggered during transportation and installation.

Finally, during the operational phase, the FMEA is now as SFMEA (System Failure Mode and Effect Analysis).

Mostly, FMEA focuses on component failure modes which impact on operational availability, but it's also important to take into account unsafe failures which lead to accident with personal or environmental consequences.

Depends on process consequences, FMEA is carried out based on risk assessment and evaluation and use risk matrixes which takes into account asset, safety, environmental and reputation consequences.

By the other way round, there's a variation of FMEA is known as FMECA which criticality index is defined based on frequency, consequence and detection. The criticality index is also defined as RPN number. Such index is defined by multiplying frequency, consequence and detection. Some example of FMECA will be given in the next section in detail and the advantages and drawback perform such methodology will be clear.

In fact, the higher advantages to applied FMECA is to take into account the failure detection when RPN number is calculated. In doing so, the hide failures are well assessed having an inspection program as well as detection devices to monitor them implemented.

When applying FMEA analysis is important to be aware about the different types of failure that are basically:

⇒ Failure on demands;

⇒ Hide failure;

⇒ Common cause failure;

⇒ Unsafe failure.

"Failure on demands" the equipment fails when is required to operate. A good example is standby equipment such as pumps, valves or compressors as well as safety functions which are required to work when unsafe condition happens.

In case of standby equipments, it's important to test them to certify it will be available when the main equipment fails or need some preventive maintenance. The maintenance team is responsible to monitor and test such standby equipment. The same approach is addressed to safety function equipment.

The **"hide failure"** occurs when it's not possible to detect easily equipment failure. That typical failure of safety equipment like SIF (Safety Instrumented Function) which there is more than one sensor to trigger logic element in order to final element (valve) be activated. In some cases, for example, there must be two of three signals (2003) to trigger the safety function. In this case if one sensor has spurred failure, there will be another two or three working properties which send the correct signal to logic element. Another example of hidden failure occurs in the tube and shell heat exchangers. In some cases, occurs obstruction in some tubes due bad water quality, but the equipment keeps their performance. During inspection, it's possible to detect some tubes obstructed.

The "**common cause failure**" is when one or more equipment has the similar cause of failure. In the case of oil field in refinery plants, when such oil has different specification, corrosion may occur in different equipments such as vessel, pipelines, towers. The common cause of corrosion in such different equipment is the oil with different specification that is more corrosive than usual and affect all equipment material.

The "**unsafe failure**" cause an unsafe condition in equipment and process. In industrial plants, some valves fail to close and can be bypassed without any problem. When the relief valve fails to close, it's an unsafe condition because it's not possible to release the pressure in a vessel. Some accident happens because the plants operate under unsafe condition.

In terms of safety and risk management, the unsafe failures are the most important type of failure despite many of them does not shutdown system but system operate in unsafe condition. Because many of unsafe failures do not shutdown system is very hard task for safety engineers to convince maintenance engineer to priories such inspection and maintenance on a layer of protection or protection component. Some unsafe failures are easier due the catastrophic consequence, but others are not because in many processes there are more than one layer of protection and operation and maintenance protection have not a clear risk perception of unsafe failure in one specific layer of protection or protection component.

In order to exemplify an unsafe failure in one specific system, there will be carried out an example of a pipeline. Basically, the toxic product is pumped through the pipeline. The Table **7** shows FMEA of pipeline. In red are highlighted unsafe failures.

In general terms, we can say that FMEA Disadvantages are:

- As qualitative analysis do not provide information about equipment reliability;

- In some cases, unsafe failures are not taking into account as it would to be because operational and maintenance characteristic of this analysis;

Table 7: Pipeline FMEA

Oil Company			**FMEA**		**Management: Process**	
System: Hydrogen Plant			**Subsystem: Nafta pipeline**		**Date: 30/07/2009**	
Draw Number: DE-22323-03			**Team: xxxxxxxxxxxxxxxxxxxxxxxxxxxxxxxxxxxxxxxxxxxxxxxx**			
Component	**Failure Mode**	**Causes**	**Effect to System**	**Other Effects**	**Detection**	**Recommendation**
Pipeline	- Weld fatigue	- High vibration	- Low flow	- Huge Leakage	- Visual inspection and low performance	R001 – Operate system under project specified conditions. Action by: Operation Group
	- Erosion	- High flow velocity	- Low flow	- Huge Leakage	- Visual inspection and low performance	R002 – Operate system under project specified conditions. Action by: Operation Group
	- External Corrosion	- Chemical attack	- Low flow	- Huge Leakage	- Visual inspection and low performance	R003 – Define correct material during the project phase. Action by: Design team
Joint	- Worn out	- Incorrect installation	- Flow reduction	- Small leakage	- Visual inspection	R004 – Follow up installation and apply procedures: Action by: Operation group
		- High vibration	- Flow reduction	- Small leakage	- Visual inspection	R005 – Operate system under project specified conditions. Action by: Operation Group
Gasket	- Worn out	- Incorrect installation	- Flow reduction	- Medium leakage	- Visual inspection	R006 – Follow up installation and apply procedures: Action by: Operation group
		- Age relate degradation	- Flow reduction	- Medium leakage	- Visual inspection	R007 – Follow RCM and RBI recommendations: Action by: Maintenance Group

The FMEA advantages are:

- FMEA is a very well defined Risk analysis and covers all equipment, component failure mode.
- Is possible to understand the failures to supply designers with improvement information as well as operational and maintenance professional with criticality of failure modes.
- FMEA provides information to maintenance and inspection plan that can be defined by RCM (Reliability Centered Maintenance).

2.7. RISK BASED INSPECTION (RBI)

The American Petroleum Institute (API) initiated a project named Risk-Based Inspection (RBI) in 1983. As a risk methodology, RBI is used as the basis for prioritizing and managing the efforts of an inspection program (American Petroleum Institute, API Practices 580 and 2002).

Some industries such as oil and gas, nuclear, chemistry face risk which poses in equipments such as vessels, heat exchangers, pipelines and valve which must be monitored and managed throughout their life cycle.

In order to maintain the risk under acceptable level during the operational phase, it's necessary to implement an inspection and preventive maintenance program which assures such equipments integrity.

An RBI program allows to define the prior inspection and maintenance as well as resources to maintain such risk at acceptable level based on preventive action which is implemented based on inspection, preventive and predictive maintenance.

The main goal of RBI is to maintain equipment integrity based on specific inspection and maintenance program. In fact, it's necessary to have a dynamic and flexible program to attend new modifications and demands to assure equipment integrity.

Traditional practices are based on rules and standard methods which defines specified interval of time to perform inspections throughout the whole equipment lifecycle. The condition based maintenance is the best approach because enables to predict the integrity of equipment and update the future inspections and NDT. Even though, this strategy must take into account the risk as index to prior action and equipments. In fact, different inspection strategies were developed in the last decades (Lee, 2006) as shown in Fig. **6**.

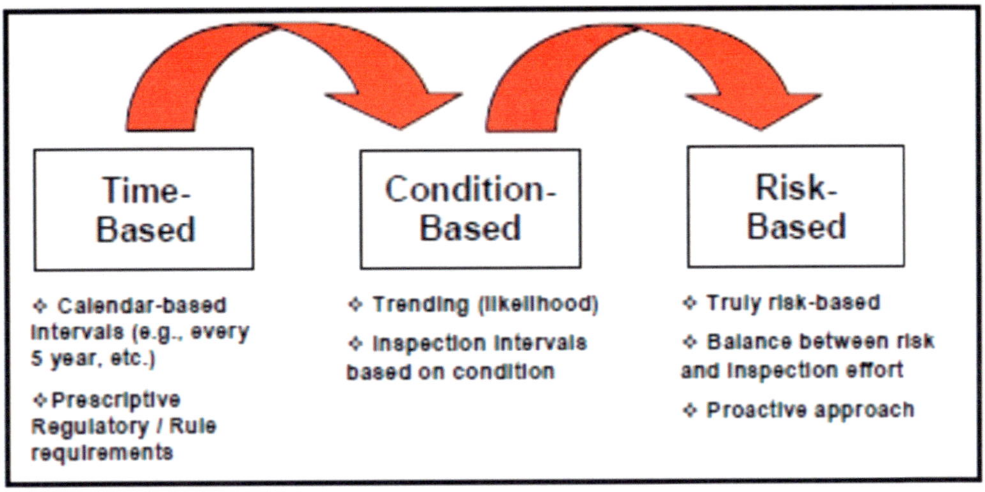

Figure 6: Evolution of Inspection and Maintenance Plan Strategies Author – Lee, 2006.

In fact, inspection is the first task to detect potential equipment failure. Therefore, it's very necessary to define the interval time to inspect equipments as well as the type of inspection. Depends on the situation, some visual inspection will be enough equipment to detect the failure. Mostly, when it's happening such equipment is degraded and the functional failure may happen in a relative short period of time. Therefore, additional approaches such as non destructive test (NDT) such as ultrasound and radiography must be carried out to define failures such as corrosion and erosion. Even though, such methods have some limitation and in some cases other methods are required.

In addition to defining the interval of inspection and NDT it's also important to establish the priority of such preventive action based on risk posed in such equipments such as vessels, pipelines, tanks, towers and valves.

The catastrophic accident which occurs by toxic product release, which pose on such equipments. Therefore, it's very important to carry out the RBI program for such critical equipment in order to maintain the acceptable risk level based on inspection and predictive maintenance tasks.

The RBI method can be applied based on qualitative and quantitative risk analysis methods. In the first case, in order to define the risk, the risk matrix which combines the frequency of failure with a measure of the consequences of such failure is applied. In the second case, quantitative methods can be applied to define the frequency of failures and it's consequence. Such methods will be discussed in Chapter 3 and 4.

The qualitative risk analysis methods which identify clear the hazards are FMEA and PHA. The quantitative methods which predict the frequency of failures and its consequence are Fault three analyses (FTA), Event Three Analysis (ETA) and Quantitative Risk analysis (QRA). In addition, methods which define preventive maintenance such as Reliability Centred Maintenance (RCM) and the legislation must also to be taken into account when RBI is performed. Fig. **7** shows the RBI information flow.

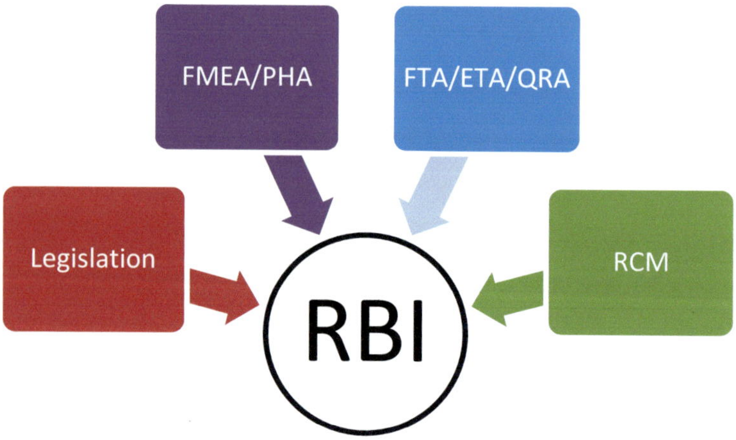

Figure 7: RBI information flow.

Based on qualitative approach, after hazard identification, inspection interval is defined to mitigate the risk based on such risk assessment. The Fig. **8** and Tables **8** and **9** below show an example of a risk matrix, frequency and severity rank used in RBI analysis.

Frequency	A	B	C	D	E
5	5A	5B	5C	5D	5E
4	4A	4B	4C	4D	4E
3	3A	3B	3C	3D	3E
2	2A	2B	2C	2D	2E
1	1A	1B	1C	1D	1E
	Severity				

Figure 8: RBI Risk Matrix.

The color references are:

The risk on the matrix is a combination of frequency and severity categories as shows on Table **8** and 2-9.

Table 8: Frequency Rank

Frequecy Qualification	Frequency	Rank
Very High: Failure is almost inevitable	< 1 Year	1
High: repeated Failures	1 in 3 years	2
Moderate: Occasional Failures	1 in 5 years	3
Low:Relatively few failures	1 in 10 years	4
Remote: Failure is unlikely	1 in 30 years	5

Table 9: Severity Rank

Severity Level	Severity Description	Rank
Very High	Fatalities or multi severe injuries	E
High	One fatality and/or severe injuries	D
Moderate	Severe Injuries	C
Low	Minor Injuries	B
Very Low	Insignificant Injuries	A

An example of RBI would be applied to define the inspection tasks for Distillation Plant. The first step is to define the failure mode for each system equipment such as distillation tower, vessels and pipelines. In order to define the failure modes, cause and consequence the FMEA analysis will be applied.

In addition, the qualitative risk assessment based on risk matrix will evaluate the risk for each failure mode.

In cases which the risk posses an unacceptable level a specific inspection task must be defined as shows Fig. **9**.

The risk policy defines action based on criterion below.

1. Red Color (High Risk – unacceptable = is obliged to reduce risk

2. Pink Color (Medium High Risk – Tolerable = is a advisable reduce risk if possible

3. Yellow Color → Medium Risk –Tolerable = maintain risk level

4. Green Color → Low Risk – Minor = monitor risk

Oil and Gas Company	**RBI (Risk based Inspection)**	**Management: Project Engineer**
System: Distillation System	**Subsystem: Distillation**	**Date: 20/01/2007**
Draw Number: DE-001.12.110.1	**Team: xxxxxxxxxxxxxxxxxxxxxxxxxxxxxxxxxxxxxxxxxxxxxxxxx**	

Component	**Failure Mode**	**Causes**	**F**	**Effect to System**	**S**	**R**	**Recommendation**
Distillation Tower	- Internal Corrosion	- Material out of specification	4	- Loss of performance in Tower	D	**4D**	R001– Perform inspection and preventive maintenance in each 5 years Action by: Maintenance
	-External Corrosion		4	- Toxic product spills - Damage to employee health	D	**4D**	
Vessel	- Internal Corrosion	- Material out of specification	4	- Loss of performance in Tower	D	**4D**	R002 – Perform inspection and hydrostatic test and NDT in each 5 years. Action by: Maintenance
	-External Corrosion	- Material out of specification	5	- Toxic product spills - Damage to employee health	E	**5E**	
Pipelines	- External Corrosion	- Material out of specification	5	- Toxic product spills - Damage to employee health	E	**5E**	R003 – Perform inspection and NDT each 5 years. Action by: Maintenance

Figure 9: Propylene System (RBI)

It's necessary to update the RBI during the asset life cycle and also review the inspection interval based on NDT results. Basically, The RBI methods must be applied during all asset life cycle starting in project phase by a qualitative or quantitative approach. In order to apply a quantitative approach, it's necessary reliable historical failure data. Whenever such reliable data is not available, a qualitative analysis is the further option. In fact, a qualitative analysis performed by experience technicians has more values than a quantitative analysis performed based on unreliable data. The important issue is to involve specialist even when quantitative methods are applied and review the RBI during the asset life cycle.

In order to perform RBI is necessary to have a team of specialist with technical knowledge and experience in the following areas:

- Risk Analysis Methods.
- Safety Process Hazards.
- Project, maintenance, operation and inspection.
- Legislation related to inspections and maintenance.

In addition, in order to perform the RBI analysis is also necessary resources such as:

- Room to carry out the meeting.
- Reliable information about historical failures and accident.
- Time available to perform analysis.
- Managers support.
- Risk analysis softwares licenses when it's necessary.

Likewise, others risk analysis methods, it's necessary to manage the RBI recommendation during the asset life cycle and it's requires managers support. In many cases, the RBI recommendation is in conflict of investment or priority with others recommendation and actions which must be clarified and eliminate. In fact, safety must be a priority during all asset life cycle. In order to avoid such conflict and facilitate RBI recommendation, their benefit must be communicated throughout different managerial organization levels. The final product of RBI are inspection and predictive maintenance tasks which achieve the main benefits such as:

- Define clear timing of inspection to mitigate the risk and prevent unsafe failures.
- Better identification of deteriorating process.
- Operational availability increasing by eliminating unnecessary inspections.

- Reduce insurance premiums.
- Reduce costs of legal action.
- In general terms, we can say that RBI disadvantages are:
- When not properly performed can lead to a high number of inspections which increase the operating cost without mitigating the risk.
- When not integrated with another program such as Maintenance and Integrity management can increase operational costs.

REFERENCES

American Petroleum Institute, API Recommended Practice 580. (2002). First edition, Risk-based Inspection.

Australian Standard, AS/NZS 4360. (2004). Risk Management.

A. Vozella, G. Gigante, L. Travascio & M. Compare. RAMS for aerospace: Better early or late than never. ESREL 2006.Safety and Reliability for Managing Risk –Taylor & Francis Group, London, ISBN 0-415-41620-5

Calixto, E. (2007). Sensitivity analysis in critical equipments: The distillation plant study case in the Brazilian oil and gas industry". ESREL 2007, Stavanger.

Calixto, E. (2007). The safety integrity level as hazop risk consistence. the Brazilian risk analysis case study. ESREL 2007, Stavanger.

Calixto, Eduardo. (2007). the integrated preliminary hazard analysis methodology regarding environment, safety and social issues. The platform risk analysis study. ESREL 2007, Stavanger.

Calixto, Eduardo. (2012). Gas and Oil Reliability Engineer: Modeling and Analysis. Imprint: Gulf Professional Publishing, ISBN: 9780123919144.

C. Ericson. (1999). Fault tree Analysis-A history. 17 ° International System Safety Conference, 1999, EUA.

Carson,Carl S. (2005). Fazendo da FMEA uma ferramenta de Confiabilidade Poderosa. SIC 2005.

Carson,Carl S. (2005). FMEA mais eficazes a partir das lições aprendidas. SIC 2006.

Crawl, A; Louvar, J. F. (2002).Chemical Process Safety Fundamentals with Applications. Prentice Hall PTR Upper Saddle River, New Jersey: 07458.

Duarte, M. (2002).Riscos industriais: Etapas para investigação e a prevenção de acidentes. Rio de janeiro: Funenseg.

Guidance for Risk Based Inspection. (1998). TWI/RSAE Proposal RP/SID/6306.

J. Dunjó, J.A. Vílchez & J. Arnaldos. (2006).Thirty years after the first HAZOP guideline publication. Considerations. Safety, Reliability and Risk Analysis: Theory, Methods and Applications Taylor & Francis Group, London, ISBN 978-0-415-48513-5.

K. Lee, C. Serratella, G. Wang R. Basu & R. Spong. (2006). Flexible Approaches to Risk-Based Inspection of FPSOs. 2006 Offshore Technology Conference held in Houston, Texas: Energo Engineering Inc.OTC 18364.

Marzal, Edward. M; Scharpf, Eric. (2002). Safety integration Level selection. Systematics methods including layer of protection Analysis. The Instrumentation, systems and Automation Society.

OHSAS 18001. (1999). Especificação do Sistema de Gestão e Saúde Ocupacional. Healthand and Safety Assessment Series.BSI.

Control of Major Accident Hazard Regulations. (1999). published by The Stationary Office. ISBN 01108 21920.

M J. Sobral; L.A. Ferreira. (2010). Development of a new approach to establish inspection frequency in a RBI assessment. Reliability, Risk and Safety. Taylor & Francis Group. London: ISBN 978-0-415-60427-7.

Risk Based Inspection (RBI): A Risk Based Approach to Planned Plant Inspection. (1999). Health and Safety Executive. Hazardous Installations Division. CC/TECH/SAFETY/8.
Risk Based Inspection: Development of Guidelines. (1991). General Document. The American Society of Mechanical Engineers (ASME), CRTD, Vol. 1. ISBN 0 7918 0618 9.
Risk Based Inspection Base Resource Document. (2000). API Publication 581, Preliminary Draft, American Petroleum Institute.
Simpson J. (2007).The Application of Risk Based Inspection to Pressure Vessels and Aboveground Storage Tanks in Petroleum Fuel Refineries. 5th Australasian Congress on Applied Mechanics, 12 December 2007, Brisbane, Australia.
Straub Daniel and Faber Michael Havbro. (2004). Computational Aspects of Generic Risk Based Inspection Planning. Swiss Federal Institute of Technology, ETH Zürich. ASRANet Colloquium, Barcelona.

Chapter 3: Quantitative Risk Analysis: Concepts and Methods

CHAPTER 3

Quantitative Risk Analysis: Concepts and Methods

Abstract: Quantitative risk analysis encompasses different types of methods to be applied during risk assessment. These include Fault Tree Analysis, Event Tree Analysis, Bow Tie, Safety Integrity Level, and Consequence and Effect Analysis. These methods are principally applied in situations where the level of risk is defined as intolerable by the application of qualitative methods. A quantitative approach is, therefore, required, in order to predict precisely the frequency of hazard occurrence, as well as the consequence and effect of accidents. So that the level of risk can be confirmed, and decisions that will mitigate such intolerable levels of risk to a tolerable level can be taken. Those quantitative methods are based on mathematical constructs such as Bayes theory, statistics, and engineering reliability concepts that are applied to equipment with safety functions. Despite the successful achievements obtained by applying such methods, improvement is still necessary in order that more realistic results in risk assessment can be achieved. Indeed, in many cases, the quantitative risk analysis methods are not fully applied to new projects or to real operational situations due to alleged lack of time and information. This chapter aims to describe the best quantitative risk analysis methods with examples, in order to clarify the advantages of applying such methods in the risk management process.

Keywords: Risk, ISO-Risk, Societal Risk, Fault Tree Analysis, Event Tree Analysis, Bow Tie Analysis.

3.1. INTRODUCTION

As already mentioned, quantitative risk analysis methods can predict the frequency and consequence of accident scenarios. The application of such methods requires a mathematical background, and in many cases may be performed by software packages because of their complexity and time requirement. The most usual quantitative risk analysis methods are described below.

3.2. FAULT TREE ANALYSIS (FTA)

This is a quantitative deductive method which identifies top events, and incidents or accidents or the combination of events that trigger top events. Such combinations are defined by logic gates, based on Boolean Logic, which define the combination of basic events. Such combinations can be represented by logic gate OR, or by logic gate AND. The main advantage of FTA is that a combination of events enables the cut sets to be defined. The cut sets are combinations of

Eduardo Calixto

events which trigger the top event. From a risk management point of view, once such cut sets are identified, it is possible to follow this with the prevention of accidents during a long life cycle of an enterprise.

3.3. EVENT TREE ANALYSIS (ETA)

This is a quantitative inductive method which identifies the initiating event, incident or hazard, and assesses the sequence of events that can trigger one or more accident scenarios. Such a combination is usually calculated by multiplying the initiating event frequency by the sequence event probability, resulting in the frequency of accident scenarios. The ETA does not calculate risk, It is therefore necessary to combine each frequency of the accident scenario result with the severity of accident consequence (*e.g.* The number of deaths), in order to calculate the risk. In some cases, Consequence and Effect Analysis (CEA) is used to calculate the numbers of deaths for each accident scenario, and such a number is combined with the frequency defined by ETA. The ETA is principally applied at the project phase in order to calculate the frequency of accident combined with the consequence of an accident.

3.4. LAYER OF PROTECTION ANALYSIS (LOPA)

This is a quantitative inductive method which identifies an initiating event, incident or hazard, and identifies the sequence of layers of protection that can be taken in order that the accident may be avoided. The main objective of LOPA is that the effect of the layers of protection is checked in order that accidents are avoided and a tolerable risk level is achieved. The concept of layers of protection is very important in terms of devices that can prevent accidents without human intervention being put in place. In some cases, human actions are considered as a layer of protection. However, whenever accident consequences are catastrophic, other layers of protection against such accidents are projected. Such a combination is usually calculated by multiplying the initiating event frequency by the sequences of layers of protection probability, resulting in frequency. The LOPA method as well as ETA does not calculate risk. It is therefore necessary to combine the frequency of accident scenario with the severity of accident consequence (*e.g.* The number of deaths) in order to calculate the risk.

3.5. THE SAFETY INTEGRITY LEVEL (SIL)

This is a quantitative deductive method which identifies the probability of failure on demand (PFD) that must be applied to one specific Safety Instrumented

Function (SIF) in order to mitigate risk to an acceptable level. SIF comprises at least one sensor, one logic element and one final element. In this way, such SIF can monitor the process condition and whenever an unsafe condition occurs, it is possible to bring the system to a safe condition. The PFD is associated with SIL numbers which vary from 1 to 4. The PFD should be lowered as much as possible in order to mitigate risk. In order to define the SIL level for SIF, there are four methodologies: Risk Matrix, Risk Graphic, Risk Reduction and Individual or Societal Risk. The Risk Matrix, which defines SIL is based on a combination of probability and consequence, each combination having a SIL number. The Risk Graphic defines risk based on the following criteria: consequence, presence of operational ground, the probability of the accident being avoided, and the demand for SIF. Such analysis is presented in a graphic from left to right with each criterion being defined according to a pre-established definition. The Risk Reduction method considers the relation between accident frequency and tolerable frequency. The Individual and Societal risk method also consider the relation between accident frequency and tolerable frequency, but in this case tolerable frequency takes into account individual or societal risk criteria as well as the probable number of deaths.

3.6. THE BOW TIE ANALYSIS (BTA)

This is a quantitative (or qualitative) deductive method which identifies the causes and consequences of incidents as well as the control and recovery measures. This method defines the combination of causes of accidents as well as the sequence of events that results in accident scenarios. As a concept it can be understood as a combination of the FTA on the left with ETA on the right. It is possible to calculate mathematically the accident frequency or probability of each event and their combination. Such solutions are easier to obtain when supported by the software. BTA is a good tool to aid understanding of incident analysis and is also a very good tool with which to manage risk once the combination of incident causes, control and recovery measures as well as consequences has been established. This method can be applied in all the phases of an enterprise.

3.7. CONSEQUENCE AND EFFECT ANALYSIS (CEA)

This is a quantitative deductive method which defines the vulnerable area based on the consequence of accident and the expected number of deaths. This method is based on the mathematic methods of calculating gas and liquid leakage from pipelines, tanks, vessels and takes into consideration the environment conditions around the accident scenarios. In most cases software is used to calculate the vulnerable accident area, and some software is able to calculate individual risk, the ISO-RISK curve, and societal risk when the frequency of accident is input. This methodology is principally applied at the project phase, but can also be applied at the operational phase in order to support decisions.

Before applying risk analysis methods, it is necessary to understand well the different concepts of risk. In general terms, risk concepts can be understood by a combination of the possibility of a specific hazard occurrence, and its consequences expressed by injuries or deaths in a specified time. Risk can also be assessed by a quantitative approach and in this case Individual and Societal Risk is the most appropriate and frequently used method.

Individual risk is frequency of death per year for people located in a vulnerable area. The individual risk index is represented by the ISO-Risk curve or the As Low as Reasonably Practicable (ALARP) limits.

The ISO Risk curve is a graphic representation of a vulnerable area which an individual or population are exposure to some accident consequence considering the value of individual risk. In some countries, there is a risk criterion to project which depends on an individual risk value, for example $1x10^{-6}$. The contour curve cannot achieve an unacceptable region in the presence of community as shown in Fig. **1**.

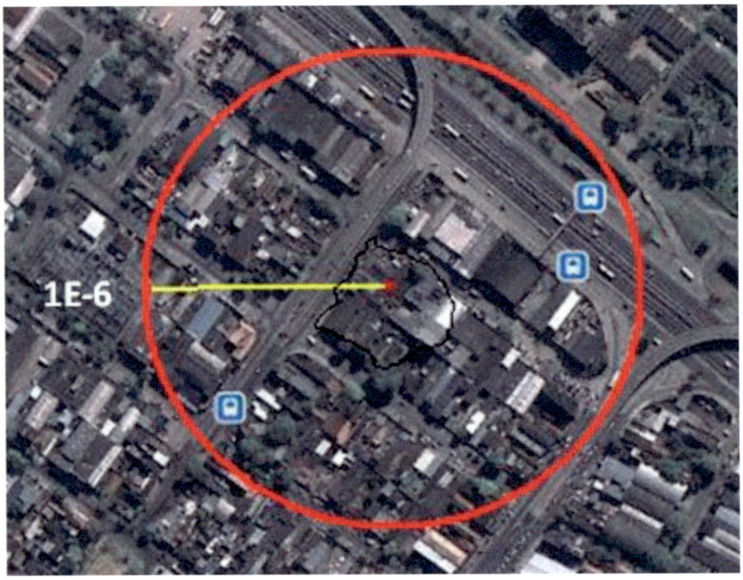

Figure 1: ISO-Risk Curve (Source: Calixto, 2011).

The ALARP factor is also principally used in project analysis as a risk criterion. A crucial point in this criterion is the consideration of how much a risk must be mitigated below an acceptable region. There is an investment implication and in most cases it is not clear how much return such additional investment gives in term of safety. Fig. **2** shows the ALARP risk tolerability.

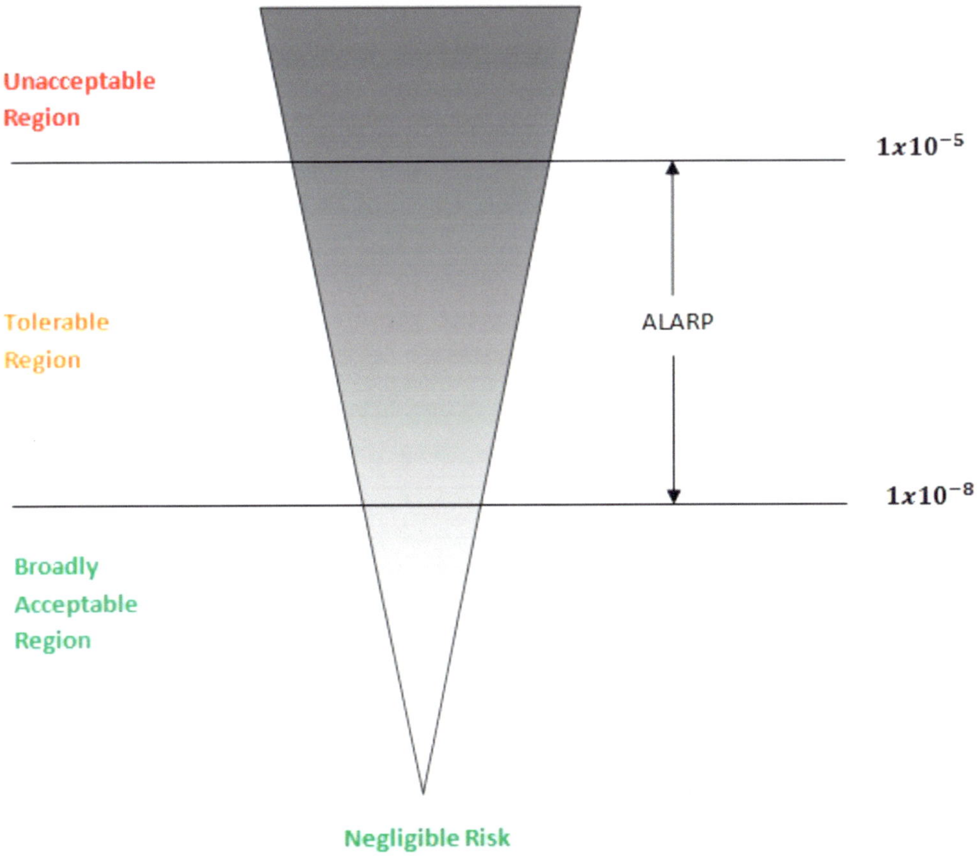

Figure 2: ALARP Individual Risk (Netherland).

Individual risk is calculated by the sum of all risks of each accident scenario in a plant facility, and is expressed in terms of the number of deaths per year. In order to define the number of deaths in each accident scenario, it is necessary to carry out a CEA in order to predict such a number based on the effect on employees in a vulnerable area of the accident. Such a calculation considers the consequences (radiation, toxic level, pressure wave) and tolerance that is defined by PROBIT equations. This will be discussed in more detail with examples of a CEA.

Remarkably, in some countries only societal risk is a decision criterion for acceptance of new projects; individual risk is not. Regardless of how many deaths occur during a plant installation, if such an accident does not affect the community outside the plant, the project is accepted. From a safety point a view, this makes no sense and means that projects with a low level of safety are accepted by authorities when they are located in places where no community is present, or where there will be no significant effect on the community.

Furthermore, in individual risk calculation in most of cases, such risk is considered independent of time. That means the calculated risk remains constant for a long period of time. This is unrealistic because initiating events are mostly equipment failures that are dependent on time and are better represented by cumulative density function distribution. Consequently, the probability of failure increases over a long period of time, and this results in the associated risk increasing over a period of time. In order to keep risk at an acceptable level, inspection and preventive maintenance are required so that failure in layers of protection and equipment with possible unsafe failures may be detected.

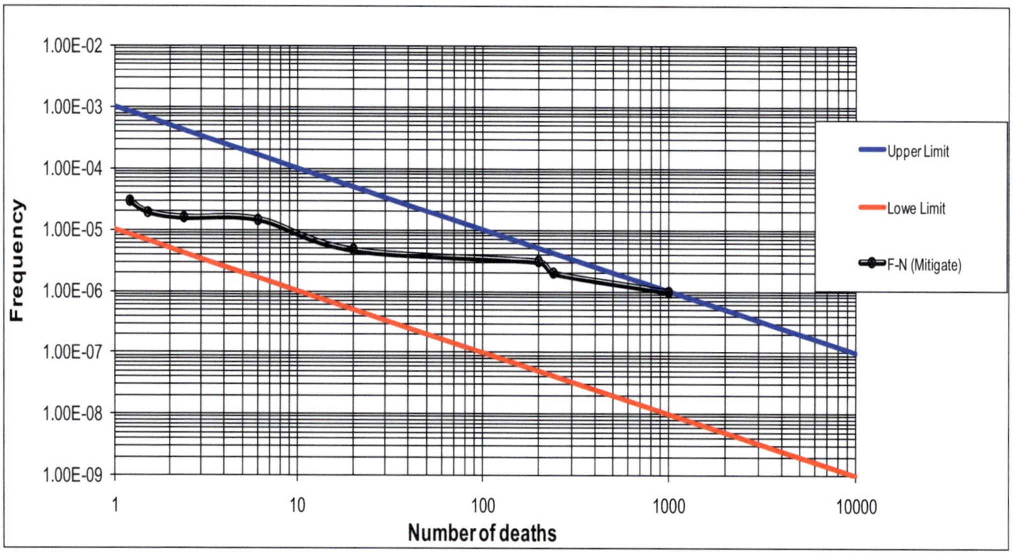

Figure 3: Societal Risk.

Societal risk is the frequency of death per year, which a community outside the industrial area are exposed to accident consequence. Societal risk is usually represented by the F-N curve that shows the cumulative expected number of fatalities on each frequency level. Such a curve represents the combination of the expected number of deaths and the frequency, and is thus a cumulative curve, which takes into account all the hazard scenarios from one or more specific hazard sources in the plant facility which may affect the community outside the plant facility area. The ALARP is also defined in F-N curve in order to define when the societal risk is acceptable or not. A high level of reliability on layers of protection and equipment can also help to mitigate risk as well as the implementation of preventive maintenance in order to keep a high level of reliability and availability of such devices. Fig. **3** shows an example of an F-N curve within the ALARP region.

Two of the most important concepts in quantitative risk analysis are the probability and frequency of an event. The probability of failure is the inverse of reliability. Reliability is the probability of equipment, products or services operating successfully over a specific period of time, and is the mathematical complement of the probability of cumulative failure. Thus the equation below represents the relation between cumulative failure and reliability, which are complementary. In other words, if the two values are added together, the result is 100% (or 1).

$$F(T) = 1 - R(t)$$

Fig. **4** shows an example of a cumulative probability density function which represents a 31% chance of an accident over 3.8 years.

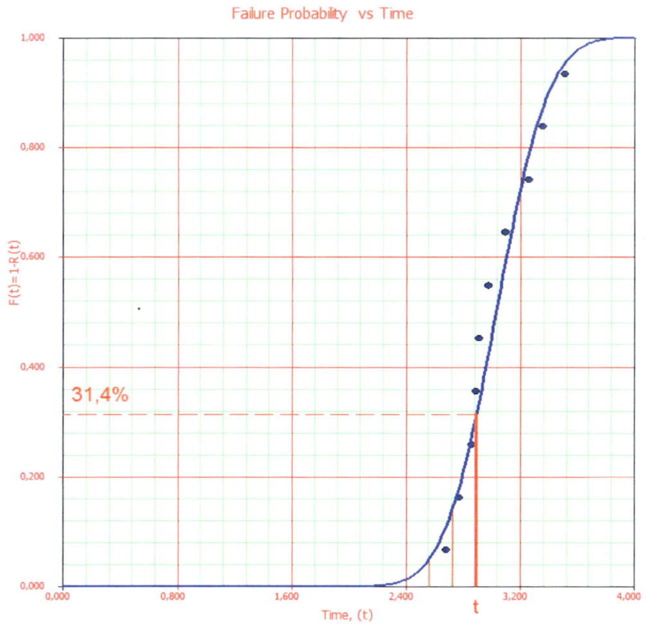

Figure 4: Accident probability of time.

The most simple probability density function that is applied in reliability and risk analysis, and is exponential and based on such functions as the cumulative probability of failure, is:

$$F(T) = 1 - e^{-\lambda t}$$

where

λ = constant failure rate

Indeed, the constant failure rate is only one assumption because the probability density function can represent different functions such as normal, Gumbel, Lognormal, Weibull, Gama, logistic, Generalized Gama, *etc.* In fact, the probability density function definition depends on historical failures, and to define which one of them fits better in such data it is necessary to perform a lifetime data analysis which is not the objective of this book. However, it is important to have such information in mind because it affects the risk calculation. In addition, it is necessary to understand the concept of failure rate that is defined by the relation between PDF and reliability functions as shown in the equation below.

$$\lambda(t) = \frac{f(t)}{R(t)}$$

The failure rate will be constant only in an exponential PDF case as shown in the equation below

$$\lambda(t) = \frac{f(t)}{R(t)} = \frac{\lambda e^{\lambda t}}{e^{\lambda t}} = \lambda$$

In addition, in an exponential PDF case, the MTTF will be the inverse of the failure rate as shown in the equation below.

$$MTTF = \int_0^t t.f(x)dx = t\int_0^t \lambda e^{\lambda t} dt$$

$$MTTF = t.\frac{\lambda}{\lambda^2 t} = \frac{1}{\lambda}$$

3.8. FAULT TREE ANALYSIS (FTA)

Fault Tree Analysis (FTA) was originally developed in 1962 at Bell Laboratories by H.A. Watson, under a U.S. Air Force Ballistics Systems Division contract to evaluate the Minuteman I Intercontinental Ballistic Missile (ICBM) Launch Control System.

This method is mainly used in the fields of safety engineering and reliability engineering to understand the events combination which leads to systems equipment and components fail.

FTA is a deductive quantitative risk analysis method which defines the combination of events which trigger a top event. The first step is to define top events, and then the main event (intermediary and basic) and logic gates necessary in order to calculate top event probability. Top events usually consist of an accident or equipment failure, and the combination of events from the top event down to basic events is assessed. In order to calculate the top event probability based on the combination of intermediary and basic events, it is necessary to consider Boolean Logic as follows:

$P(A) \cup P(B) = P(A) + P(B) - P(A)*P(B)$. That is graphically represented by the whole areas in the figure below:

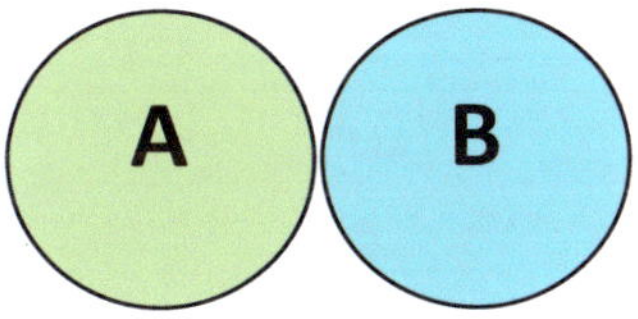

OR

$P(A) \cap P(B) = P(A)*P(B)$, That is graphically represented by the interception area in the figure below:

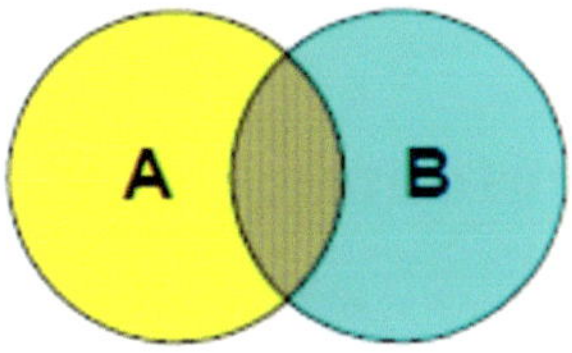

The Fault Tree can comprise a combination of more than two events and it is advisable to calculate two by two, *i.e.* a probability resultant of two combinations combined with another event and so on. The combinations of such events are represented by logic gates as follows:

The Top Event triggered is necessary to satisfy the combination of all the events below the top event.

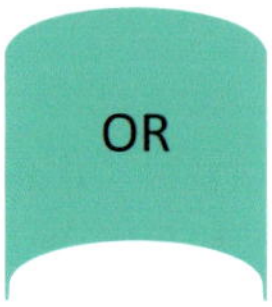

Logic Gate OR: One of the events below this gate must happen to trigger the logic gate OR.

Logic Gate AND: All of the events below this gate must happen to trigger the logic gate AND

Stand By Event: All active and passive events must happen to trigger the fault in the standby event.

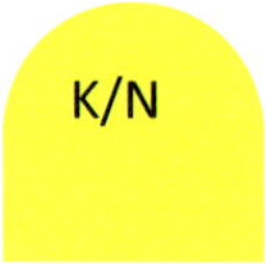

K/N Event: This gate is triggered when more than K of N events happen.

Basic Event: That is the last event on FTA.

Exclusive OR Gate: This gate is triggered when one of the events happen.

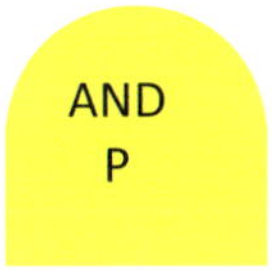

Priority AND Gate: This gate is triggered if all the events below happen in a defined sequence.

The FTA is called static when it comprises simple logic gates like OR and AND. The FTA is called dynamic when it comprises more complex logic events like Exclusive OR, Priority AND, and Inhibit.

The logic gates represented in a dynamic FTA in most cases are very rare and consequently there are few data available. Such gates can be represented at a higher level that turns a dynamic FTA into a static FTA.

An additional and more important issue concerning FTA is the relationship between the basic event and time. When probabilities of the basic event are described by constant probability the FTA is independent of time. By contrast, when probabilities of the basic event are described by cumulative density function, the FTA is dependent on time.

FTA that is dependent on time are more realistic than FTA that are independent of time, but it is harder to model them because it is necessary to define the cumulative density function for each basic event.

In terms of safety, the time dependent FTA is more accurate because it demonstrates how top event probability increases with time, *i.e.* the chance of an accident happening during a specific time. Because such a calculation may be quite complex an alternative approach is the use of software to model and simulate time dependent FTA.

3.8.1. Time Independent Fault Tree Analysis

Time independent FTA is when probabilities of basic events are constant. Whatever the probabilistic characteristics, the Fault Tree is built from the top event to basic events and takes account of event combinations. FTA is simpler than Reliability Diagram Block (RDB) in terms of representation of the top event analysis, but both representations show the same result regarding opposite logic. A simple example of the FTA and RDB is represented by a simple SIF which comprises the initial element (sensor), the logical element and the final element (valve) as shown in Fig. **5** A. Fig. **5** B represents SIF RDB that is inverse logic from FTA and shows similar results.

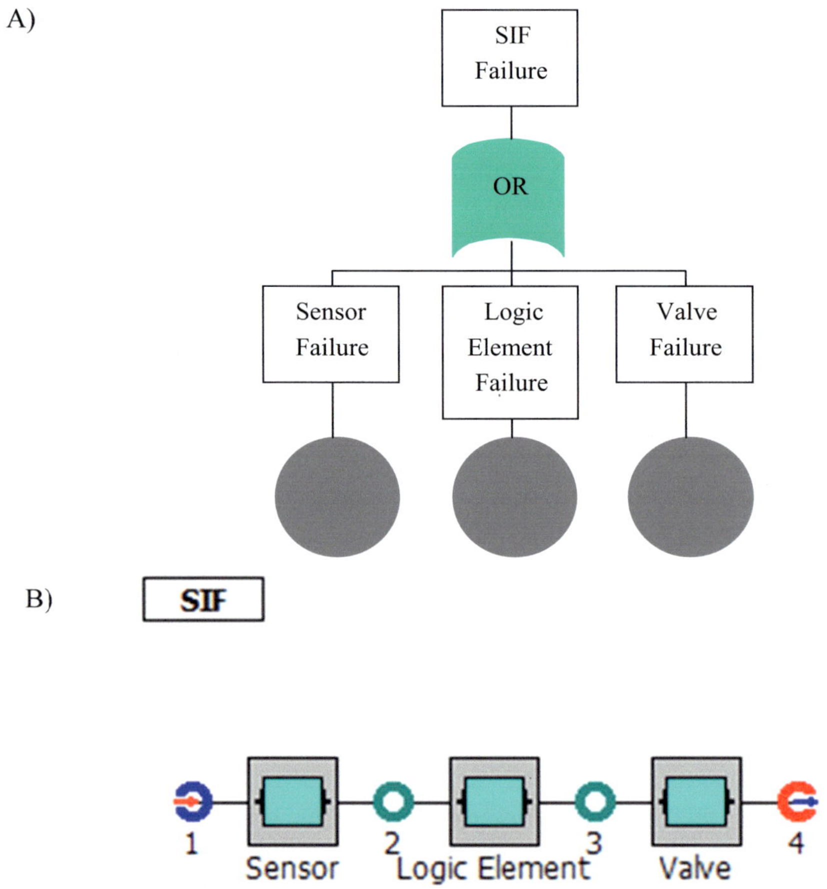

Figure 5: Fault Tree x Reliability Diagram Block (FTA/RBD -BQR Software).

Thus, when regarding the probabilities of failures, we have:

The Probability of SIF failure on the Fault Tree diagram:

P(Sensor) = 0, 3

P(Logic Element) = 0, 2

P(Valve) = 0, 1

P(SIF Failure)= P(Sensor) $\cup$ P(Logic Element) $\cup$ P(Valve)

R1 = P(Sensor) $\cup$ P(Logic Element) = (P(Sensor) + P(Logic Element)) – (P (Sensor) x P (Logic Element)) = (0, 3+0, 2) -(0, 3x0, 2) =0, 5-0, 06=0, 44

R1$\cup$P(Valve) = (R1 + P(Valve)) – (R1 x P(Valve))=(0, 44+0, 1)-(0, 44x0, 1)=0, 54-0, 044=0, 496

3.8.2. Time Dependent Fault Tree Analysis

The time dependent FTA considers the cumulative probability function for basic event value, thereby, the probability will increase with time. Therefore, in most cases, as long as the time pass, higher is the chance of failure unless some preventive maintenance takes place. The Fig. **6** shows the similar safety instrumented function demonstrated in Fig. **5** A. The top event is the safety instrumented function failure considering time dependent FTA.

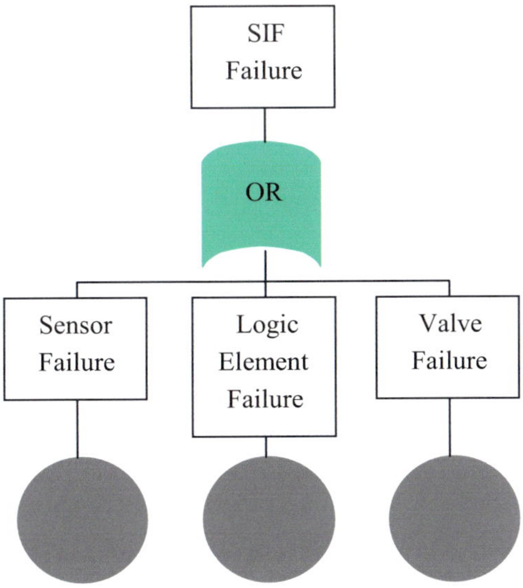

Figure 6: Safety Instrumented Function (FTA).

In terms of the failure rate of each basic event, the probability of each safety instrumented function varies with time. Thus, it is possible to calculate, for example, what is the probability of a safety instrumented function within 1.5 years. The exponential CDF for all events takes into consideration:

E0 = Sensor failure

E1 = Logic Element Failure

E2 = Actuator failure

The probability cumulative functions for each event are:

$$P(E0)(t)=1-e^{-\lambda t}=1-e^{-0{,}0002\,t}$$

$$P(E1)(t)=1-e^{-\lambda t}=1-e^{-0{,}00004\,t}$$

$$P(E2)(t)=1-e^{-\lambda t}=1-e^{-0{,}0002\,t}$$

After 1.5 years (13140 h) the probability values are:

$$P(E2)(t)=1-e^{-\lambda t}=1-e^{-0{,}0002\,t}=1-e^{-0{,}0002\,(13140)}=0{,}91$$

$$P(E1)(t)=1-e^{-\lambda t}=1-e^{-0{,}00004\,t}=1-e^{-0{,}00004\,(13140)}=0{,}37$$

$$P(E0)(t)=1-e^{-\lambda t}=1-e^{-0{,}0002\,t}=1-e^{-0{,}0002\,(13140)}=0{,}91$$

Thus, calculating the gates' resultant probability we have:

P(SIF failure) = P(E0) $\cap$ P(E1) $\cap$ P(E2) = (P(E0) + P(E1) + P(E2) – P(E0 $\cap$E1) – P(E0 $\cap$E2) – P(E1 $\cap$E2) + P(E0$\cap$ E1$\cap$E2= (0.91 + 0.37 + 0.91) – (0.91 x 0.37) – (0.91 x 0.91) – (0.37 x 0.91) + (0.91 + 0.37 + 0.91) = (2.19) – (0.336) – (0.828) – (0.336) + (0.306)=0.99

3.9. EVENT TREE ANALYSIS (ETA)

The Event Tree Analysis main objective is to predict the frequency or probability of an accident considering a sequence of events. Basically, such prediction is calculated by multiplying the frequency of the initial event by the probability of failure of a layer of protection and control measures. In ETA, the model is built up from left to the right, beginning from the initial event to sequence events. A good example is a blowout accident at an oil well. Basically, the blowout initiates due to loss of circulation, no equipment supply and overpressure. Indeed, the kick phenomenon happens when the pressure information is higher than inside well. Whether the kick is not controlled the blowout accident will happen. The Event Tree in Shown in Fig. **7** represents such sequence of event considering also the combination of cause which trigger a kick event and also the combination of event which lead to a loss of kick control. Both combinations are represented by FTA. In fact the ETA encompass only the kick and kick control events having as result the well under control or the Blowout. In some cases, hybrid analysis encompasses more than one method to give a better estimations. In Fig. **7** the hybrid model encompasses the ETA and ETA.

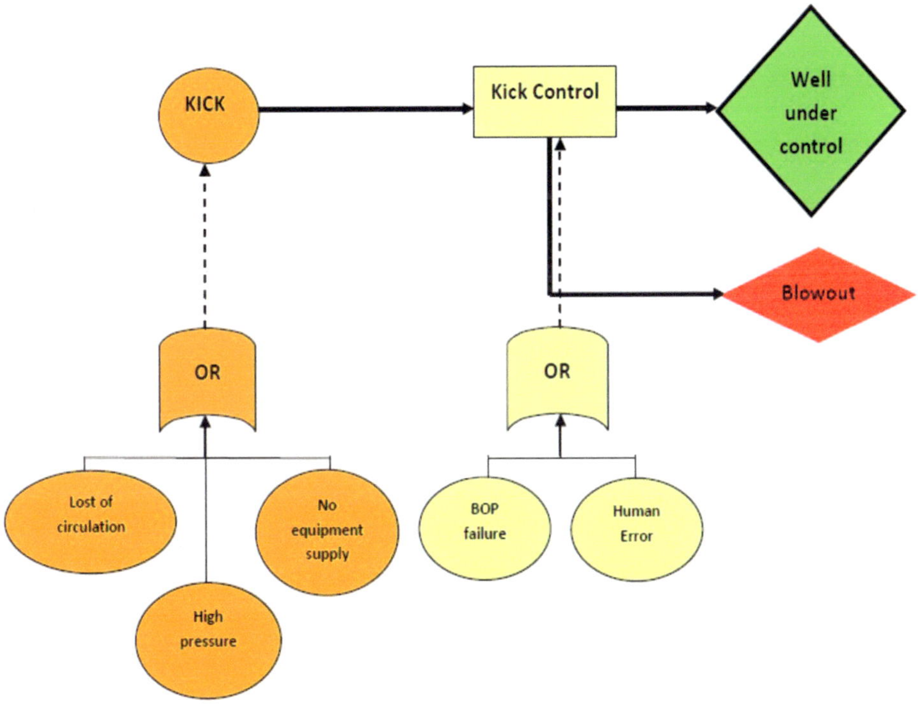

Figure 7: Event Tree (Blowout).

3.9.1. Independent of Time ETA

The independent of time ETA considers the values of probability and frequencies constant, and that is applicable in most risk analyses. This does not, however, take into account the degradation of equipment. Based on Blowout ETA demonstrated in Fig. **8** the frequency of the results of accidents is calculated by multiplying the initiating event frequency by each branch event probability.

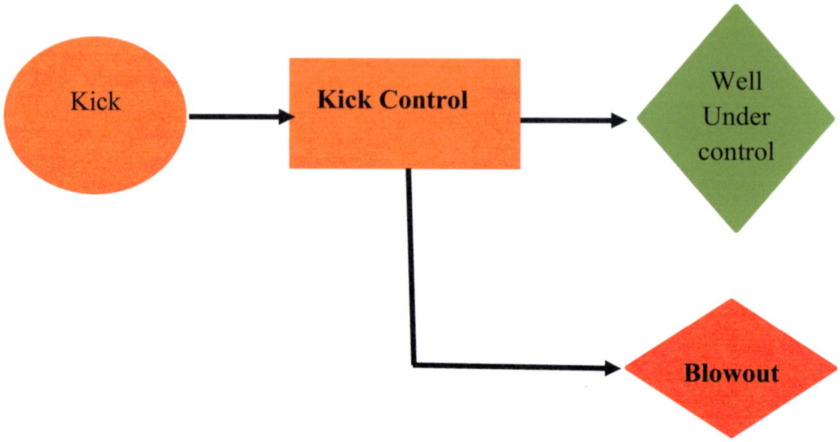

Figure 8: Event Tree (Blowout).

Thus the frequencies of Blowout are:

f(Blowout) = f(Kick) x P(Kick control) = (1x10-5) x (0,8) = **9,6 x 10-7**

In fact, it's also possible to calculate the probability of Blowout considering the FTA results represented in Fig. **7**.

The Kick occurrence depends on a loss of circulation or high pressure or no equipment supply on a well. Such an event combination is represented by the equation:

P(Kick) = P(Loss of Circulation) U P(High Pressure) U P(No Equipment Supply) = P(Loss of Circulation) + P(High Pressure) + P(No Equipment Supply) – (P(Loss of Circulation) x P(High Pressure)) – (P(Loss of Circulation) x P(No Equipment Supply)) – (P(High Pressure) x P(No Equipment Supply)).

To have no kick control, it is necessary for BOP failure or human error. This is represented by the equation:

P(Kick Control) = P(BOP Failure) U P(Human Error) = P(BOP Failure) + P(Human Error) – ((P(BOP Failure) x P(Human Error)).

The probability of the wellbeing under control is calculated by event three and is mathematically represented by the equation:

P(Well under Control) = P(Kick) x P(Kick Control)

The complementary event is the probability of a blowout occurring.

P(Blowout) = P(Kick) x (1-P(Kick Control))

Such type of calculation was demonstrated in item 3.2 considering the SIF example. In fact the frequency is a more used index rather probability because enables to calculate risk considering the frequency of deaths per year.

3.9.2. Dependent on Time ETA

Time dependent ETA considers considers the cumulative probability function for probability and frequency value calculation, thereby, the probability will increase with time and frequency may remain constant or increase with time. It is a more realistic probability model because it represents equipment degradation. Based on the example of Blowout, the Time dependent ETA takes into account the frequency of Kick on time as well as the probability to kick control. The Kick event can be represented by a Gumbel PFD (μ=20, σ=2). Fig. **9** shows kick event rates increasing along a time axis.

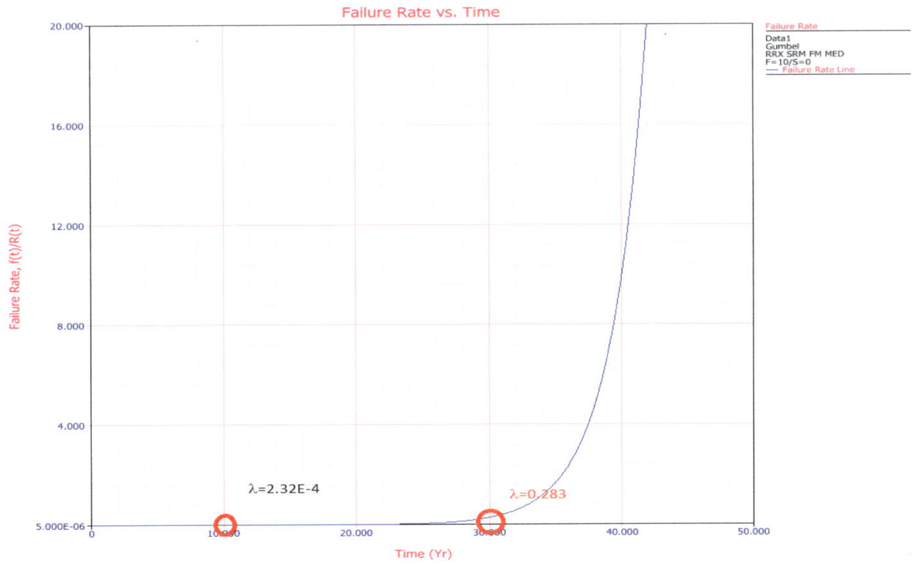

Figure 9: Kick Event Failure Rate Function.

Based on Fig. **9** above, there are two values of Kick event rate. In 10 years the kick event rate is $2.32x10^{-4}$ and in 25 years, it is 0.283. Applying those values shown in Fig. **9**, to an ETA we have two cases:

Case 1 (10 years) - Frequency of Kick = λ(10) = $\mathbf{2.32x10^{-4}}$

f(Blowout) = f(Kick) x P(Kick control) = $(2.32\text{x}10^{-4})$ x (0,8) = 1,85 x 10^{-4}

Case 2 (25 years) - Frequency of Kick = λ(30) = 0,283

f(Blowout) = f(Kick) x P(Kick control) = (0.283) x (0,8) = **0.2264**

In 10 years, the frequency of the blowout is very remote, based on the risk matrix frequency classification shown in Fig. **10**, and the risk of accidents is moderate, but after 30 years, the risk level of the blowout is not tolerable.

		Frequency Category					
		A (Extremely Remote)	**B (Remote)**	**C (Little Frequent)**	**D (Frequent)**	**E (Very frequent)**	**F (Extremely frequent)**
		At least 1 between from 1000 to 100.000 years	At least 1 between from 50 to 1000 years	At least 1 between from 30 to 50 years	At least 1 between from 5 to 30 years	At least 1 between from 1 to 5 years	At least 1 in 1 years
Severity Category	**IV**	M	NT	NT	NT	NT	NT
	III	M	**Blowout**	NT	NT	**Blowout**	NT
	II	T	T	M	M	M	M
	I	T	T	T	M	M	M

Figure 10: Frequency of Blowout from 10 years to 25 years on Risk Matrix.

The main advantage of a time dependent ETA approach is that it enables an assessment of risk levels as they vary in time, and this supports the performance of preventive actions in order to avoid accidents.

The reliability of safety function equipment such as BOP (Blowout preventer) as well as preventive maintenance and inspection has a very important hole to prevent accident. In fact, asset integrity management, which encompasses reliability, maintenance, risk and human factor must be implemented a long critical asset life cycle in order to maintain the risk under acceptable level and avoid accidents.

The challenge to achieve high reliability begins during concept phase and go through the whole asset life cycle.

3.10. LAYER OF PROTECTION ANALYSIS (LOPA)

The LOPA main objective is to predict the frequency or probability of an accident considering a sequence of events. Basically, such prediction is calculated by multiplying the frequency of the initial event by the probability of failure of a layer of protection.

Like many other risk analysis methods, the primary purpose of LOPA is to determine if there are sufficient layers of protection against an accident scenario, in other words, to keep risk at a tolerable level. In order to keep risk at an acceptable level, more than one layer of protection must be put in place in order to achieve the risk target and to reduce vulnerability. From a preventive point of view, whenever it is possible, it is advisable to put in place layers of protection which achieve a tolerable level of risk and not to rely on layers of protection which minimize the accident effect and consequently minimize the risk.

Some examples of preventive layers of protection are rupture disks, relief valves, SIFs and even operator action. There are some layers of protection which minimize accident effects like blowout Preventers, areas around tanks to contain oil in case of spill, walls around operational areas to contain toxic product releases and even windows which support pressure waves in case of explosion. Fig. **11** below shows the layer of protection concept as a means of preventing accidents or reducing accident consequences.

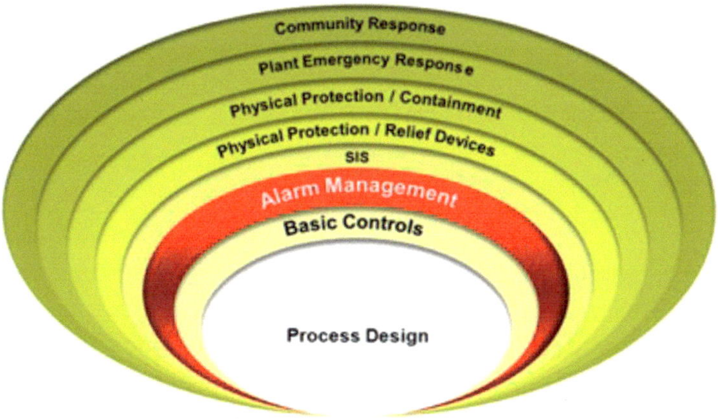

Figure 11: Layer of Protection (Source: www.isa.org, 2014).

Similar to previous quantitative risk analysis methods, LOPA can also take into account the constant probability for layers of protection and constant failure rate values of the initiating event, or cumulative density function values of the layer of protection and the rate function or the cumulative density function of the initiating event having different values of failure rate or probability over time. In the first case, the constant failure rate and probability values are carried out by static layer protection analysis. In the second case, the failure rate and failure probability vary over time and are assessed by dynamic LOPA, as will be shown in the next section.

3.10.1. Independent of Time Layer of Protection Analysis

The independent of time LOPA considers the values of layers of protection probability and initiate event frequencies (or probability) constant.

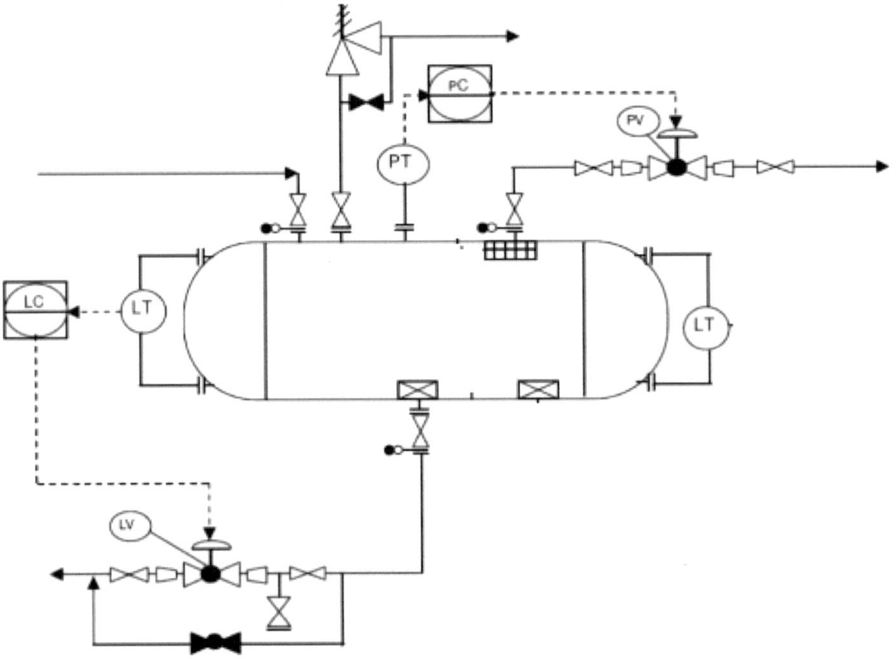

Figure 12: Layer of Protection of Vessel.

Therefore, the final accident will have a constant frequency or probability value independent of time. An example of LOPA is an overpressure inside a horizontal vessel shown in Fig. **12**. In this case, a high flow of product is sent to the horizontal vessel and the pressure must be controlled in order to avoid leakage. In this way, there are three layers of protection. These are a pressure control, safety instrumented function (SIF 1 = PT Sensor + PC Logic element + PV actuator + Valve), the level control, safety instrumented function (SIF 2 = LT Sensor + LC Logic element + LV actuator + Valve) and the relief valve. In order to calculate the frequency of accident, it's necessary to define the values of high flow frequency and layers of protection probability that are:

- High flow (f=1);
- SIF 1failure (P=$1x10^{-1}$);
- SIF 2 failure (P=$1x10^{-1}$);
- Relieve Valve (P=$1x10^{-3}$).

In case of failure of all the layers of protection, the overpressure will trigger a leakage or even an explosion. Based on the values of the initiating event rate and layers of protection failure probability, we calculate the frequency of furnace explosion as follows:

f(vessel overpressure) = f(High Flow) x P(SIF 1) x P(SIF 2) x P(Relieve Valve)

f(vessel overpressure) = (1) x ($1x10^{-1}$) x ($1x10^{-1}$) x ($1x10^{-3}$) = $1x10^{-5}$

The LOPA can be performed based on a different approach such as a template which each value of layer of protection is recorded and the vessel overpressure failure is calculated as shown in Fig. **13**.

3.10.2. Dependent on Time Layer of Protection Analysis

The dependent of time LOPA considers the values of layers of protection probability, not constant, based on the cumulative density function values for each specific time. The frequency of initiating event is based on event rate function which may be constant or not depend on event characteristics.

Therefore, the final accident frequency will depends on time. In fact, the layer probability of failure may also vary over time because such devices getting older increasing the chance of failure. An example of Dependent on time LOPA is horizontal vessel overpressure. Let's consider the similar layer protections defined in item 3.4.1 such as:

- High flow (PDF exponential: MTTF=1)
- SIF 1 Failure (PFD Normal:μ=5;σ=0,5)
- SIF 2 failure (PFD Normal:μ=5;σ=0,5)
- Relieve Valve (PDF Gumbel: μ=10;σ=2)

Accident Scenario Number	**Equipment**		
1	Vessel		
Date			
04-02-15			
Consequence description	Leakage of toxic product		
Tolerable Risk Criterion	**Tolerable Risk**	**Individual risk**	
		1x E-4≥IR≥1x E-6	
Trigger Event	High product flow		
LOPA events	**Event**	**Probability**	**Frequency**
Accident Condition	N/A	N/A	N/A
Initiate Event	High flow	N/A	1
Layer of protection	SIF 1	$1x10^{-1}$	N/A
	SIF 2	$1x10^{-1}$	N/A
	Relieve Valve	$1x10^{-3}$	N/A
Total Frequency or probability of accident		N/A	$1x10^{-5}$
Risk IS tolerable ?		Yes	No
		X	
Other layer of protection is required ?		What ?	Prob or Freq
Previous frequency or probability of accident		**Probability**	**Frequency**
		N/A	N/A
New Layer of protection		N/A	N/A
New total Layer of protection		N/A	N/A
Risk IS tolerable ?		Yes	No
		X	

Figure 13: Template Layer of Protection Analysis (Vessel).

In order to demonstrate the difference of the risk on time, two cases will demonstrated considering 2.5 years and 4.5 years such as case 1 and 2 respectively.

Case 1 – 2.5 years

f(vessel overpressure) = f(High Flow) x P(SIF 1) x P(SIF 2) x P(Relieve Valve)

f(Vessel Disrupt) = (1) x (0,008)x (0,008) x (0,0105) = $6.72x10^{-7}$

Case 2 – 5 years

f(vessel overpressure) = f(High Flow) x P(SIF 1) x P(SIF 2) x P(Relieve Valve)

f(Vessel Disrupt) = (1) x (0,95) x (0,95) x (0,046) =$4{,}15x10^{-2}$

The difference in risk between case 1 and case 2 is significant. In the first case, the risk is moderate and on the second case is not tolerable based on the risk matrix defined in 3.10. In order to mitigate the risk its necessary the asset integrity management program which encompasses reliability, maintenance, risk management and human factor issues. The Asset integrity management will be discussed in chapters 7 and 9.

3.11. SAFETY INTEGRITY LEVEL (SIL)

There are three principal standards that are used worldwide: ISA S84.01, IEC 61508 and IEC 61511. The standard ANSI/ISA S84.01 "Application of Safety Instrumented Systems for the Process Industry" was the first published by the three (ANSI/ISA [1997]) in 1996, although the committee in charge of writing ICE 61508 was already working in that direction during the same year. This standard appeared for addressing the application of Safety Instrumented Systems, with an approach that puts attention on the field devices as well as the logic solver as integral components of the SIS, specifically in the realm of the process industry. That means that this document is an industry-specific standard (Echeveria, 2003).

SIL analysis is a semi-quantitative methodology used to define whether it is necessary to implement SIF as a layer of protection in a process in order to achieve an acceptable risk level (Calixto, 2012).

In SIL analysis, the failure on demand is represented by a SIL number from 1 to 4. The higher the SIL number, the lower the probability of failure on demand as shows Table **1**.

Table 1: SIL Classification (Source: Schwartz, 2002)

SAFETY CLASS	PFD	SIL
I	≥ 10-1	0
II	≥ 10-1	0
III	≥ 10-2 - < 10-1	1
IV	≥ 10-3 - < 10-2	2
V	≥ 10-4 - < 10-3	3
VI	≥ 10-4 - < 10-3	3
X	≥ 10-5 - < 10-4	4

Each SIL number is related to one SIF, and each SIF comprises a minimum of one initiating element (sensor), one logical element and one final element (valve). In fact, SIF can comprise more than one such element. At a higher level, there is the Safety Instrumented System (SIS) which comprises more than one SIF. The Fig. **14** show an example of two SIF in horizontal vessel.

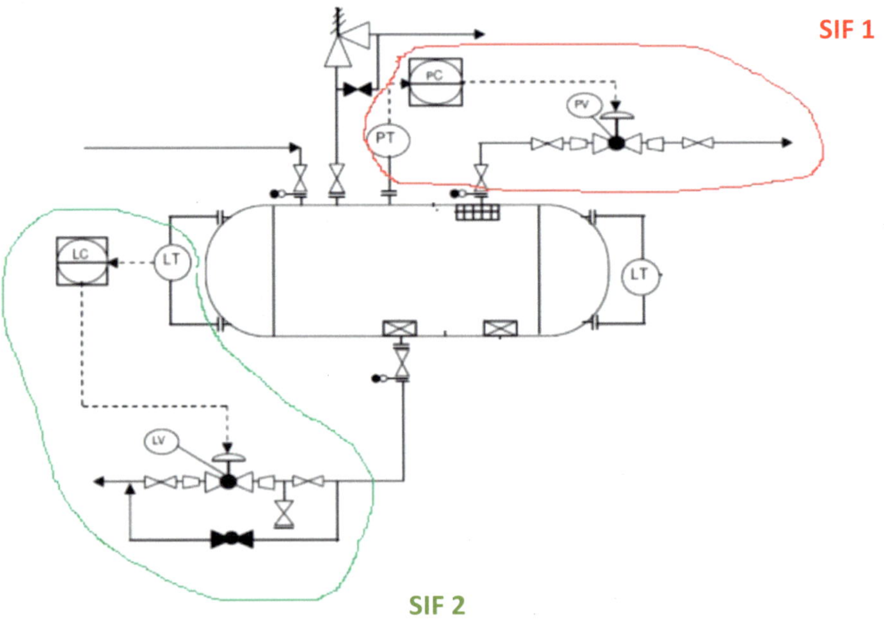

Figure 14: Safety Instrumented System.

The SIS project configuration depends on each project requires and characteristic, but most be always to be taken into account safety and cost issues. The safety Instrumented function is associated to a specific hazard and in order to identify such hazard some qualitative risk methods are applied such as HAZOP, PHA and FMEA. In order to define the SIL required to mitigate the risk associated with hazards four methodologies are proposed such as:

- Hazard Matrix,
- Risk Graph,
- Frequency Target,
- Individual or Societal Risk.

In fact, the SIL definition is only one first step to proceed the SIS configuration. In addition, it's necessary go thought different tasks toward Safety Life Cycle as shown in Fig. **15**.

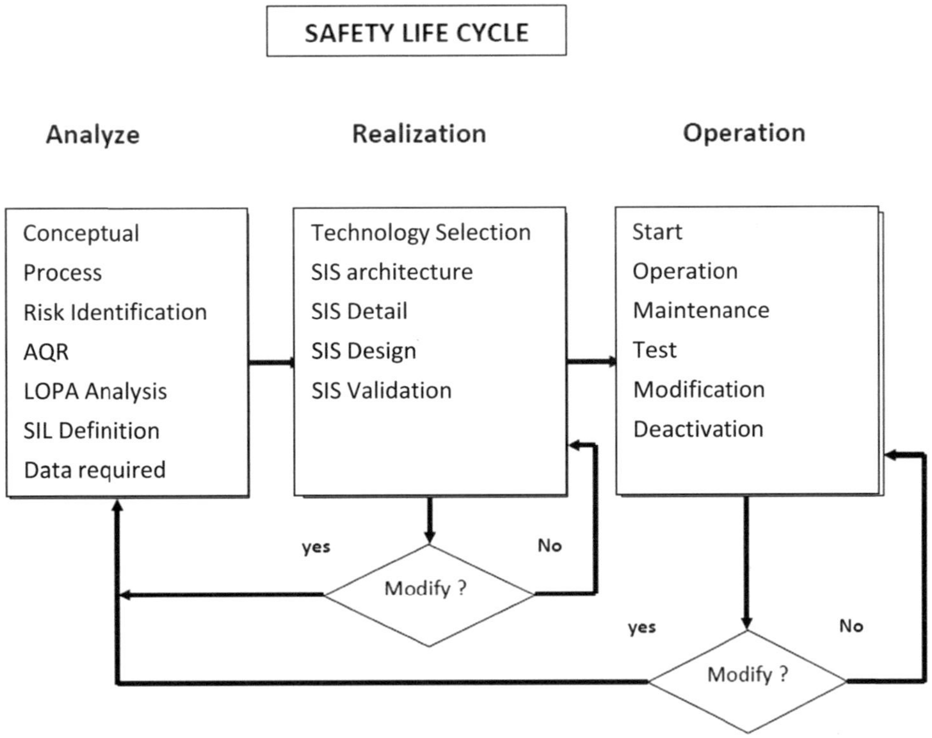

Figure 15: Safety Life Cycle (Source: Marzal, 2002).

3.11.1. Hazard Matrix

This method is based on the knowledge of the likelihood and consequences of failure associated with the process and the independent protection layers in place. It provides some guidance for determining which protective measure can be considered as an independent protection layer. It requires of:

- Establishing a safety target
- Developing a risk analysis (consequences and frequency) and
- Determining the number of independent protective functions.

The Hazard Matrix is a qualitative SIL methodology which considers a risk matrix to select SIL for one specific SIF. The combination of the frequency of the hazard and the severity of the consequence defines SIL required for SIF, as shown in Fig. **16**.

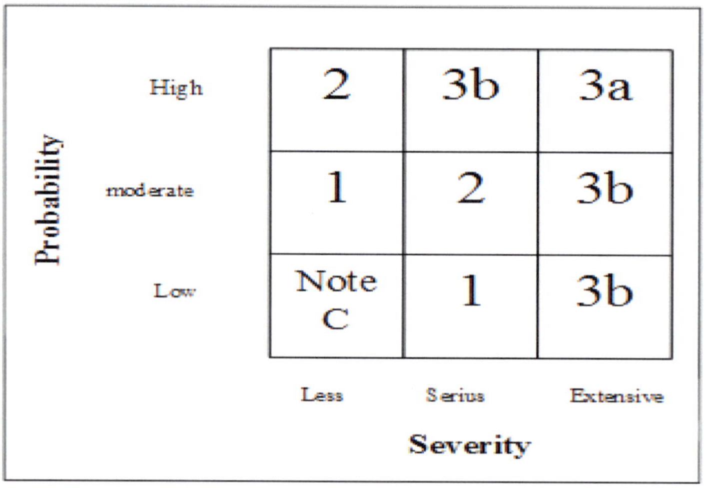

Figure 16: Hazard Matrix (Source: Schwartz, 2002).

Points to note about the Hazard Matrix are:

1. In level 3, if SIF is not provided, a Risk Reduction is necessary. Modifications are required.
2. In level 3 if SIF is not provided, a Risk Reduction is necessary. Careful assessment is necessary.

3. SIF probability does not require layers of protection.

4. This is not sufficient for SIL 4 conditions.

As usual, the risk matrix has a category for probability and consequence, and such criteria have qualitative definitions as shown in Tables **2** and **3** respectively.

Table 2: Consequence Categories (Source: Schwartz, 2002)

Consequence Categories	
Severity Category	**Description**
Minor	Impact initially limited to local area of the event with potential for broader consequence if corrective action is not taken
Serious	One that could cause any serious injury or fatality on site or offsite or property damage of $1 million off site or $5 million on site
Extensive	A failure can reasonably be expected within the lifetime of the plant

Table 3: Frequency Categories (Source: Schwartz, 2002)

Likelihood Categories		
Likelihood Categories	**Frequency (per year)**	**Description**
Low	< 10-4	A failure or series of failure with a very low probability that is not expected to occur within the lifetime of the plant
Moderate	10-2 to 10-4	A failure or series of failure with a low probability that is not expected to occur within the lifetime of the plant
High	>10-2	A failure can reasonably be expected within the lifetime of the plant

An example of a toxic product leakage on a vessel can be assessed based on Risk Matrix in order to implement SIF to prevent an incident. The incident frequency is once in every 1000 years, having approximately one hundred fatalities. There's only one alarm to alert the operator. In addition, there is only one layer of protection which is the relief valve. Based on Hazard Matrix, SIL 3 would be selected as shown in Fig. **17**.

In fact, the layer of protection must be considered when hazard is assessed in order to define the correct risk. In fact, the layers of protection are not clearly defined in the Hazard Matrix in Fig. **17**. Therefore, the group of specialists who perform SIL selection must take that into account the layer of protection when accessing the risk. By the other hand, there are some matrixes that consider the number of layers of protection in place during SIL definition as shown in Fig. **18**.

Considering the similar example, observing the consequence category in Table **2**, the risk analysis group classified the consequence as serious, and based on Table **3**, they classified the likelihood as moderate. SIL 2 is selected as shown in Fig. **18**.

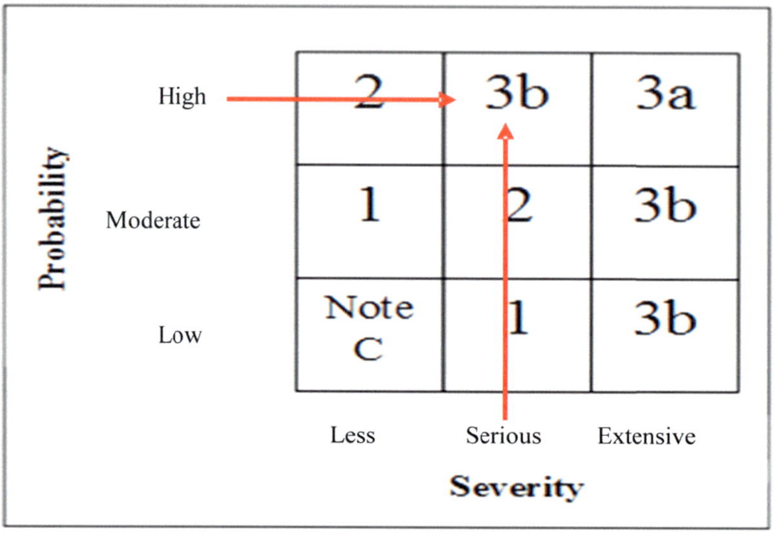

Figure 17: Hazard Matrix without Numbers of Layers of Protection (Vessel Example).

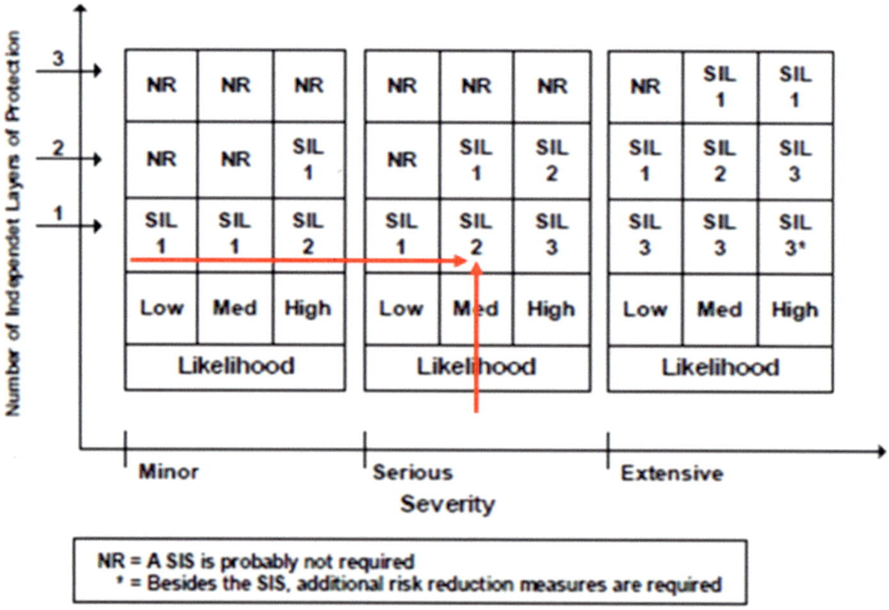

Figure 18: Hazard Matrix with a Number of Layers of Protection (Vessel Example).

The importance of considering the layer of protection when performing SIL definition is to avoid mistakes causes by conservative risk assessment. Such conservative occurs because most specialists, when facing uncertainty, are conservative.

3.11.2. Risk Graph

The Risk graph method is more complete than matrix method because will consider additional factors which influence on risk rather than frequency and consequence such as occupancy and avoidance.

Basically, this methodology selects SIL based on different accident scenarios which take into account different criterion classification based on specialist opinion which defines the graph configuration.

The consequence criterion takes into account the severity of health damage caused by accident. The probable loss of life is also a parameter to specify the consequence of the probable loss of life (PLL). The consequence is classified into four levels, such as:

- Ca (Minor Injury),
- Cb ($0.01 < PLL < 0,1$),
- Cc ($0.1 < PLL < 1$),
- Cd ($PLL > 1$).

Usually, PLL is better defined by a consequence analysis and, even when qualitative analyses like PHA are carried out, similar studies can be consulted in order to obtain a more accurate estimation of PLL number.

The second criterion to be assessed is occupancy and such criterion is related to the level of occupancy in the accident area. Therefore, such index represents the employee vulnerability to accident scenarios. The occupancies are classified into four levels, such as:

- Fa (Rare exposure to an accident in a vulnerable area. The vulnerable area is occupied less than 10% of the time);
- Fb (Frequent or permanent exposure to an accident in a vulnerable area. The vulnerable area is occupied more than 10% of the time).

The third Criterion to be assessed is avoidance and represents how much is feasible for the operator to avoid the accident. Basically, the avoidance is classified into two levels such as:

- Pa (The operator will be alerted if SIF has failed. Facilities with resources to prevent an accident are provided, and they are independent, enabling the operator to escape from the vulnerable area. There is sufficient time for the operator to be alerted about the incident and to take action to avoid it);

- Pb (If one of the conditions above is not satisfied).

The fourth criterion is the demand rate, which defines the frequency of initiates' event. The demand rate is classified in three levels such as:

- W1 (less than 0.03 times per year),

- W2 (0.3<W2<0.03 times per year),

- W3 (3<PLL<0.3 times per year).

In order to perform the SIL selection based on Graph method the first step to define well the accident scenario considering all criteria defined in Graph method. The next step is follow the graph from the left to the right by defining each criterion level in order to select SIL. Thus, considering the similar example applied in item 3.5.1 the Graph criterion classification are Cd, Fa, Pa and W2, the SIL selected is 2 as shown in Fig. **19**.

3.11.3. Frequency Target

Frequency target methodology is based on risk reduction that can be described by the equation:

$$RRF = \frac{Fac}{Ft}$$

RRF = Risk reduction factor

Fac = Frequency of accident

Ft = Frequency toleration.

So RRF is based on accident frequency and frequency toleration. Table **4** below shows the RRF and SIL required and Table **5** defines the frequency toleration which depends on the severity of the accident.

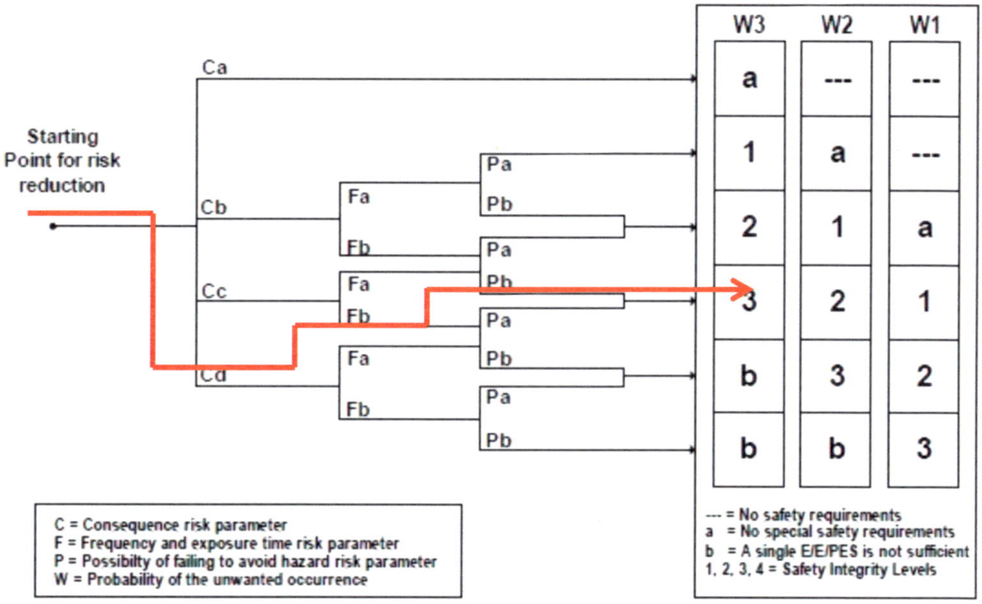

Figure 19: Risk Graph (Vessel).

Table 4: Risk Reduction Factor (Source: Schwartz, 2002)

Safety Integrity Level (SIL)	Average Probability of Failure on Demand (PFD)	Availability* in %	Risk Reduction Factor
1	10^{-2} to $<10^{-1}$	>90 to 99	>10 to 100
2	10^{-3} to $<10^{-2}$	>99 to 99.9	>100 to 1000
3	10^{-4} to $<10^{-3}$	>99.9 to 99.99	>1000 to 10,000
4	10^{-5} to $<10^{-4}$	>99.99 to 99.999	>10,000 to 100,000

In order to exemplify the frequency target method, let's consider the similar example applied to item 3.5.1. Therefore, the frequency of incident is once every 1000 years and the consequence is one hundred fatalities. Based on Table **5**, the severity is classified as serious, and the frequency tolerance is $1x10^{-4}$. The RRF will be:

$$RRF = \frac{1x10^{-3}}{1x10^{-4}} = 10$$

Based on RRF, the SIL classification is SIL 1 (Table **4**). In fact, it's advisable to define a higher SIL, it means, one level higher than the one selected to be more conservative. Therefore, in the vessel analysis case, SIL 2 must be selected.

Table 5: Frequency Target (Source: Schwartz, 2002)

Severity Rank	Impact	Target Frequency
Less	Low health disturb and environment impact. No process losses.	$1x10^{-3}$
Serious	Equipment damages. Process shutdown. High environment impact.	$1x10^{-4}$
Extensive	High equipment damage. Long process shutdown and catastrophic health and environment impact.	$1x10^{-6}$

3.11.4. Individual and Societal Risk

The Individual and Societal risk method is the most accurate because will consider the individual or societal risk criterion to define the SIL. In fact, the individual and societal risk is calculated based on consequence analysis result which is a robust approach to predict the probable loss of life. The Individual risk method define also the RRF which is calculated as:

$$\boldsymbol{RRF = \frac{Fac}{Ft}}$$

RRF = Risk reduction factor,

Fac = Frequency of accident,

Ft = Frequency toleration.

$$\boldsymbol{Ft = \frac{Fc}{PLL^{\alpha}}}$$

Fc = Frequency criterion for individual or societal risk limit (frequency toleration of deaths),

PLL = Probable loss of life,

α = Risk aversion value (α>0).

The Specialist can also define a weigh to PLL that must be included in the Ft calculation.

Considering the similar examples of item 3.5.1, the occurrence of such an incident is expected once every 1000 years, and if that were to happen approximately one hundred fatalities would be expected. The individual risk is$1x10^{-4}$. The probable number of deaths is 100 and the risk aversion value is 1 based on specialist team opinion. Based on such values the Ft will be:

$$\boldsymbol{Ft} = \frac{\boldsymbol{Fc}}{\boldsymbol{PLL^{\alpha}}} = \frac{1x10^{-4}}{100^{1}} = 1x10^{-6}$$

The next step in to calculate the risk reduction factor based on equation below:

$$\boldsymbol{RRF} = \frac{\boldsymbol{Fac}}{\boldsymbol{Ft}} = \frac{1x10^{-3}}{1x10^{-6}} = 1x10^{3}$$

Thus, based on Table **4**, SIL 2 is selected for the SIF in this case.

One important remark about SIL analysis is that the necessity to be on mind the complete life safety cycle in order to achieve high performance, such as reliability and operational availability for each safety instrumented function.

In order to achieve such high performance, it's necessary to apply the best reliability engineering methods during the realization phase such as HALT, ALT, FMEA, RCM and RAM analysis and implement the test and inspection tasks during the operation phase.

It's important to have in mind that reliability is achieved by assurance rather than definition and calculation.

3.12. BOW TIE ANALYSIS (BTA)

BTA is the newest quantitative risk analysis method, and has been in use since the 1970s. It has been incorporated into the Hazards Effects Management Plan methodology used by the Shell Oil Company in the 1990s.

BTA methods catch up the other quantitative risk concepts and methods such as FTA, ETA, and LOPA but enables a better graphic presentation which encompass all problem causes and consequences taking into account the layers of protection that prevent accidents and mitigate the consequences. An example of BTA, referring to an incident of pipeline leakage, is shown in Fig. **20**. On the left side are the threats of the incident and on the right side are the consequences.

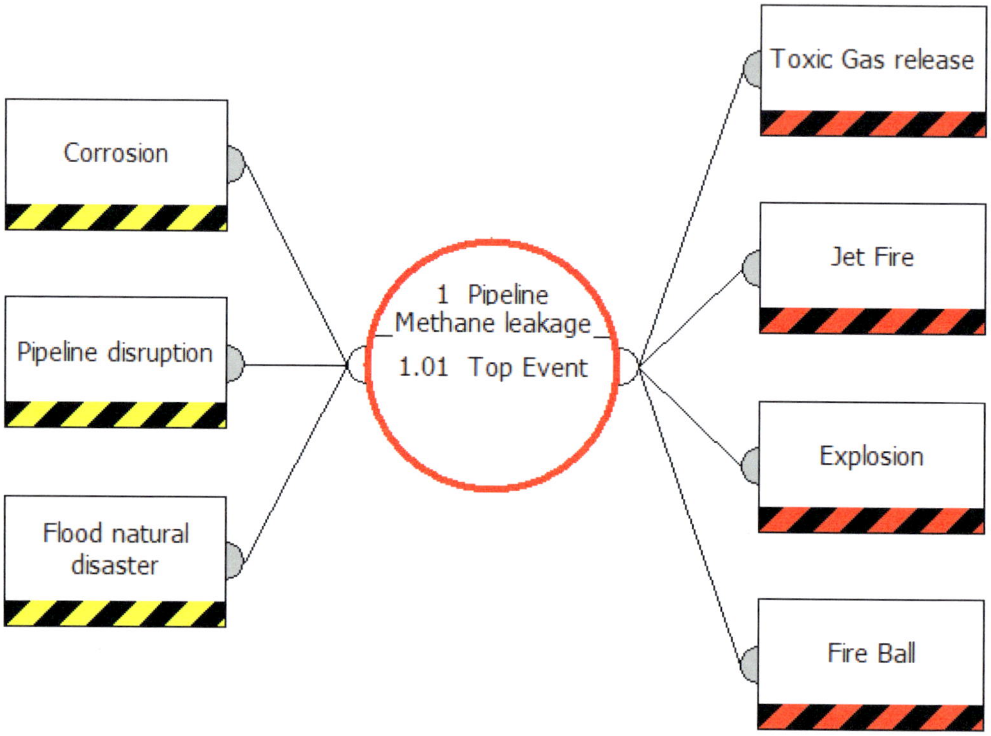

Figure 20: Bow Tie Analysis: Pipeline Methane Leakage.

Whether pipeline methane leakage occur the different consequences such as Toxic gas release, Jet fire, Explosion and fireball may occur. Some actions such as Emergency response and the layers of protection might mitigate such accident consequence. These layers of protection can also be represented in the Bow Tie analysis as a control or recovery actions as shows Fig. **21**.

The control measures that must take place to avoid corrosion are pipeline reliability specifications and inspection and preventive maintenance. Considering that the pipeline disruption is caused by sabotage, the security control must be implemented. In case of seismic events, the storm forecast must be implemented to predict the possible rain storms.

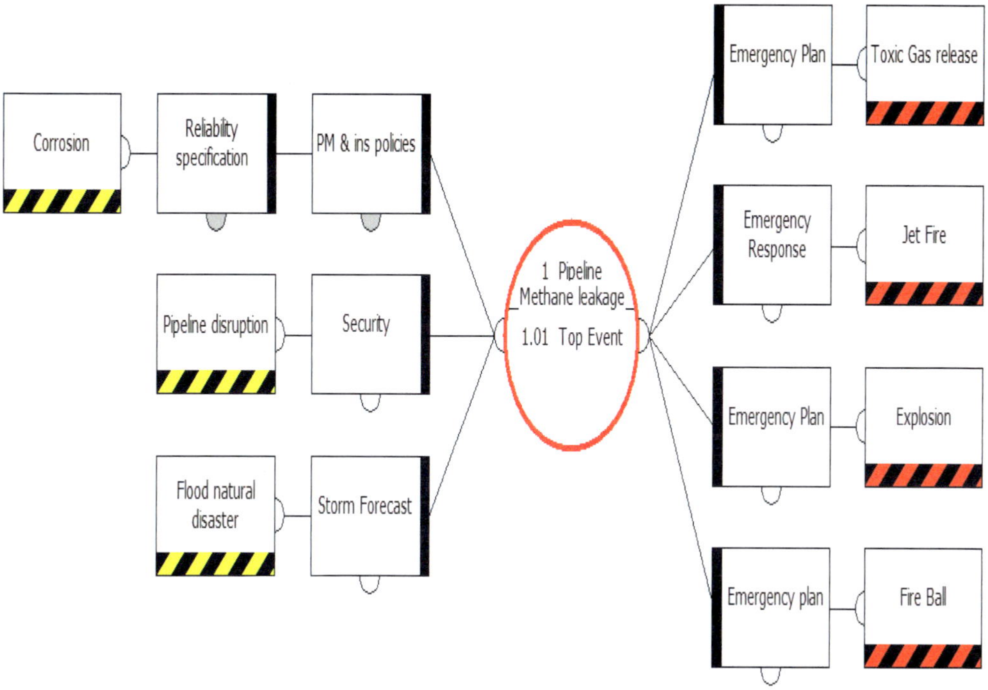

Figure 21: Bow Tie Analysis: Pipeline Methane Leakage (control and recovery measures).

3.12.1. Independent of Time Bow Tie Analysis

In most of the cases the BOW TIE analysis is only a graphical representation to support a decision or an independent time model to perform the risk prediction. In the second case, the frequency of Initiate event and the probability of control and recovery measure are independent of time, in other words, constant.

Such type of approach is easier and faster, but in some cases such as equipment failures or accidents may not represent the operational reality. In fact, some initiates events as well as control and recovery measures might be better represented by dependent on time event.

In fact, let's consider the independent of time approach first. An example of a usual threat for subsea equipment such as riser is the effect of extreme weather conditions. The Fig. **22** represents the BOW TIE model for such event considering the robust design as a control measure, Riser damaged as top event, ROV inspection and emergency response as recovery measures and Loss of containment as a consequence.

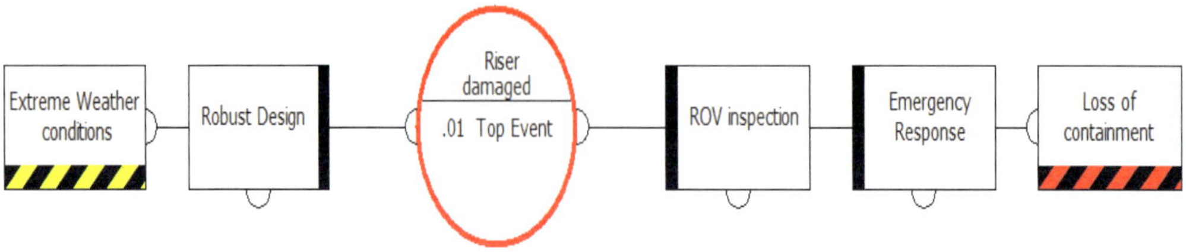

Figure 22: Bow Tie Analysis: Riser damaged.

In order to define the frequency of loss of containment, it's necessary first to define the other events frequency and probabilities. The extreme weather frequency is 2.83 E-2. Therefore, the robust design is able to support extreme weather conditions and the probability of failure in robust design is 0.826 E-13. Consequently the frequency of Riser damage is 2.34 E-15. Considering the probability of failure in ROV inspection is 0.1 and emergency response is 0.2. Finally, the frequency of loss of containment will be:

f(Loss of containment) = f(Extreme weather) x P(Robust design failure) x P(ROV inspection error) x P(Emergency response fail) = 2.83 E-2 x 0.826 E-3 x 0.1 x 0.2 = **0.0468 E-5**

The main discussion concerning the extreme weather constant failure rate applied in independent time approach which gives the same frequency of accident independent of time.

The next item will change the constant failure rate for a dependent time failure rate in order to highlight the difference in the final result.

3.12.2. Dependent on Time Bow Tie Analysis

The Dependent on time approach, consider the increase of frequency and probability of failure related to time. Basically, probability density functions are used to predict the frequency in different stage of the asset life cycle. Therefore, it will be possible to see the different rate of accident when comparing different periods of time. In fact, in order to maintain the risk at an acceptable level, inspections, test and maintenance are required for safety functions equipment.

Considering the Riser damaged example discussed previously, the probability density function applied now is normal distribution (μ=15, σ=5).

As mentioned above the dependent on time Bow Tie is established for two cases. Case 1 considers 10 years' time and case 2 considers 15 years' time.

Case 1: 10 year

f(Loss of containment) = f(Extreme weather) x P(Robust design) x P(ROV inspection) x P(Emergency response) = 0.057 x 0.826 E-3 x 0.1 x 0.2 = 0.94 E-6

Case 2: 15 year

f(Loss of containment) = f(Extreme weather) x P(Robust design) x P(ROV inspection) x P(Emergency response) = (0.15) x (0.826 E-3 x 0.1 x 0.2 = 2. 47 E-6

The Bow Tie analysis is an excellent tool to support risk management and asset integrity management. Such approach will be developed in chapter 9 in more details.

The principal objective of this chapter was to present the major concepts concerning risk and risk management, and to focus on risk analysis methods which were presented with applied examples from different industries. Thus, we can conclude that such methods are extremely important in a risk management context as well as in planning, in implementing the recommendations, in communicating risk and in updating risk during the long life cycle of an enterprise.

DISCLOSURE

"Part of this chapter has been previously published in Gas and Oil Reliability Engineering Chapter 6 – Reliability and Safety Processes 2013, Pages 421–496".

REFERENCES

A. Vozella, G. Gigante, L. Travascio & M. (2006). Compare. RAMS for aerospace: Better early or late than never. ESREL 2006.Safety and Reliability for Managing Risk –Taylor & Francis Group, London, ISBN 0-415-41620-5

Calixto, Eduardo. (2007). Sensitivity analysis in critical equipments: The distillation plant study case in the Brazilian oil and gas industry. ESREL 2007, Stavanger.

Calixto, Eduardo.(2010). RAMS Analysis Methodology: Regarding Safety Process Effects in System Availability. ARS 2010 Amsterdam, Netherlands.

Calixto, Eduardo. (2012). Gas and Oil Reliability Engineer: Modeling and Analysis. Imprint: Gulf Professional Publishing, ISBN: 9780123919144, 07 Nov 2012.

C.Ericson. (1999).Fault tree Analysis - A history. 17 ° International System Safety Conference, 1999, EUA.

Crowl, Daniel A; Louvar, Joseph F. (2002).Chemical Process Safety Fundamentals with Applications. Prentice Hall PTR, Upper Saddle River, New Jersey 07458.

Echeverria, Alejandro C Torres. (2003). Practical application of a SIL Analysis to a process plant. The University of Sheffield Faculty of Engineering. Department Of Chemical and Process Engineering.

La Rovere, Emilio Lébre. (2005). Notas de aula. Poluição ambiental. PPE, Rio de Janeiro.

Layer of Protection Analysis: Simplified Process Risk Assessment.(2001). ISBN: 978-0-8169-0811-0, published by Center for Chemical Process Safety (CCPS)

Marzal, Edward M; Scharpf, Eric. (2002).Safety integration Level selection. Systematic methods including layer of protection Analysis. The Instrumentation, Systems and Automation Society.

OHSAS 18001. (1999). Especificação do Sistema de Gestão e Saúde Ocupacional. Health and Safety Assessment Series.BSI, 1999.

Control of Major Accident Hazard Regulations. (1999). (SI-1999-743), ISBN 01108 21920, published by The Stationary Office.

M J. Sobral and L.A. Ferreira. (2010). Development of a new approach to establish inspection frequency in a RBI assessment. Reliability, Risk and Safety – Ale, Papazoglou & Zio (eds)© 2010 Taylor & Francis Group, London, ISBN 978-0-415-60427-7. Esrel 2010,Rhodos.

Summers, Angela E. (2010).Introduction To Layer Of Protection Analysis. October 2002. Published in Journal of Hazardous Materials.

CHAPTER 4

Consequence and Effect Analysis

Abstract: The consequence and effect analysis addresses to analyze the major accident like a toxic product release, explosion, jet fire, fire ball and BLEVE. Those accident scenarios are first identified and assessed by qualitative risk analysis methods and then quantitative methods are applied to have a more precise assessment.

Indeed, the main objective of consequence and effect analysis is to define the vulnerable are which employees are exposed to such major accident and also define the effect of such accident in employees' health. In order to perform its required to collect some information from the accident scenario like weather conditions, type of product and their physical and chemical characteristic, the volume and storage conditions. In most of the cases due complexity, such analysis is performed by different software packages. In order to support decisions about risk mitigation to reduce the consequence of major accident the individual risk and societal risk criterion is defined as a baseline to such mitigation action. In addition, based on the consequence and effect analysis, it is also possible to take decision about facilities, location into industrial area as well as location of industrial facilities surrounded by population.

Keywords: Consequence and effect analysis, individual risk, societal risk, ISO-Risk, jet fire explosion, fire ball, BLEVE.

4.1. INTRODUCTION AND CONCEPTS

The most of accident that happens in the workplace are "Minor accident". The Minor accidents are those unexpected events that cause damage to employees, but they do not need to be out of the workplace for a specific period of time to recover their health.

Major accident causes severe damage to employee's health and they need a specific period of time out of workplace to recover their health. In addition, the major accident consequence can cause death for one or a group of people as well as environmental damage. Such events are rare to occur in the workplace compared to minor accidents.

Because of consequences of Major accident, most of people are more sensitive to this type of accident rather than minor accident. In fact, whenever a major accident happen the risk perception of the whole society increase. After some time, the risk perception tends to reduce the whole society. That happens mostly because such events are rare happening. Therefore, the risk communication is

Eduardo Calixto

always important to be aware about such events. The Fig. **1** shows the risk perception over a long period of time.

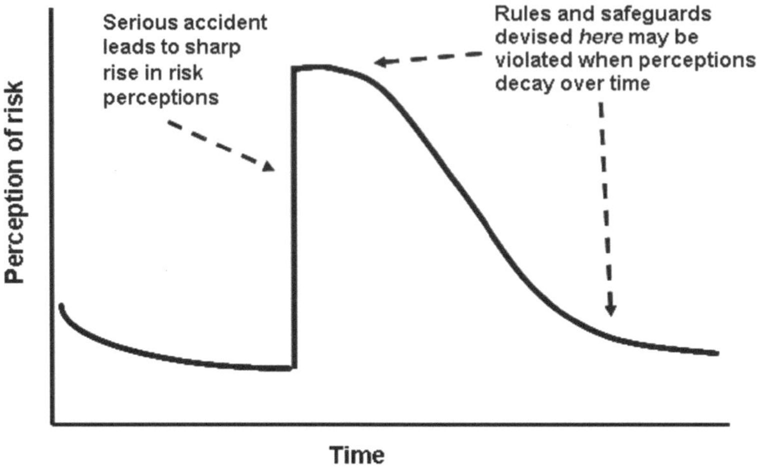

Figure 1: Risk perception (Source: Booth, R.T., 2000).

A recent example of a major accident is "Fukoshima Nuclear Power Plant accident" after a tsunami that trigger very incident on a plant that culminate in a nuclear accident. After such accident, society all over the world were more sensitive to nuclear accidents and in many countries like Germany start a discussion to limit the number of nuclear plants and even close them. In this specific case some points are very important to be arisen:

- The probabilistic risk nature of events;
- The vulnerability of plant to external factor;
- The necessity to be very well prepared for emergency response.

In the context of Risk management, the accident is always possible to happen no matter how good and effective is the risk process management. That happens due to the probabilistic risk nature and the human factors influence on technological systems. As much as the consequence severity involved in a specific industry like aviation, Nuclear, Metallurgy, Oil and gas, Transportation, more reliability are required for layers of protections and equipment. That happens in order to avoid unsafe failures and guarantee that even when incident are triggered the layers of protections are available and reliable to control the incident and keep risk under acceptable level, in other words avoid major accidents.

The reliability by definition is a chance for one device carries on their function property by a period of time. By this point of view, systems that are comprised of a group of equipment have a chance to operate a property as well as to failure on time. No matter how good is equipment quality, how high it is the reliability on time there will be always a chance for failure. In addition, equipment on system, that can be repairable or replaced, require a human intervention to keep system in a high reliability and availability level. In case of no repairable equipment it's necessary to replace them and in case of repairable equipment is necessary to define a maintenance policy. The first challenge in this case is defined well the period of replacing or maintenance in no repairable or repairable equipment. The second challenge is to carry out such replace and maintenance, property to reestablish most of equipment reliability. Whenever maintenance is not carrying out property the equipment may degrade more than expected and the system operates at a reliability level lower than expected.

Nevertheless, even when reliable equipment and property maintenance are carried out operational factors take influence on equipment reliability because in many cases the equipment are operated under condition that was not specified and such conditions may change over time or even for misuse.

Even though those entire factor are taken into account, most of the cases the reliability analysis are dedicated to assessing the product and plant in order to guarantee high performance and not high safety. In other words, layers of protections are not assessed as frequently as other devices to keep high availability and reliability. Most of the cases, Reliability Engineers dedicate more time to assess system and equipment to keep a high production level than high safety level. The safety professional most of the time assesses layer of protections based on risk analysis (ETA, FTA, LOPA, SIL) but mostly qualitative or semi quantitatively. All those issues (probabilistic and human) can explain why the major accident happens even in high reliability systems. But even though all such factors are taken into account and are very well manage along enterprises lifecycle, there are other important issue that is external factors.

The vulnerability of plant to the external factor is not under Organizations control and is necessary to be aware about them to be prepared for the response.

Unfortunately, in many cases the risk of major accident was underestimated, because of that society has been face and learning with major accident a long last decade. In order to have a more effective emergency response in case of a major accident is necessary to follow the steps below:

- Define major accident scenarios;

- Define and establish an emergency plan ;

- Integration between local authorities and companies;

- Trainee and carry on the simulate response exercise.

- Update and improve emergency plan.

The first step to be carried out before defining an emergency plan defines the major accident scenarios. Such scenarios came from company and industry historical data and risk analysis and addition, it is necessary to understand and define the major accident scenarios intensity as well as the vulnerable area affected for such accident. The best practice in these terms has carried on a Consequence and effect analysis that will predict the consequences, vulnerable areas and probable number of deaths. Based on such prediction, is possible to see which are the population into and external plant can be affected by a major accident scenarios as well as the locations and that is important information to evacuation routes.

In addition to consequence and effect analysis an evacuation exercise response as well as simulation of virtual reality can help too much to configure an emergency plan.

It is very important to understand that all such tools to support decision have limitations. Even though, those are the best practice and technology that we have nowadays. The uncertain of such methods exist and to face such fact, many industry consider the worse scenarios and try to be prepared for the response such worst case, but even though, in some cases there will be a worse scenario that was not considered. That was happening on Gulf of Mexico oil spill accident in 2011 for example.

Once a major accident scenarios are defined and quantified, the second step is preparing an emergency plan based on such scenarios. The Emergency Plan can be understood as a set up actions, resources, responsibilities, communications and organizational framework which will be deployed in emergency case. All such aspect is important to have an effective emergency plan. The organization frameworks define all specialists necessary in emergency response and their responsibilities. Those specialists are from different subjects like law, logistics, safety, environment, security and all of them work on different levels. The required resources are technical devices for emergency response as well as communication devices, food, transportation, signalization and economic recourses.

4.2. FIRE EMERGENCY AND FIRE PROTECTION DEVICES

One common emergency situation for industry, business builds and civil location is fire. The fire definition is the oxidation of a material by the combustion process, releasing mainly heat, light, and smoke with different products.

Basically, for a fire to exist, it's necessary the presence of oxygen, heat and fuel combustion.

Such condition defines the known fire triangle which these three components exist in balance condition as shows Fig. **2**. Without a proper balance between heat, oxygen and fuel fire will not exist.

In case of fire, the fire protection devices are essential to avoid or mitigate fire accident. Such devices are classified as Active or Passive.

The Active Fire Protection are systems, which require to be triggered by some condition or manually activate in order to work. The Active Fire protection devices are:

- Fire suppression;
- Sprinkler systems;
- Fire detection;
- Hypoxic air fire prevention;

Fire Suppression can be controlled manually or even automatically depends on the project. The fire extinguisher or a Standpipe system are examples of manual fire suppression. By the other way round, the fire sprinkler system, a gaseous clean agent, or firefighting foam system are examples of automatic fire suppression. Automatic suppression systems would usually be found in large commercial kitchens or other high-risk area.

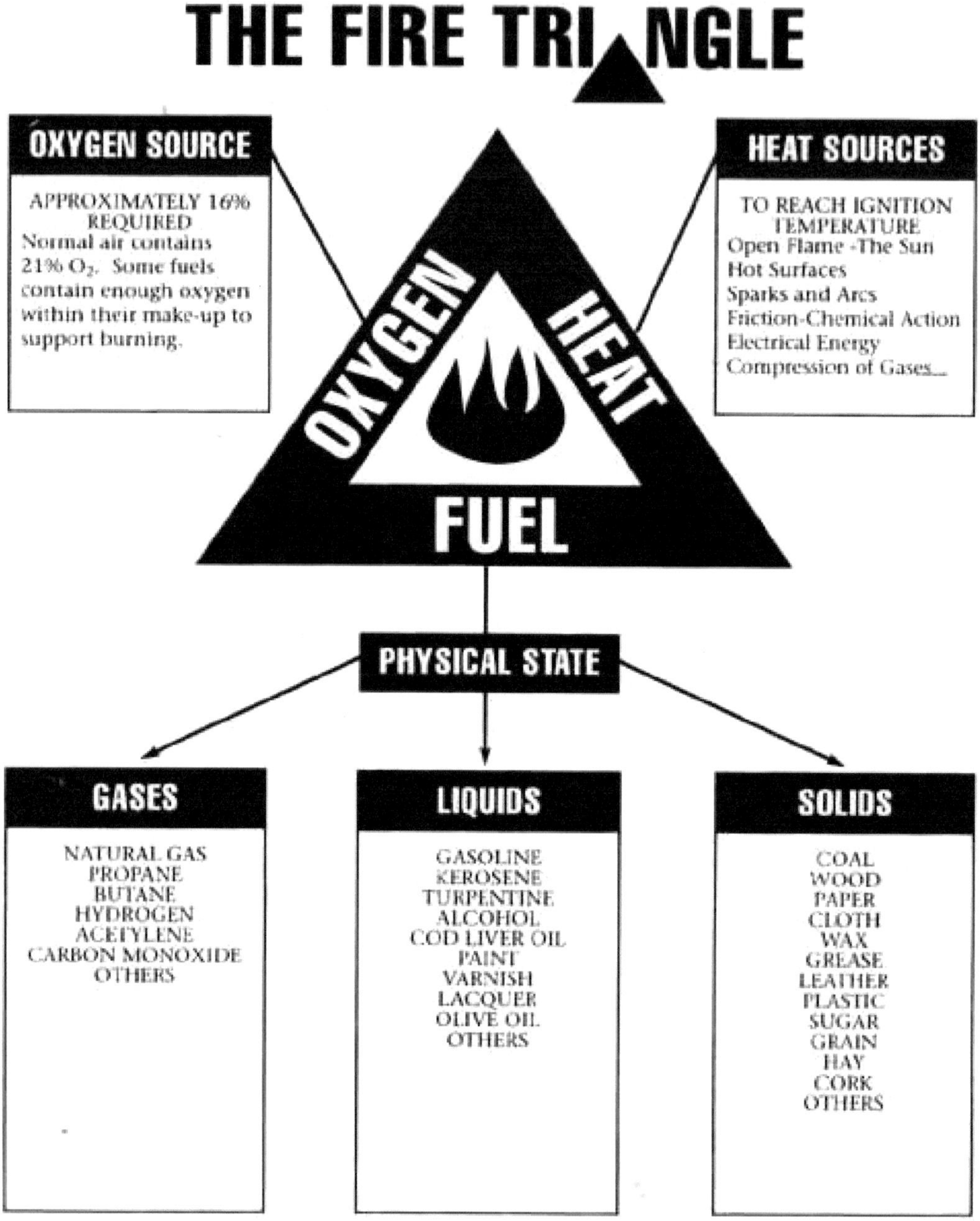

Figure 2: Fire triangle (Source: http://www.self-sufficient-blog.com).

The sprinkler systems are located at ceiling level and are connected to a water source being installed in different systems and locations such as commercial and residential buildings, trains and aircraft. A sprinkler system operates when heat at the site of a fire causes a glass component in the sprinkler head to fail, thereby releasing the water from the sprinkler head.

The fire detection system detects smoke, flame and heat and trigger the alarm to alert the hazard and consequently starts the evacuation emergency response. Such system also includes de-energizing magnetic hold open devices on Fire doors and opening servo-actuated vents in stairways. The Fig. **3** shows examples of active fire protection devices that are from left to right side, the fire suppression, Splinker, fire detector and alarm.

Figure 3: Active Fire Protections Systems (Source: http://www.firesuppresion.co.uk).

The other type of fire protection is the hypoxic system which reduces the oxygen concentration inside the protected area and consequently the ignition can not occur. In fact, it's necessary to have more than 16% of oxygen in total air volume of combustion take place. The advantages of such system are the oxygen regulation inside and specific area and low operational cost (installation and maintenance). The Fig. **4** shows an example of Hypoxic air fire system.

Figure 4: Hypoxic air fire prevention System (Source: http://www.firesuppresion.co.uk).

The Passive Fire Protection (PFP) such as fire-resistant (walls, glass, floor and doors), dampers, fire dampers, fire stop, cable coating, fire cladding and microtherm enclosures have the main objective to reduce the fire spread out through the facilities.

In general terms, the passive fire protection systems effectiveness is assured of carrying out fire testing at 140 °C (for walls, floors and electrical circuits) or even at 550 °C, for structural steel.

In some projects when firewalls are applied as protection, it's not necessary another active fire protections such as sprinklers. In addition, it's necessary to take into account the collapse effect of part of the wall structure to assure the firewall effectiveness.

The Fire-resistant glass using multi-layer as a barrier for the fire separation. The tumescent technology or wire mesh embedded within are used in the fabrication of fire-resistance glass.

The fire-resistance rated floors are also barriers applied to separate parts of facilities with different use in order to mitigate the fire effect in one of them.

The Fire dampers are devices used in heating, ventilation, and air conditioning (HVAC) ducts to prevent the fire inside the duct spread out through facilities. The Fig. **5** shows an example of passive fire protection systems.

Figure 5: Passive Fire Protection System (Source: blog.armchairbuilder.com).

There are different types of passive fire protection methods shown in Fig. **5**. The first, is the, heat resistant foam in orange color applied to fill the voids around the electrical and HVAC lines. The resistant foam slows down the fire spread out.

The second one is the metal connected to the furnace vent pipe on the top. The heat resistant caulk (red) to tightly seal up the top of the wall is also applied to slow down the fire spread out. The Fire stop is designed to seal walls and services holes against fire and smoke the Fig. **6** shows an example of the fire stop system.

Figure 6: Fire stop Protection System (Source: blog.armchairbuilder.com).

The cable coating is also applied to reduce flame and smoke spread out.

The Fire cladding is another example of protection against fire. The material for such cladding include perlite, vermiculite, calcium silicate, gypsum, intumescent epoxy and Durasteel.

The MicroTherm are boxes or wraps made of fire resistent materials, including tapes to protect equipments.

4.3. CONSEQUENCE AND EFFECT ANALYSIS

The first step to define the major accident scenarios is to identify them. Such information may come from historical accident and mainly for risk analysis. That is the most important link of information between risk analysis and accident scenarios.

In fact, even the most simple risk analysis like a Preliminary Risk Analysis may have major accident scenarios identified and that is the first step not the last one.

That means Preliminary Risk Analysis is a good practice to identify major accident scenarios, but a step forward must be done that is carried on consequence and effect Analysis based on major accident scenarios identified in Preliminary Risk analysis. One remarkable aspect is when a risk qualification is carried on in preliminary Hazard Analysis thus we have a Preliminary Risk Analysis and mostly, the major accident scenarios are those with unacceptable risk level or even those with high severity consequences. Such approach is logical and save time on consequence and effect analysis, but is necessary to be careful because in some cases, some accident scenarios identified in Preliminary Risk Analysis can cause some death and are not modeled or its sub estimated in Preliminary Risk Analysis. The best practice to avoid that is considers that there will be at least one employee exposure to such accident scenarios. Despite in some cases, even if one death is related to accident scenarios do not characterize a major accident, such number of death have an influence on the individual and societal risk as well as having an influence on Emergency Plan. The Fig. **7** shows an example of major accident hypothesis on Ammonia vessel identified in Preliminary Risk Analysis.

Preliminary Risk Analysis								
System: Cooling			**Subsystem: Ammonia Cooling (Vessel, valves and pipelines.)**				**Draw: xx-xxx-xxx**	
Hazard	**Cause**	**Consequence**	**Detection**	**P**	**S**	**R**	**Recommendations**	**AH**
Small Ammonia leakage	Valve corrosion	Small quantity of Toxic gas leakage	Gas detection sensor	**D**	**I**	**M**	Carry on inspection and preventive maintenance on valves. Action: Maintenance	1
	Vessel corrosion	Small quantity of Toxic gas leakage	Gas detection sensor	**D**	**II**	**M**	Carry on inspection and preventive maintenance on vessel. Action: Maintenance	2
Huge Ammonia leakage	Valve corrosion	Small quantity of Toxic gas leakage	Gas detection sensor	**B**	**III**	**M**	Carry on inspection and preventive maintenance on valves. Action: Maintenance	3
	Vessel corrosion	Small quantity of Toxic gas leakage	Gas detection sensor	**B**	**IV**	**NT**	Carry on inspection and preventive maintenance on vessel. Action: Maintenance	4
	Pipeline abruption	Small quantity of Toxic gas leakage	Gas detection sensor	**B**	**IV**	**NT**	Carry on inspection and preventive maintenance on vessel. Action: Maintenance	5

Figure 7: Accident Hypothesis from PRA (Ammonia Cooling System).

Based on Accident Hypotheses shown in Fig. **7**, is necessary to define the features of whether, process, and leakage to define the accident scenario as show Table **1** and start to perform consequence and effect analysis.

Table 1: Accident Scenario

Accident hypotheses: Huge Ammonia leakage caused to vessel corrosion.	
Weather and environment	
Latitude	22° 45' 37 63"
Longitude	42° 53' 26 24"
Altitude of Wind Measurement	6,0 m
Air temperature	23,8^0 C
Ground temperature	23,8^0 C
Atmospheric Pressure	1 atm
Air humidity	83%
Atmospheric Class	D
Wind velocity	1,5 m/s
Wind direction	Vulnerable area
Ground type	Concrete
Weather formation	Open sky
Temperature (day / night)	23,8 °C / 22,3°C
Process Characteristics	
Type of vessel	horizontal
Volume	50%
Vessel hole diameter	6 in
Valve hole diameter	xx in
Pipeline hole diameter	xx in
Internal pressure	14,60 kgf/cm^2
Internal temperature	-33°C
Product Characteristics (Ammonia)	
Molecular Weigh	17 g/mol
Freeze Point	-77°C
Ebolution point	35,3°C
Explosion Limit	16% - 25%
IDLH	300ppm
Dangerous concentration	20 ppm

Based on such information presented in Table **1** is possible calculate or simulate the accident scenario consequence. It is very important to have in mind the importance of such information because that influence of consequence results. Mostly the process and product characteristic are much known, but the atmospheric class not and such value takes high influence on accident scenario, I mean, in toxic product leakage. The atmospheric class depends on wind speed, solar radiation and cloud cover basically. Such atmospheric class is the result of such factor combination and varies from A to F that means from more unstable to more stable as shown in Table **2**.

Table 2: Accident Scenario (Source: ALOHA MANUAL, 2009)

Wind Speed*			Day: Incoming Solar Radiation			Night: Cloud Cover	
Meters per second	Knots	Miles per hour	Strong**	Moderate	Slight***	>50%	<50%
<2	<3.9	<4.5	A	A-B	B	E	F
2-3	3.9-5.8	4.5-6.7	A-B	B	C	E	F
3-5	5.8-9.7	6.7-11.2	B	B-C	C	D	E
5-6	9.7-11.7	11.2-13.4	C	C-D	D	D	D
>6	>11.7	>13.4	C	D	D	D	D

Note: Stability is D for completely overcast conditions during day or night.

Note: This table is for releases over land. If the release occurs over water, the stability class will be either D or E.

* Wind reference height is 10 meters.

** "Strong" solar radiation corresponds to clear skies with the sun high in the sky (solar angle greater than 60 degrees).

*** "Slight" solar radiation corresponds to clear skies with the sun low in the sky (solar angle between 15 and 35 degrees).

As much as wind speed, solar radiation and lower are cloud cover more unstable is atmosphere and in accident scenarios with toxic product leakage that means is it harder to achieve explosive limit and toxic concentration. By the other way round as much as stable the atmospheric class the explosive limit as well as toxic concentration is achieved easily. That also depends on quantity of release on accident scenario.

The toxic product dispersion characteristic has also an influence on dispersion and most of dispersion models that regard toxic product release consider the higher Gaussian concentration close to source release and as far as from release source the concentration reduce. According to this model wind and atmospheric turbulence is the force that moves the toxic cloud crosswind and upwards as shows Fig. **8**. That happens with neutral buoyant gas that has a density similar to atmospheric air.

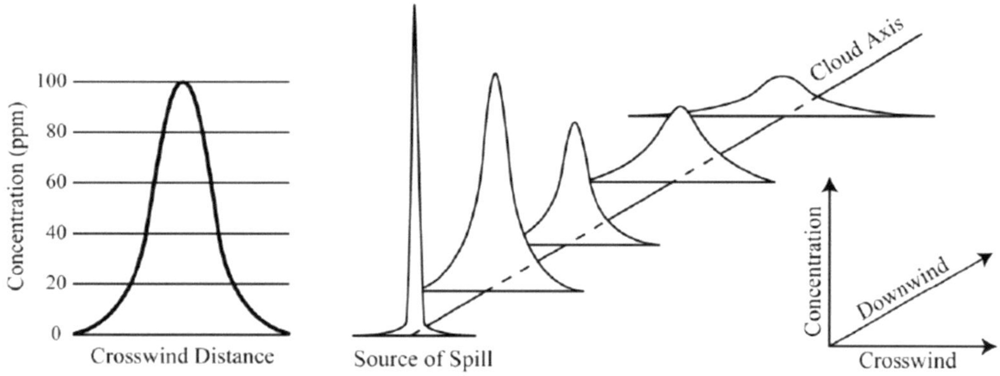

Figure 8: Gaussian Dispersion (Buoyant Gas) (Source: ALOHA MANUAL, 2007).

In some cases the dispersion may occur differently depends on characteristic of gas. The Heavy gas as the instance is denser than air around and consequent dispersion goes down first and then goes downwind direction. The Fig. **9** shows the heavy gas dispersion.

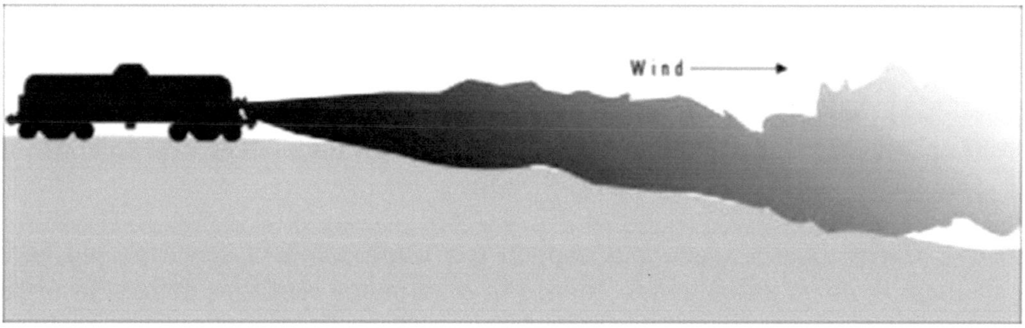

Figure 9: Gaussian Dispersion (Heavy Gas) (Source: ALOHA MANUAL, 2007).

Such gas characteristic is very important to model dispersion as well as to know in which position will be installed the gas detection. In case of pure product is easier to predict such characteristics, but when there is a product combination is harder to predict dispersion as well as other physical and chemical characteristics to model dispersion. Mostly in this case is considered the worse product parameters based on combinations, but even though the dispersion characteristic may occur differently that is expected.

Thus, taking into account all such characteristics, the consequence and vulnerable area are demonstrated when modeling accident scenario by software. In ALOHA software, for example, considering ammonia puddle released is possible to see different ammonia concentration along cloud dispersion direction and distance as shown in Fig. **10**.

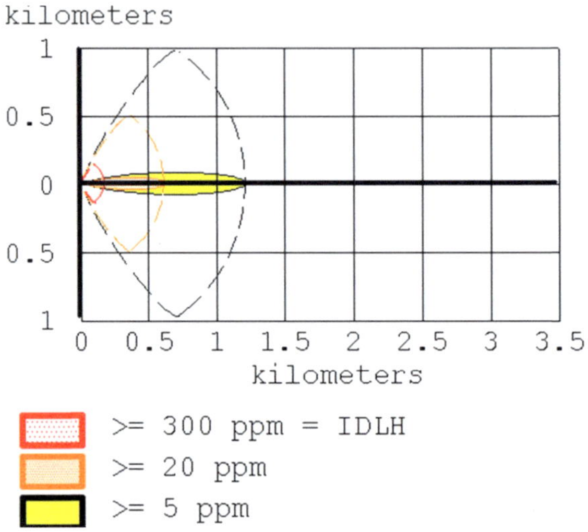

Figure 10: Ammonia Puddle cloud dispersion.

With such solution is possible to estimate which is the most probable area to be affected in an accident scenario as well as define population affected. In doing so, is possible to define the emergency route and the required time to do dislocation as well as which are the population that must be communicated first to start the emergency dislocation to safety place.

In fact, different consequence can happen from one accident scenario and each one of them requires a different response in emergency response action. In order to know which consequences are probable to happen from one unique accident scenario is a good practice to establish all consequence possibilities. A good risk analysis tool to describe consequences is Event Tree Analysis and is also possible to define the frequency of occurrence for each one. In case of ammonia leakage for example, is possible to have as consequential jet fire, cloud explosion, flash fire or only toxic release. The Fig. **11** shows an example of such consequences and their frequency calculated by Event Tree Analysis. The historical data on accident help to define the event probabilities. With such frequency values and consequence and effect results is possible to calculate the individual and societal risk.

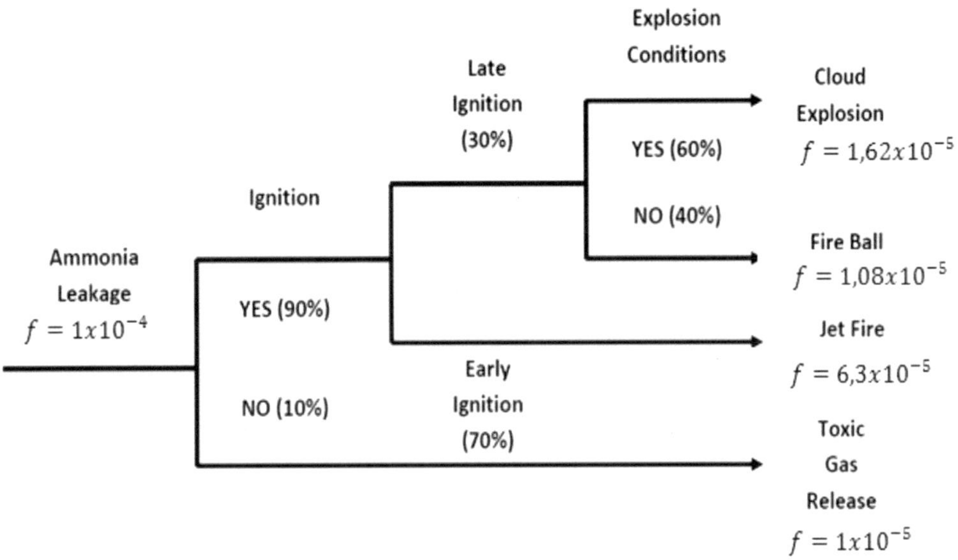

Figure 11: Event Tree Analysis.

In fact, the consequence analysis shows the vulnerable area and consequence that means the level of radiation, toxic concentration and pressure of shock waves in case of explosion. Based on such values is possible to calculate the probability of deaths that is defined by Probit function. There are different Probit equations to calculate the effect of the pressure wave, radiation and lethal toxic concentration. The lethal dose concentration depends on product characteristic, but radiation and overpressure effect have a similar Probit Function for all products. However, depends on the product and the accident scenario different overpressure wave and radiation will be released. The probability of death varies from 1% to 100% and most of cases is defined three values of probability between this range of probability value when is necessary to calculate individual and societal risk.

Despite some reference values is also necessary to calculate the expected number of deaths for population exposure to other levels of consequences. If the look on Fig. **10** for example on the first red area, population exposure to 300 ppm of ammonia concentration represent Immediate Danger to Life or Health (IDLH), that means the concentration level limit for people be exposed to ammonia within 30 minutes without healthy damage. In case of radiation effect on human health the Probit Functions are represented by equations:

$$\Pr = A + B \ln\left(I^{\frac{3}{4}} \times t \right)$$

where:

I=radiation (kw/m^2)

t=exposure time

Pr=probit value

A=-14,9

B=2,56

In order to know the radiation level of 1% of chance of death exposure during 30 seconds, for example we can rewrite the probit equation as:

$$I = \left(\frac{e^{\frac{\Pr - A}{B}}}{t} \right)^{\frac{4}{3}}$$

The probit value of 1% is 2,67. Thus the radiation value will be:

$$I = \left(\frac{e^{\frac{2,67-(-14,9)}{2,56}}}{0,5} \right)^{\frac{4}{3}}$$

I=15 kw/m^2

That means that 1% of population exposure to this radiation (15 kW/m^2) during 30seg will die. In consequence analysis, we can look at an area that will be affect for such radiation and estimate the number of people exposed to such consequence. In fact, the consequence (jet fire, for example) make different radiation value in different distance from the accident, such as show Table **3**.

Table 3: Thermal radiation effect (Source: ALOHA Manual. (2007))

Thermal Radiation (kW/m^2)	Effect
35	Significant chance of fatality
20	Fatality unless rescue is effected quickly
12.5	Extreme pain when exposure more than 20 s
Lesser than 5	Will cause pain in 15 to 20 seconds and injury after a 30 second exposure

In case of overpressure wave the Probit equation will be represented by the equation:

$$Pr = -77{,}1 + 6{,}91 \ln(P)$$

where:

P is overpressure in Pascal (Pa).

Pr = probit to fatality value

Considering 1% of chance of death the probit value is 2,45 (Pr = 2,45). Thus, applying the equation above, we have.

2,45 = -77,1 + 6,91 ln P

$$P = e^{\frac{2{,}45-(-77{,}1)}{6{,}91}}$$

P = 0,1 bar ≈ 1.45 psi

That means 1% of population exposure to overpressure wave with 1psi value in case of explosion. The Table **4** shows the expected effect of different overpressure values.

Considering the toxic cloud effect in case of toxic level the profit equation is represented mathematically by the equation:

$$Pr = A + B \ln(C^{n} \times t)$$

where,

A, B e n = parâmetros which depends on toxic substance

C = concentration, [ppm]

t = exposure time, [min]

Considering the toxic level, the profit function parameters are A, B en and depends on the product such parameter have different values. In amonia case the parameters are A=-16,5; B=1, n=2. Thus, take place the values on Probit function and considering 10 minutes of release and 1% as the chance of death to population exposure to ammonia cloud we have as concentration C=4599 ppm.

$$2,67 = -16,5 + 1\ln\left(C^2 \times t\right) \quad C = \left(\frac{e^{\frac{2,67-(-16,5)}{1}}}{10}\right)^{\frac{1}{2}}$$

C= 4599 ppm

It's also important to define the distance in which toxic concentration level achieve in order to define vulnerability and effect as well as to calculate Individual and Societal risk values. In this example, we have 4599 ppm of ammonia achieve 357 m in 10 minutes. The Fig. **12** below shows the concentration over 60 minutes.

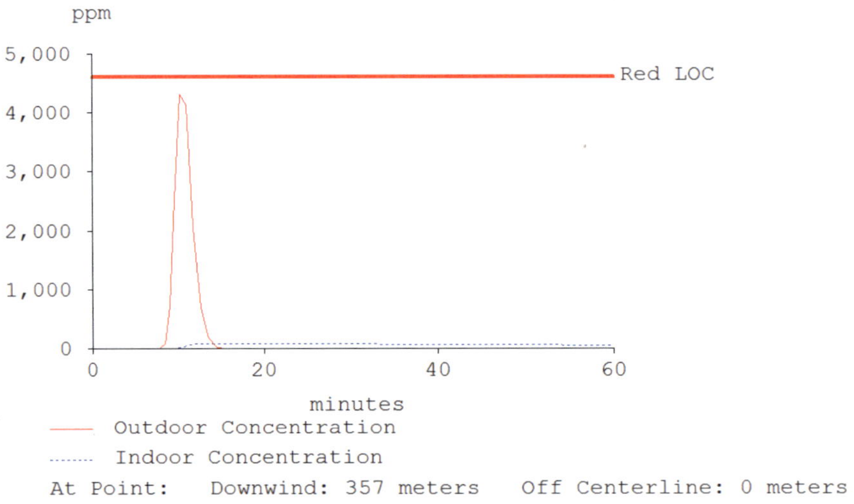

Figure 12: Ammonia concentration (ppm) x Time (minutes).

In such an accident scenario, considering 1% (of chance of death to exposure population we have as an effect:

→ 15 kW/m^2 – 30 s of exposure (thermic radiation);

→ 0,1 bar (overpressure);

→ 4599 ppm – 10 min of exposure (toxic level).

The different probability of deaths gives different values of radiation, the overpressure and toxic level. Tables **3** and **4** are good guidelines to carry on vulnerability and effect analysis.

Table 4: Overpressure effect (Source: ALOHA Manual. (2007))

Overpressure (bar)	Effect
1.01	Loud noise (143 dB); sonic boom glass failure.
1.0	Typical pressure for glass failure.
1.05-1.08	Windows usually shattered; some window frame damage.
1.06	Minor damage to house structures.
1.08	Partial demolition of houses; made uninhabitable.
1.08-1.58	Range from slight to serious laceration injuries from flying glass and other missiles.
1.15	Partial collapse of walls and roofs of houses.
1.15-1.22	Non-reinforced concrete or cinder block walls shattered.
1.17-1.85	Range from 1-90% eardrum rupture among exposed populations.
1.18	50% destruction of home brickwork.
1.22	Steel frame building, distorted and pulled away from the foundation.
1.35	Wooden utility poles snapped.
1.35-1.49	Nearly complete destruction of houses.
1.70	Probable total building destruction.
201-3.01	Range for the 1-99% fatalities among exposed populations caused by direct blast effects.

These are peak pressures formed in excess of normal atmospheric pressure by blast and shock waves.

4.4. INDIVIDUAL AND SOCIETAL RISK

Based on consequence and effect analysis and probit function calculation is possible to estimate the number of dead to person's exposure to accident scenario in different distance from the accident source. As soon as such accident scenario consequence and effect are calculating as well as estimate the exposure population is possible to calculate individual and societal risk. One additional information is the frequency of each accident scenario as shown in Fig. **6** (Event Tree) for example.

The risk concept is a combination (multiplication) between frequency and consequence. The Individual Risk is a measure of frequency (or probability) of death into the operational area considering all accident scenarios. Such individual risk is mathematically represented by the equation:

$$RI = \sum_{i=1}^{n} C_i X f_i$$

where:

f = frequency of accident scenario (year)

C = consequence of accident scenario (deaths in plant area)

As example of individual risk is an example of ammonia leakage represented by an event tree in Fig. **6**. Considering consequence and effect analysis result simulation (Aloha Software) for each accident scenario we have:

⇒ Jet fire risk (10 death in 6,3 x 10-5 years);

⇒ Fireball (10 death in 1,08 x 10-5 years);

⇒ Toxic cloud (17 death in 1 x 10-5 years);

⇒ BLEVE (10 death in 1 x 10-5 years);

Thus the individual risk will be:

$$RI = \sum_{i=1}^{n} C_i X f_i$$

$$RI = [10\,X\,(6{,}3\,x\,10^{-5}\,)] + [10\,X\,(1{,}08\,x\,10^{-5}\,)] + [17\,X\,(1\,x\,10^{-5}\,)] + [10\,X\,(1\,x\,10^{-5}\,)] = 1{,}008\,x\,10^{-3} \approx \mathbf{1x\,10^{-3}}$$

Thus, depending on the individual risk criterion as shows figure 8 such individual risk value is accepted or not and if not, mitigations risk actions are required to be implemented in order to have acceptable individual risk level. Considering the example above, the individual risk value of $\mathbf{1x\ 10^{-3}}$ Is not acceptable based on criteria established on Fig. **13**. In fact, different countries have different individual risk tolerance levels as well as in some countries different states have different individual risk criteria. That means one project accepts in one specific country may not be accepted in other due more strict individual risk tolerance and to do so, more mitigation actions are required. Such mitigation is required no matter how effective the emergency plan can respond to a major accident.

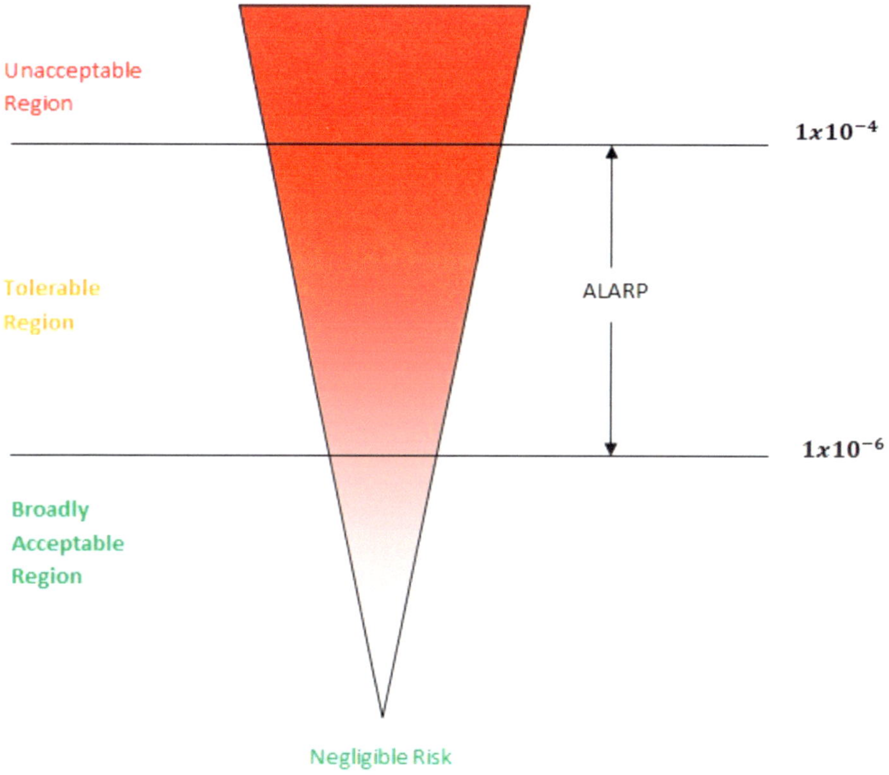

Figure 13: ALARP (As Low As Reasonable Practicable).

The example above did not use the Probit function and regards only population exposure to scenarios to the effect that has 100% of chance of death. In fact, the Probit Function can be used to include in the individual risk calculation with different values (1%, 50% and 75%) as usual.

The other risk criterion is FAR (Fatal Accident Rate) that also consider the individual risk to calculate the number of fatal accidents in label life time that employees are exposed in an operational area. The FAR is calculated by equation:

$$FAR = RI_{med} x T_{ex}$$

where

$$RI_{med} = average\ individual\ risk$$

$$T_{ex} = Total\ exposure\ hour$$

The average individual risk is the individual risk divided by number of employee's exposure to such risk scenarios and the total exposure hour regards total labor hour of one employee. The total exposure time that is:

$$T_{ex} = \frac{10^8}{(24\ h\ x\ 365\ days)} = 1{,}14x10^4$$

Thus, considering the example of individual risk is an example of ammonia leakage to calculate individual risk and considering that ten employees are exposed to accident scenario in case of ammonia leaked the FAR will be:

$$RI_{med} = \frac{1x10^{-3}}{10} = 1x10^{-4}$$

$$FAR = RI_{med} x T_{ex}$$

$$FAR = 1x10^{-4}\ x\ 1{,}14x10^4 = 1{,}14$$

The FAR means is expected 1.14 deaths in labor life in such operational Plant that have ammonia in their process. The question is such number is acceptable or not. In order to compare such number is necessary to check some FAR references as shown in Table **5**.

Table 5: FAR (Nordic Countries for the period 1980–1989) (Source: Rausand, 2004)

FAR at Nordic Countries for the period 1980–1989.	
Industry	**FAR (Fatalities per 10^8 Working hours)**
Agriculture, forestry, fishing and hunting	6.1
Raw material extraction	10.5

Table 5: contd...

Industry, manufacturing	2.0
Electric, gas and water supply	5.0
Building and construction	5.0
Trade, restaurant and hotel	1.1
Transport, post and telecommunication	3.5
Banking and insurance	0.7
Private and public services, defense, *etc.*	0.6

Another Individual Risk Criterion is "ISO-CURVE" that shows graphically the individual risk that society is exposure considering the different distance and population affected. Mostly is establishing one specific individual risk level (1E-6 for example) that society cannot not be exposed and to calculate such is estimated by different distance and individual risk. There are software which calculates "ISO-CURVE" automatically when input data like population and distance as well as accident scenarios and their frequency. One example of "ISO-CURVE" is shown in Fig. **14**. In some cases other ISO-Curve Values are shown on the same map, but that is not good because even when ISO-Curve with values lowers than 1E-6, for example 1E-10, such risk is questioned by society. In fact, Individual and Societal risk are questioned because the ideal situation is when the population is not vulnerable to any kind of major accident scenario. By the other way round there are many cases where the community comes to leave after industrial Plant start their operation. In this case, individual and societal risks are a fair criterion to balance society and company's interest.

The other important risk index is societal risk and that is the most important and used risk criterion to approve projects which may have a major accident with environmental impacts. The societal risk has a different concept of individual risk because regards the cumulative number of deaths outside operational plant area that means an effect on society population. The societal risk is represented graphically with upper and lower risk limit. The societal risk calculated based on accident scenarios consequence and effect considering also population affected must be between such risk limits. Thus the first step to define Societal Risk graph is define the expected number of population death per scenario and then listed in crescent sequence to calculate the cumulative value points. In order to clarify such procedure, we will consider as an example the follow result from ammonia vessel consequence and effect analysis of external population (outside plant). Thus the accident scenarios with respect effects are listed below:

⇒ Explosion - Probit 1% (20 deaths in 1,62 x 10^{-5} years);

⇒ Fire Ball (200 deaths in 1,08 x 10^{-5} years);

⇒ Toxic Cloud (1000 deaths in 1 x 10^{-5} years);

⇒ Toxic Cloud - Probit 1% (2,4 deaths in 1 x 10^{-5} years);

⇒ BLEVE radiation - LC 1% (6 deaths in 1 x 10^{-5} years);

⇒ BLEVE overpressure wave - Probit 1% (1,2 deaths in 1 x 10^{-5} years);

⇒ Fire Ball Toxic puddle evaporation - Probit 1% (1,5 deaths in 3,6 x 10^{-6} years);

⇒ Toxic puddle evaporation (240 deaths in 1 x 10^{-6} years);

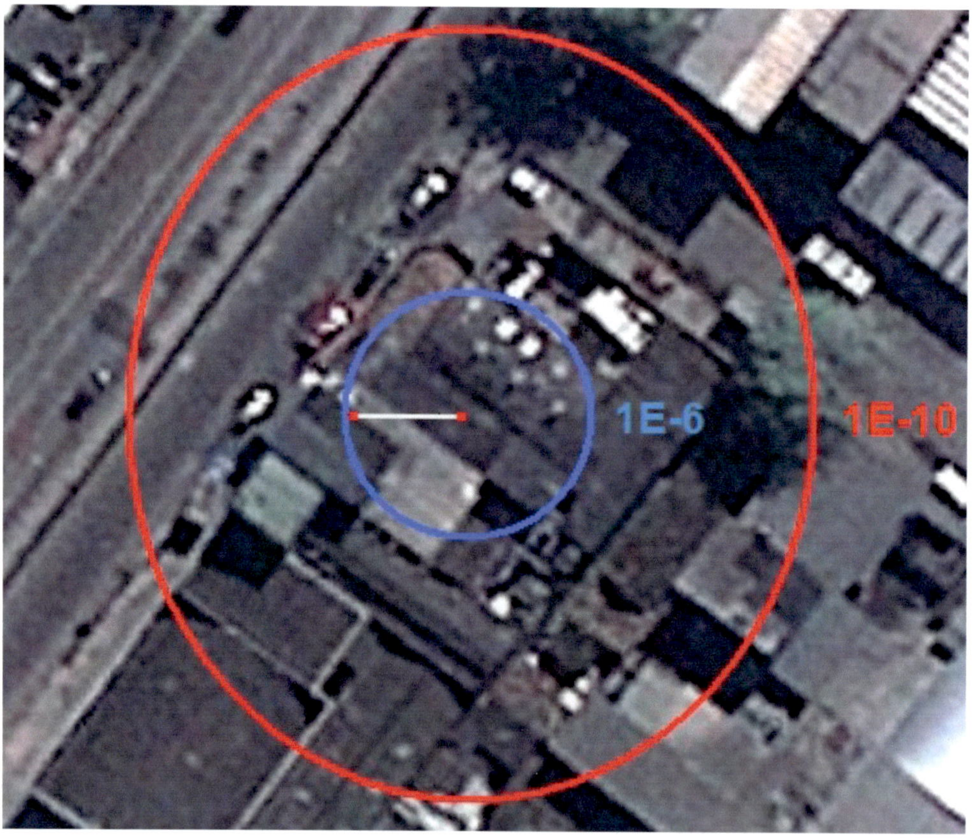

Figure 14: ISO-CURVE (1E-6).

If the risks are listed in crescent order from the lower to higher risk value and the cumulative frequency is calculated based on the frequency of each scenario and subsequence frequencies from following accident scenarios. So for example the value of frequency 2.93E-5 = (1E-5)+(3.6E-6)+(1.0E-6)+(1.0E-5)+(1.62E-6)+(1.08E-6)+(1.0E-6)+(1.0E-6).

The frequency value of 1.93E-5= (3.6E-6)+(1.0E-6)+(1.0E-5)+(1.62E-6)+(1.08E-6)+(1.0E-6)+(1.0E-6) and so on. Thus we are able to define the F-N points of the F-N curve. Such point is the expected number of deaths (first column) and the cumulative number of deaths (third column). The Table **6** show the F-N points plotted in Fig. **10** (F-N Curve).

Table 6: F-N curve points

Expected Number of Deaths	Frequency of Accident Scenario	Cumulative Frequency of Accident Scenario
1.2	1.00E-05	2.93E-05
1.5	3.60E-06	1.93E-05
2.4	1.00E-06	1.57E-05
6	1.00E-05	1.47E-05
20	1.62E-06	4.70E-06
200	1.08E-06	3.08E-06
240	1.00E-06	2.00E-06
1000	1.00E-06	1.00E-06

On Fig. **15** is possible to see the F-N curve (black line) between the upper and lower limit. The ideal situation is when F-N Curve is between the upper and lower limit. When F-N curve has some point over the upper limit (red line) the societal risk is considered intolerable. By the other way round mitigate risk below lower limits is not an advantage once the risk is under control and such effort required investment. Such investment can be applied to maintain the societal risk under tolerable level.

The importance of individual and societal risk criterion to emergency plan is that such risk is the baseline to the emergency plan in order to define which are the prior major accident scenarios and which are the population more vulnerable to such scenario into and outside Plant Facilities.

The consequence analyses as well as individual and societal risk criterion are very important and baseline for emergency plans for some industries like Oil and Gas,

Chemical, Petrochemical, Nuclear and other facilities that deal with dangerous substances. Other industries like Construction, Metallurgic and another must be worried about with fire and other treats. Even though, on next item the emergency plan configuration will be common to all industries.

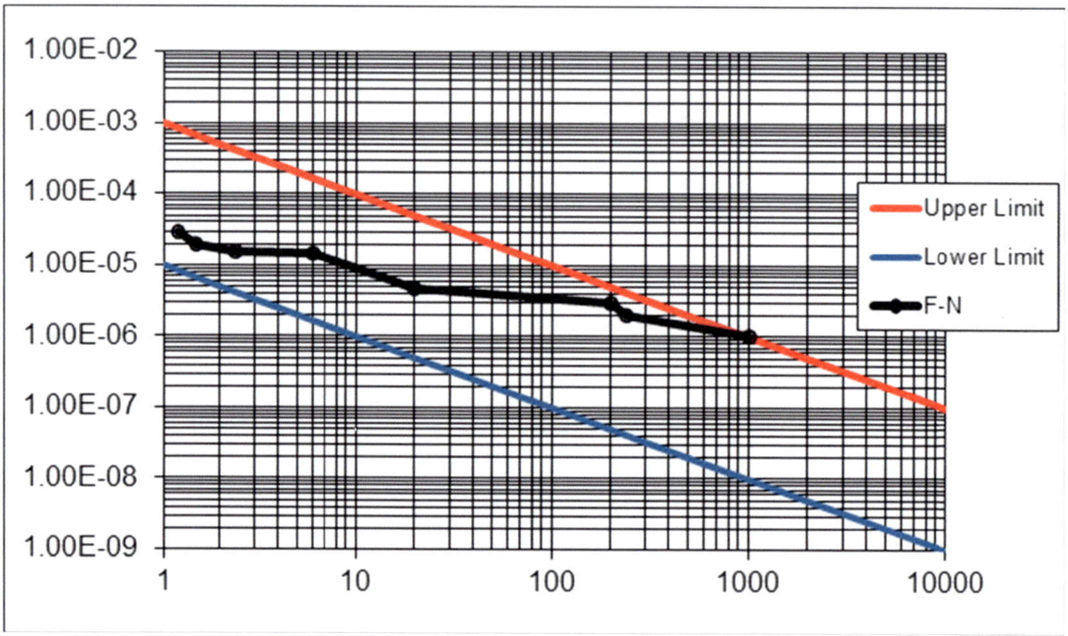

Figure 15: F-N curve.

REFERENCES

Active Fire protections Systems. Retrieve from: http://www.firesuppresion.co.uk

ALOHA Manual. (2007). Environmental Protection Agency, Retrieve from: http://www2.epa.gov/cameo/cameo-downloading-installing-and-running-aloha EUA.

Booth, RT. (2000). An iconoclastic critique of the Sacred Cows of Health & Safety. An analysis of the conventional wisdoms of health & safety management. Inter-alia, Sampson Gamgee Memorial Lecture, Birmingham Medical Institute, 22 November (unpublished)

Fire triangle. Retrieve from: http://www.self-sufficient-blog.com

Fires top Protection System. Retrieve from: blog.armchairbuilder.com

Hypoxic air fire prevention System. Retrieve from: http://www.firesuppresion.co.uk

Passive fire protection System. Retrieve from: blog.armchairbuilder.com

Rausand, M. (2004). Some Basic Risk Concepts. System Reliability Theory (2nd ed.).Wiley. ISBN: 978-0-471-47133-2.

CHAPTER 5

Emergency Response Planning

Abstract: The emergency response is the final defense line in case of a major accident case and it's also a possibility to mitigate the effects of such major accident. The main input to carry out an effective emergency response plan is the risk analysis. Based on risk analysis results, operational procedures and logistic of the emergency response are defined. Indeed the first step to verify the emergency response plan is to carry out the emergency exercise to train employees on emergency procedures as well as to improve such procedures.

Depends on major accident consequences the emergency response requires a huge response capability which involves more than one company and also local, state and national authorities support.

This chapter aims to discuss the main issues involved in emergency response like communication, emergency response framework functions, and organization integrations as well as giving some examples of different emergency response approaches.

Keywords: Emergency response, contingency response, emergency communication, virtual reality, vulnerability.

5.1. INTRODUCTION

The emergency is a situation that is not expected on usual process or workplace context and requires larger resources in response because is related to major accidents or unusual conditions.

The most of accident that happens in the workplace are "Minor accident". The Minor accidents are those unexpected events that cause damage to employees, but they do not need to be out of the workplace for a specific period of time to recover their health.

Major accident causes severe damage to employee's health and they need a specific period of time out of workplace to recover their health. In addition, the major accident consequence can cause death for one or a group of people inside or outside company as well as environmental and societal damage. Such events are rare to occur in the workplace compared to minor accidents.

Because of the consequences of a major accident, the most of people are more sensitive to this type of accident than minor accident. In fact, whenever a major

Eduardo Calixto

accident happens the risk perception of the whole society increases. After some time risk perception tends to reduce for the whole society. That happens mostly because such event is rare to happen, thus risk communication is always important to be aware about such events.

A recent example of a major accident is "Fukoshima Nuclear Power Plant accident" after a tsunami that triggers very incident on a plant that culminates in a nuclear accident. After such accident society all over the world were more sensitive to nuclear accidents and in many countries like Germany started a discussion to limit the number of nuclear plants and even close them. In this specific case some points are very important to be arisen:

- The probabilistic risk nature of events;
- The vulnerability of plant to external factors;
- The necessity to be very well prepared for emergency response.

In the context of risk management, the accident is always possible to happen no matter how good and effective is a risk management process. That happens due to the probabilistic risk nature and the human factors influence on technological systems. As much as consequence severity involved in a specific industry like aviation, nuclear, metallurgy, oil and gas, transportation, more reliability are required for layers of protections and equipment's in order to avoid unsafe failures and guarantee that even when incident are triggered the layers of protections are available and reliable to control the incident and keep risk under acceptable level, in other words avoid major accidents.

The reliability by definition is a chance for one device carries on their function property during a period of time. By this point of view, systems that are comprised of a group of equipment have a chance to operate a property as well as to failure on time. No matter how good equipment quality is, how high it is the reliability on time there will be always a chance for failure. In addition, equipment on system, that can be repairable or replaced, require a human intervention to keep system in a high reliability and availability level. In case of not repairable equipment is necessary to replace them and in case of repairable equipment is necessary to define a maintenance policy. The first challenge in this case is defined well the period of replacing or maintenance in not repairable or repairable equipment. The second challenge is to carry out such replace and maintenance properly to reestablish most of equipment reliability. Whenever maintenance is not carrying out properly the equipment may degrade more than expected and system operate at a reliability level lower than expected.

Nevertheless, even when reliable equipment and property maintenance are carried out operational factor take influence on equipment reliability because in many cases the equipment are operated under conditions that was not specified and such conditions may change a long time or even for misuse.

Even though those entire factor are taken into account, most of the cases the reliability analysis are dedicated to assessing the product and plant in order to guarantee high performance and not high safety. In other words, layers of protections are not assessed as frequently as other devices to keep high availability and reliability. Most of the cases, Reliability Engineers dedicate more time to assess system and equipment to keep a high production level than high safety level. The safety professional most of the time assesses layer of protections based on risk analysis (ETA, FTA, LOPA, SIL) but mostly qualitative or semi quantitatively. All those issues (probabilistic and human) can explain why the major accident happens even in high reliability systems. But even though all such factors are taken into account and are very well manage along enterprises life cycle, there are other important issue that is external factors.

The vulnerability of plant to the external factor is not under Organizations control and is necessary to be aware about them to be prepared for the response.

Unfortunately, in many cases the risk of major accident was underestimated, because of that society has been face and learning with major accident a long last decade. In order to have a more effective emergency response in case of a major accident it's necessary to follow the steps below:

- Define major accident scenarios;
- Define and establish an emergency plan ;
- Integration between local authorities and companies;
- Train and carry out the emergency response exercise;
- Update and improve emergency plan.

The first step is defining the major accident scenarios before defining an emergency plan. Such scenarios came from company and industry historical data and risk analysis. It's also necessary to understand and define the major accident scenarios intensity as well as the vulnerable area affected for such accident. The best practice in these terms is performing a consequence and effect analysis that will predict the consequences, vulnerable areas and probable number of deaths. Based on such prediction, is possible to see which are the population into and external plant that can be affected by a major accident scenarios as well as the locations which is important to define the evacuation routes.

In addition to consequence and effect analysis an evacuation exercise response as well as simulation of Virtual reality can help too much to configure an emergency plan.

Indeed, it´s very important to understand that all such tools to support decision have limitations. Even though, those are the best practice and technology that we have nowadays. The uncertainty of such methods exists and to face such fact, many industry consider the worst case scenarios and try to be prepared for the response such worst case, but even though, in some cases there will be a worse scenario that was not considered. That was happening on Gulf of Mexico oil spill accident in 2011 for example.

Once a major accident scenarios are defined and quantified, the second step is preparing an emergency plan based on such scenarios. The emergency plan can be understood as a set up actions, resources, responsibilities, communications and organizational framework which will be deployed in emergency case. All such aspect is important to have an effective emergency plan. The organization frameworks define all specialists necessary in emergency response and their responsibilities. Those specialists are from different subjects like law, logistics, safety, environment, security and all of them work on different levels. The required resources are technical devices for emergency response as well as communication devices, food, transportation, signalization and economic recourses.

One of the most important factors in emergency response is communication. The communication hierarch must be well defined, as well as the communication technology. When emergency involves local authorities the communication in emergency response are more complex because it's requires an interface between different organizations as well as communication with public.

The third step is to establish integration between companies and local authorities. By this way, it is necessary that companies and local authorities understand the cooperation relation importance. In some cases such cooperation is very hard to be established due to the local authorities profile as a controller and because of that companies do not send information to local authorities to avoid punishment. Even though when operation between authorities is established a big challenge is keeping such relationship over a long time. The companies must understand that local authorities have a very relevant duty that is preserving the societal interest and local authorities must understand that companies are important to society in terms of the local economy and they not wish to have accidents.

The last step to have an effective emergency response is to carry out simulate response exercise. The simulate response exercises are practical exercises which mobilize the emergency plan resources and framework in order for training and to test the emergency response plan effectiveness. As much as a complex emergency response more complex is the simulate response exercises and in some cases requires economical resources and very well planned simulate exercise. In order to simplify the simulate response exercise, such simulate can be carried on in parts. That means carry out only the emergency staff command communication exercise or only part of responses using emergency devices to specific major accident scenario.

A great alternative nowadays is to use virtual reality to train people and simulate the emergency response in order to update and improve the emergency plan. Both alternative can be used in order to reduce economical resources and save time but the reality is the simulate response exercise must be carried out periodically with all people and resources defined in the emergency plan because that is the most close situation of major accident scenarios and show how people probably react in such emergency situation. In addition, people will have a chance to put in practice what they learned in training and partial simulates emergency exercise. The frequency of emergency response exercise depends on people and resource availability. Indeed the simulate response exercise mostly are not carried on more than once a year, but is important that whenever a new accident scenario came out in the process or workplace the risk analysis must be updated as well as emergency plans and as soon as possible such scenario must be take part in to response exercise.

Whenever an emergency requires more resources than a company can provide and support is necessary to apply a contingency response. Therefore, is necessary to prepare a Contingency Plan whenever there is an accident scenario that shows the necessity to use huge quantities of resources that is not available in the company as well as is necessary to get support to higher authorities' level. A good practice is to prepare a local Contingency Plan together with companies closer to your company that are able to supply resources to help in contingency response. In many cases, companies with similar risk in their process are more sensitive to cooperate, but by the other way round some companies that have no such risk are interested in taking part into such plan in order to have also resources from other companies and local authorities to support them in case of a contingency response.

In many countries like USA, Canada, Japan, United Kingdom and Australia the government developed a huge framework in national and also at international level with all issues that must be taken into account in contingency planning. A good example of such contingency plan effectiveness was the Gulf of Mexico accident when USA government and other companies supported the response such huge oil spill. Another good example was the nuclear accident at the Fukoshima Nuclear Plant. Despite different nature and characteristic of accident scenario, both cases have one important aspect that was the dynamic change a long accident scenario and in both cases the Contingency Plan was very well developed after the accident and the response was effective.

The contingencies Plans are not well defined at all levels (Local, State and National) in many countries around the word and that is a big challenge for authorities and a big treat for the whole society. Indeed, in many cases, the company has more resource than local authorities and also high influence due to economic impact on such local society. In such cases, is necessary that companies develop their contingency plan together with local, state and national authorities depends on the accident scenario impact and be a leader in such process. Despite the importance, that didn't happen in many cases and the history shows major accident that would have been their effect minimized if there was a such Contingency Plan.

The next item will explain the Emergency and Contingency Plan as well as vulnerabilities that treat all plants and facilities all over the world.

5.2. EMERGENCY RESPONSE PLAN

The emergency plan encompasses resources, organizational framework, communication and responsibilities that will take place when a major accident occurs. The Emergency Plan comprise a minimum:

- Plant or Facility description and location;

- Accident Scenarios;
- Emergency Response Organizational Framework;
- Emergency Response Communication Flow;
- Emergency Response Procedures;
- Emergency Response Equipment List;
- Emergency Response Equipment Maintenance Plan.

The plant or facility process description, layout and location are a very important information and give a real idea about how difficult is to access or evacuate such facilities as well as the populated area into and outside facility which are closer than accident sources. In order to approve or not the plant location in some countries all over the world define the vulnerable effect area as a main criterion, but for many countries all over the world is necessary to calculate the individual and societal risk and to do that is necessary to carry out consequence and effect analysis as well as defined accident scenario frequency by specific risk analysis tools (QRA, FTA, ETA, LOPA and SIL).

After understanding how proceed to carry out consequence and effect analysis as well as calculated individual and societal risk as discussed in Chapter 4 it is time to understand The Emergency Plan configuration and framework.

Because emergency situation requires special resources like equipment, specialists and investment to keep such plan efficient along time is also required an organizational framework, communication flow and procedures to proceed in emergency case.

The emergency response organizational framework defines the hierarchy as well as the responsibilities for each professional involved in emergency response. Indeed, to define such organizational framework is essential in order to achieve proper coordination when an emergency occurs. Most companies use the proper employees and specialist and establish an emergency response framework. In companies level the emergency response must consider at minimum operational, safety, environment, communication and logistic specialist.

The operational emergency specialist group must be coordinated by professional who work on operational process and supported by safety and environmental professional. That is because there are many operational issues involved in emergency response like shutdown plant safely and that requires operational experience and mostly supervisor and operator who work on shifts are the people more prepared to carry out such action. Despite the experience of operational specialist there are issues that require support from safety specialists as well as environment specialists. In the first case, for example, when is necessary to assess risk of specific task in emergency response or assess the quality of the air in a confined place. In the second case, environment specialist helps to define which area are prioritized to protect and avoid contamination due sensible environment and species exposure as well as assess if action carried out to mitigate environment impacts are enough.

Initially on first emergency response organizational framework concept was thought that operational and planning, emergency group was enough, but after some major accidents was realized that during emergency response there are other activities related to logistic like supply equipment, transportation and food for people involved in emergency response that is so important as operational and planning response.

There, the Logistic Emergency Response group is responsible to define appropriate transport, flow and equipment stock, food and tools for all people involved in emergency response. In some cases, emergency takes more than one day for example to be priced and that involve two or more groups of Operational Emergency Response and Planning Emergency Response as well as logistic. Such people need food, transportation and equipment available and in case of logistic inefficiency the emergency response efficiency may be compromised. The equipment must be available and reliable to be operated and that is a big challenge of the Logistic Emergency Group because such equipment must be maintained a long time, even where there is not an emergency response simulation exercise or real major accident which requires such equipment.

In addition is also important to have a Communication Emergency Group, which is responsible of communicates accident status in society and authorities whenever is necessary. Even though an accident does not affect the community is also important to communicate the accident for the community because the need to know how their lives can be affected by an accident. In fact, what happens is sooner or later the accident is spread out on society and it's harder to tell the truth when you omit the truth. Considering communication is also important to have an official channel between companies and society because in many cases employees involved or not in emergency response may give wrong information which affect company image and reputation.

In most of Emergency response there is an emergency plan group and such group is comprised of specialist responsible to plan and define emergency response actions and assess the scenario constantly during an emergency situation and update response action. The weather conditions must be taken into account and the emergency response team must be prepared to take place a response even in worse weather conditions. The Emergency Planning Group is also comprised for operational, safety and environment specialist. In most of cases it is not possible to assess the emergency response situation and carry out emergency response action at the same time, but is important to both group update one each other. The common mistake doesn't update operation, an emergency response group about new action and priorities and is a good practice to have one specialist or leader of an Operational emergency response group on Emergency response Planning meeting.

Finally, the operational response group is responsible to carry out the response action in order to mitigate the emergent effects. Such group is comprised of operational, safety and environmental professional most of the time. In this case effective training as well as practice and good emergency response devices make difference in emergency response effectiveness.

The Fig. **1** below shows an example of Emergency Plan organizational framework an on top is On Scene response that are responsible for all actions carried on in emergency response.

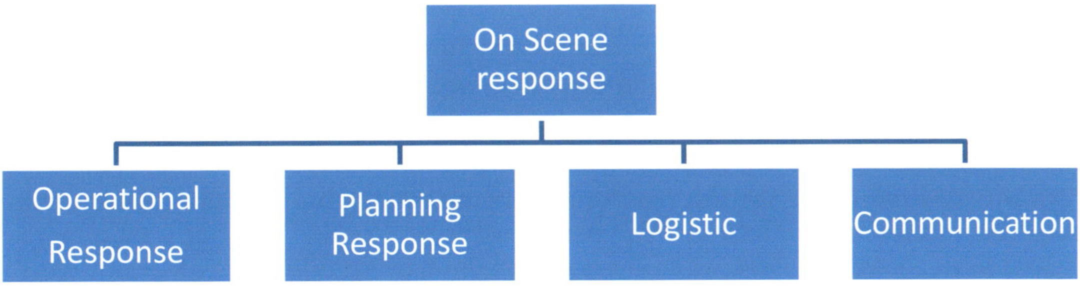

Figure 1: Emergency Plan Organizational Framework.

Most, of course, many people understand wrongly, that the On Scene response must be the high authority on plant or company but is not. In fact, the emergency

On Scene response must be someone with knowledge and authority to take decision on emergency situations and such people can be an experience, operational, safety or environment specialist who took part into emergency teams a long years and is able to take correct decision and lead specialist in case of major accident. Whenever the On Scene response is a highest manager in a company or in the plant it is required to have staff to support him and in reality the staff group must all time convince them what is the right decisions and actions and unfortunately in some cases such One Scene response do not take the best decisions. When major accident involves consequences outside company and requires support from other companies and local authorities is a different situation and that will be discussed in next item about Contingency Plan.

Once defined the organizational framework is necessary to establish the communication flow. In this case there are two main communication processes. One is to start the emergency response and the other one is communication between emergency response groups as well as employees and community. One first case, whenever one employee face one specific emergency situation he or she must communicate to emergency team in appropriate channel that mostly is defined a telephone number to do so. Another important communication aspect is how do inform emergency situation to emergency group. In this case, employees must be trained to communicate the situation and in such communication a good description of accident scenarios is essential to first emergency approach. In fact, emergency situation is not always an accident scenario, for example events like a heart attack or brain attack of an employee. Even in this case the emergency team must be communicating because they are the most prepared group to also deal with such situation.

One second step of emergency communication is triggered by emergency response team after assessing the accident scenario and define the accident dimension. In this case, the On Scene response is communicated and he or she communicates and put together the whole emergency response group. In case of external consequences on the community or environment, the local authorities must be communicated and in this case is necessary to also contact a maximum leader in the company. It's also necessary to take into account the vulnerability of communication network and provide alternatives to contact the emergence response group as well as other people as a second option.

At third hand, there is a communication between emergency response teams and On Scene Response. In this case, whenever is centralized the decision, the longer such decision is taken and in some cases is necessary that an emergency response action group has liberty to act by themselves. Such situation is usual when major accident is triggered for natural catastrophes and more than one emergency response groups must act independently because in some case there are major accident triggered on same time in different locations on same company or in a different location in city or country. Such issues will be discussed on following item that talk about vulnerabilities.

Finally, there is a communication between emergency response teams with employees and community. The big difference in such two groups is the employees are theoretically trained to follow evacuation procedures and follows the companies' rules. Outside companies, there is nothing that guarantees that community follow the evacuation procedures even when they were communicating and trained for that and in this way a support by local authorities are required to guaranties that the affected areas are isolated as well as to help to coordinate the community evacuation.

Even when the major accident is local in a company the level of communication between emergency response groups and On Scene response must be assessed because On Scene Response are the highest authority in emergency response and in some cases he or she are not available to answer a requested authorization for specific action and such situation can compromise the emergency response efficiency.

The multi-organizational model support the organizational interaction and communication assessment and taking into account the cognitive decision process (K Furuta, 2007). The Fig. **2** shows the multi-organizational model.

The multi-organizational model considers the interaction between several agents from different emergency response groups such as Action, Planning, logistic and Communication.

Depends on major accident, communication between emergency groups can be more complex because the large number of resources involved, such as logistics, equipment and technologies from different sources which require a high level of coordination and communication.

In most cases, Emergency response decision and communication are centralized. Therefore, depending on the velocity of decision required for a short period of time such emergency response is delayed. During the September 11 terrorist attacks in the US, all decisions were centralized in US President and there was a huge delay in response. That happens because simultaneous scenarios occurred during the same time and was not possible to answer all of them based on centralized decision model. Therefore, in order to test such extreme situations, is advisable to carry out some form of analysis of the emergency response framework in order to check its efficiency.

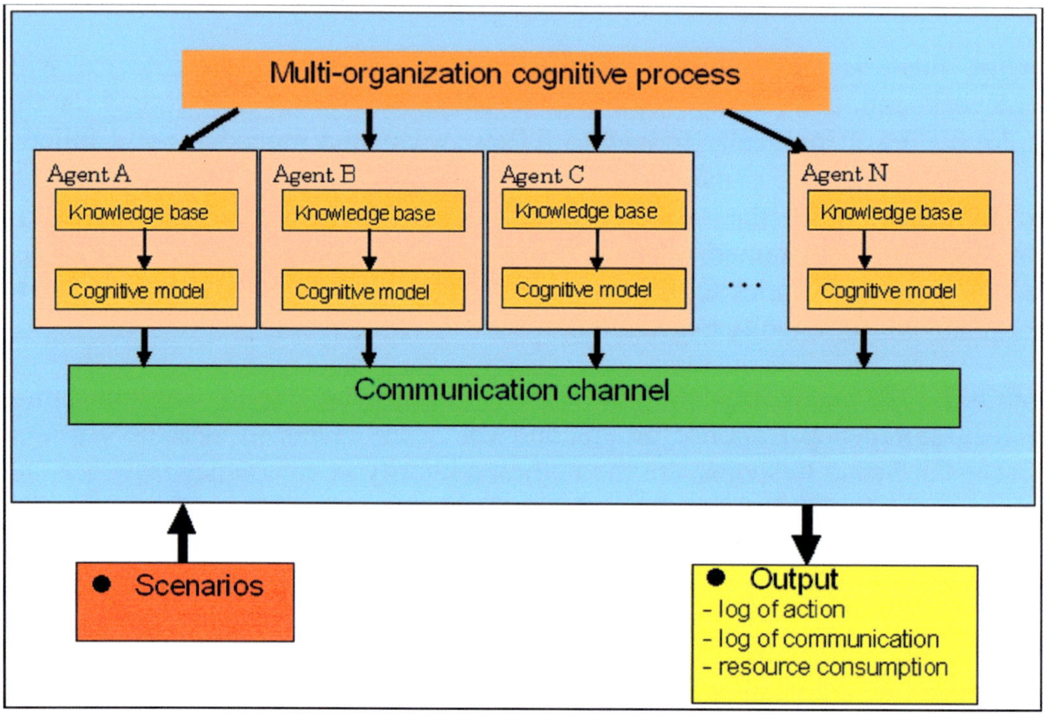

Figure 2: Multi-Organization Cognitive Process (Source: K, Furuta, 2007).

The multi-organizational simulation has the main objective to verify command availability. In Japan, natural disaster trigger different accident scenarios which require flexible and fast response. The Fig. **3** shows the Japan case study, which analyzes the availability (workload) of the emerging framework communication.

One of the main results of this model is communication workload, which is highlighted by yellow lines which shows unavailability (overload) between 537 and 604 seconds. This shows that some improvement is required in order to avoid delays in decisions communication. Therefore, it's also necessary to provide emergency communications pathways to avoid such problem which has high influence on response effectiveness.

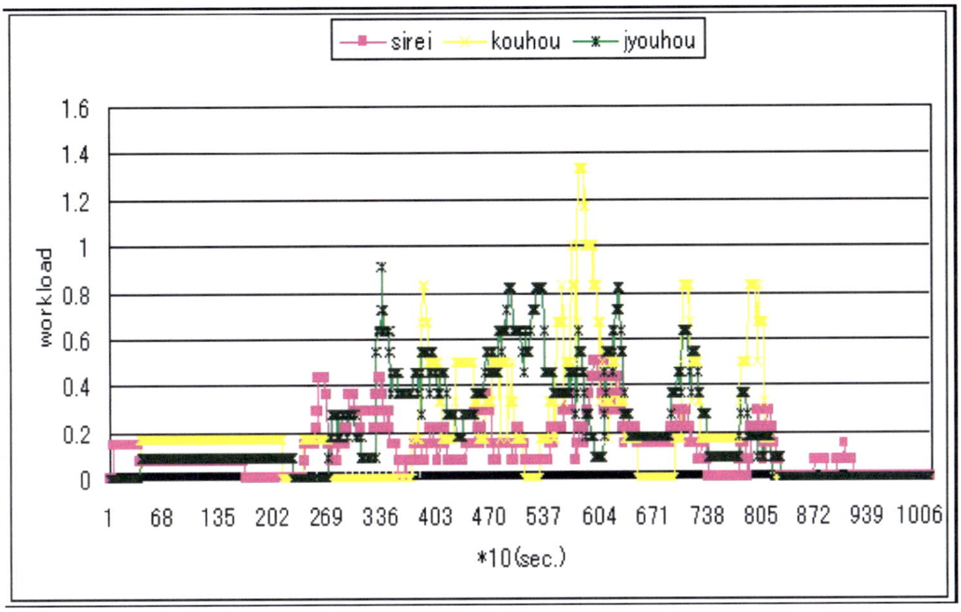

Figure 3: Communication Workload (Source: K,Furuta, 2007).

The following step in defining Emergency Plan configuration has defined all emergency procedures considering all action which is necessary to be carried on in all action from different emergency response groups. No matter how good are procedures they always need to be revised and a good practice to check if procedures are good enough is to carry out simulation exercises.

The Simulation response exercise is a practical exercise that is done to test emergency plan efficiency that means procedures, proper Emergency response organizational framework and proper emergency response communication. Such exercise means to put in practice part or all emergency response organizational framework and procedure to one specific accident scenario. Such exercises require planning because in some cases may involve many resources like logistic, time and even investment to proceed it. In fact, such procedures are carried out annually or biannually depend on necessity to test emergency plan response efficiency.

Another remarkable challenge on emergency plan is to keep equipment and devices available and reliable enough to have no failure when demanded. The first step is to make up a list of such services and have special stock in place with easy access as well as secure enough that only authorized people access them. In addition, such equipment and devices must be maintained and tested constantly in order to be available in case of emergency response. In doing so, is necessary to prepare a maintenance plan to inspect and carry out maintenance whenever is necessary for such equipment. The emergency simulation exercise helps to keep emergency response equipment and devices available and reliable because once is defined such exercise is necessary to test all equipment and devices previously. Indeed, is a big challenge to carry out inspection and maintenance in such device once there are not required demands over time. Consequently, their failure does not affect the plant availability or cause an accident. Even though, such maintenance plan is required and must be dealt with maintenance manager as well as an investment must be defined for that. Consequently, the exhausted emergency equipment must be repaired or replaced after emergency exercise when necessary and during the asset life cycle.

Considering the cost involved to maintain an emergency response plan, the most considerable cost is related to simulation emergency response exercises. Such activity cost includes planning, training and testing equipment. The simulation emergency response exercise can be reduced to one of specific emergency response groups or can be carried out by virtual reality. Even though, it is always necessary to carry out emergency response exercise for the whole Emergency Response organizational framework and that also involve employees and community. Despite the technical aspects involved in an emergency response simulation exercise what is the most important for employees who are not working on operational area as well as a community and visitors is the evacuation procedure.

The evacuation procedure is triggered by emergency response team whenever accident scenario is accessed by such group and requires an evacuation in order to prevent accidental consequences. By this way, employees and community must be trained to evacuate facilities, building our homes to a safe place which is also defined in emergency procedures. Because the accident scenario is dynamic, whenever it is possible must exist more than one evacuation route and safe place to lead employees and community. During the evacuation there are trained employees designated to lead other employees thought out evacuation routes. Those employees are called evacuation leaders and have some clothes to identify them. Mostly there are also emergency channels (Radio) to allow the evacuation leaders to communicate and follow emergency response team instruction. Usually, there is more than one evacuation leader designated to evacuate other employees in safe place. Mostly the employees stand in line and go thought out the evacuation route to a safe place. Mostly one of evacuation leader is in the beginning of the line, either at the end of line and a third one check out the facility or build to certify that anyone stood behind. In real emergency many situations can happen as people delay to stand in line, people go outside the facility without follow emergency procedures and even people who refuse to evacuate because do not believe in emergency situations. That's such a big problem for evacuation leader who coordinate evacuation and the biggest challenge for them is focused on the majority of people and try to evacuate them to a safe place as fast as possible as well as safe as possible. The evacuation efficiency depends on how much employees and population are training for such situation. In commercial build which always face fire treat such emergency response exercise is harder because risk perception for people who works in commercial builds are smaller and to proceed evacuation exercise is much more complicate because in most of cases, many companies works on the same build and even when there's only one company there are some limitation when a build is located in a city. That also happens in the hotel and in this case is more complicated to convince guests to take part in an evacuation exercise. At least they must be communicated or briefed about evacuation routes procedures. A partial solution for that is to carry out virtual reality simulation to predict how efficient is the evacuation procedures that comprise the evacuation route and safety point to lead people. Such software is able to trigger and accident scenario and define with some limitation peoples' behavior during evacuation. The simulation result shows how many people survived and how many deaths were not avoided. Such simulations do not substitute the real emergency evacuation exercise with people, but reduce their frequency. By one hand it is a good tool to test and improve the evacuation procedure, but by the other hand the real evacuation exercise enables people to increase their risk perception about the accident scenario and they will be aware about how to proceed in a real emergency situation.

5.3. VIRTUAL REALITY APPLIED TO EMERGENCY RESPONSE PLAN

The VR (Virtual Reality) technology was born in the mid 1960s. It became accessible to the industrial world (the video game and leisure field contributing to make software financially affordable) after being long restricted to major public research centers and very large company R&D divisions (J. Marc, N. Belkacem & J. Marsot, 2007).

With the rapid development of safety engineering & computer technologies and the modern management theories of safety science, they have been integrated as one and widely used in the field of industrial safety since 1970s (Giampiero *E.G.*

Beroggi, 1995; Aizhu Ren, 2006; Songbai Cheng, 2009; Judith Molka-Danielsen, 2010; Younghee Lee, 2010).

With the development of computer technology and its extensive application, computer is widely used in MIS. Modern safety management system (SMS) is mainly based on TCP/IP communication protocol and the World Wide Web technology. The existing network platforms, application technologies and safety information resources are conveniently restructured and integrated by a simple, unified browser interface, as shown in Fig. **4**.

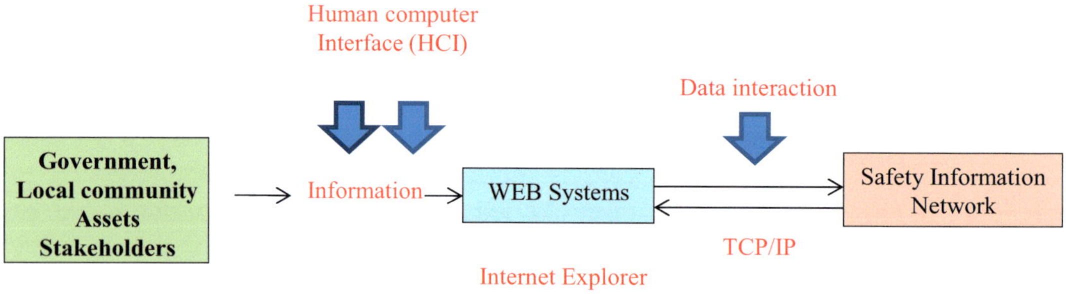

Figure 4: Web based modern safety management model.

Basically VR can be used in order to model accident scenarios as well as emergency situations and procedures like emergency evacuation.

In case of accident scenarios, to model accident scenarios in 3D is a great evolution of emergency point a view because is possible to have a better idea about contingency and fire protections position in case of toxic product leakage for example. In addition, 3D facilities evacuation models have more realist result in predicting available time to evacuate and to check if such time is higher than required time to evacuate when compared with the 2D facilities evacuation model.

One of the most important issues in emergency response is to ensure that employees will be able to reach a safe place after emergency communication. The evacuation time will depend on a number of factors such as employee's behavior, facility layout features and the accident scenario. The assumption that all persons, include disable and injurious ones, reach a safe place after emergency communication in reasonable safety. In order to model the emergency evacuation the People Movement Modeling Analysis (PeMMA) is applied. In case of fire simulation, the fire models such as zone model and CFD (Computational Fire Dynamics) when more complex simulation is required.

Considering evacuation models, under certain emergency situation conditions, such as evacuation of a crowd, some aspects such as a narrow door and reduced visibility conditions take high influence on human behaviour. Therefore, it´s very necessary to take into account human behavior in evacuation models. Such aspect should be studied from an interdisciplinary approach including physical, biological, psychological and social aspects.

In fact, in order to have a more realistic result the evacuation model should include high density of people, hard scape conditions and at the same time, behaviors related to individual survival instinct.

Thereby, computer simulations nowadays are the state of the art to study evacuation. However, given the uncertain about the real people behave in extreme conditions, computers and physical models are not properly validated. In this case, we need to take into account to the worse scenarios in evacuation model.

By the other hands, some experience with animals has arisen as a new approach in this field (Saloma, 2003).

The animals that are easy to obtain and manipulate in the laboratory are insects. In particular, ants are social insects that share certain characteristics with humans, such as communication mechanisms between individuals that is performed by pheromones, vibration, tactile contact, *etc.* (Mc Cabe, 2006; Holldobler, 1999; Detrain and Pasteel, 1987). Second, they have sensors that allow them to react to the surrounding environment. Third, they have an escaping behavior distinguishable from their normal one, and they are biological self-propelled entities. Over the last years, at least two works have been published reporting experiments with ants applied to the study of pedestrian dynamics. (Altshuler, 2005) studied the symmetry breaking in the use of two exits when ants are in a "panic" state. This herding phenomenon was first studied by computer simulation (Helbing, 2000, Shiwakoti, 2009) and investigated the influence of placing an obstacle before the exit and found that it could reduce the egress time of ants.

In such experiments is possible to investigate the so called **"faster is slower effect"**. This refers to the increase of the evacuation time when the degree of hurry of a crowd to get out through a narrow exit is high. This effect was first described by Helbing (Helbing, 2000) and it was studied by using computational simulations of the Social Force Model.

Nowadays the emergency evacuation is well modeled by virtual reality models. The TUTOR model developed by company CDC Modeling from UK is an example of emergency response which represents the evacuation as well as other issues related to emergency situation as shows Fig. **5** (Jain, 2003).

The TUTOR model allows simulation which encompasses up to 1200 entities which encompass individuals, crowds, obstacles, facilities and everything important to be accounted on simulations. A scenario can be replicated at 1 meter, 5 meter and 10 meter resolution, for up to 400 square km (Jain, 2003).

Figure 5: TUTOR virtual reality representation (Source: Jain and Mclean, 2003).

The Virtual reality is a very good tool to support and improve emergency plans and the trend nowadays is to use such tools more and more because of the improvement on the accuracy simulation result and the possibility to reduce costs as well as do not expose people in dangerous situations that can happen during emergency response exercises.

5.4. CONTINGENCY RESPONSE PLAN

The Contingency Plan is the highest level of Emergency Plan which requires resources from different organization in order to proceed the emergency response. That happens because mostly, such Contingency Plan is required to response a major accident which the consequence affect beyond company limits and in some cases have a high effect on society and the environment. Even when companies have all technical resources to apply emergency response, the local authorities are required to coordinate traffic, assess if the environment is recovered property after the accident and define community compensation. The Contingency Plan can be defined as:

⇒ Local Contingency Plan;

⇒ Regional Contingency Plan;

⇒ National Contingency Plan;

⇒ International Contingency Plan.

The Local Contingency Plan applies to emergency situations which major accident effect goes beyond company limit, but requires only resources for other local companies as well as local authorities. Local companies with similar risk or not can be integrated into the Local Contingency Plan, which the responsibility for each company is defined. The group of companies which encompass the Local Contingency Plan define regular meeting to update the Local Contingency Plan information as new risk, define Contingency emergency response exercise and propose modifications to the plan. Depends on local legislation such plan is obligatory in law, but even if it´s not been an advantage for companies which have major accident scenarios to take part in the Local Contingency Plan in order to divide resources and also experience with other companies.

The Regional Contingency Plan has the similar concept of the Local Contingency Plan but has higher scope and need more resources to be applied due to the effect of major accident consequence. Most of cases, industries with major accident scenarios such as Oil and Gas industry, Chemical industry and Nuclear Industry have such Regional Contingency plan. Mostly, the regional government authorities are responsible to define the Regional Contingency plan requirement as well as involving companies to apply and also take part in such contingency plan. In some developing countries, government in some cases has no resources to lead this process and in this case the industry must to carry out such contingency plan together with regional authorities.

The National Contingency Plan is higher than Regional Contingency Plan and integrates all resources available in one country to response a major accident. As much as complex the Contingency Plan more complicated to be applied and that required training, simulation exercise and cooperation between organizations. A good example of the national Contingency response was the major accident on EUA in 2012 in the Gulf of Mexico, when, after a blowout there was a huge oil spill for several weeks.

The International Contingency Plan required international cooperation between countries in response to major accident and both examples of major accident, I mean, Fukoshima Accident and Golf of Mexico Accident had international help despite was totally coordinated for regional and national authorities.

One of the most important procedures of Contingency Plan is the emergency communication. The emergency communication is the official contact that company performs to local authorities and community as well as communication between emergency teams. In case of emergency into company where there is not environment impact or employee's health damage, companies don't need to communicate with authorities and society. In case of employee's health damage or environmental impact it is necessary to report to Public authorities. In case of Major Accident is also possible to have an environmental impact and in this case Environment Local Authority must be communicated. That is usual on Industries like Oil and Gas, Chemical, Nuclear and Metallurgic.

In many cases in history, when companies face a major accident, they try to response accident by their selves with their own resources and when the situation is out of control and they realize that is not possible to respond the emergency they contact local authorities to apply contingency plan. That is a big mistake on major accident and that happen because companies know that in most of cases there will be some punishment for the environmental impact. In some cases, companies do not apply their emergency plan to mitigate environmental impact. That happens all over the world for example, with many ships that cause oil spill and do not repair that. Despite detected the oil spill on the ocean is not easy to determine who are responsible for.

The internal and external emergency communication is critical to mitigate major accident impact and when communication is made as soon as detected the major accident better is the chance to control the situation. Despite company responsibility for major accident response, the local, regional and national authority must have the decision power to trigger contingency emergency as well as to consider the contingency response finished. Whenever a major accident involves environmental impact, such support is really necessary because government authority is responsible to preserve environmental and community interest. The second point is that whenever a contingency plan is triggered, is necessary to have support for government organizations in order to carry out all necessary actions. For example, control vehicle traffic on the road to give priority to equipment and other resources go faster to major accident local. Another example is to isolate accident local area to avoid that community or other people become too close to the accident scene. In addition is also necessary support from the environment agency to define the priority area to protect in case of major accident as well as to assess if response actions are enough to mitigate major accident impact. The Fig. **6** shows the Contingency response communication flow. In such flow is clear that the Environment Agency has a power to trigger contingency plan in different levels based on their assessment of a major accident situation. That can happen at the beginning of a major accident or during the major accident response when the Environment agency realizes that emergency response conditions is getting worse and the available resource are not enough. Nevertheless, companies need to start the emergency response, with available resources.

Despite to Contingency communication importance is also very important to apply such contingency response efficiently the Contingency Organizational Framework with all responsibility defined. In addition is also important to train contingency plan in different levels and different groups. As much as the contingency response exercise is applied more chance to have a successful contingency response in real cases. The big challenge is as higher as contingency plan level more complex is to coordinate operational teams, logistic and even to take decisions. In addition, in case of environmental impact it's important to define the environment recovery action plan that must be followed up by authorities and implemented by companies responsible for the accident.

5.5. INTERNATIONAL APPROACHES FOR CONTINGENCY PLANS

One of the worse major accident impacts of community and environment all over the word and one of the most frequent is Oil Spill. Such major accident can be triggered for a spill in a platform, drill, refinery or even oil transportation by ship or FSO (Floating Storage and Offloading). When there is an oil spill in the ocean or in rivers the consequence can be catastrophe depends on quantity and time to respond such emergency situation. The Oil spill environment impact can last years to be mitigated and in some cases in not possible to do so. Therefore, will be given some examples of Contingency Response Plan response applied in different countries for Oil spill accident scenario.

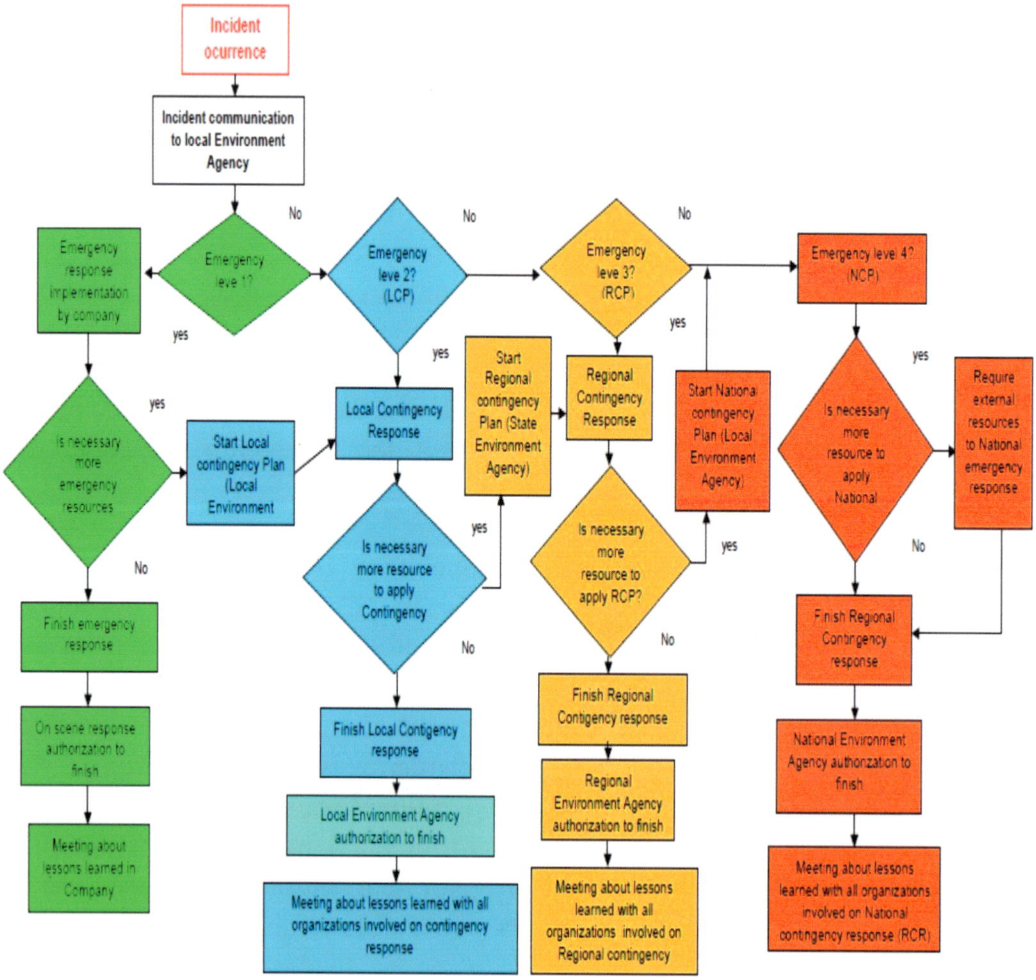

Figure 6: Contingency Communication Flow (Source: Calixto, E. 2011).

5.5.1. US Approach

The US has very successful and efficient emergency frameworks which can face different emergency situations such as natural catastrophes, industrial accidents and terrorist attacks. Such emergency response framework has achieved the highest level of cooperation between the different organizations, such as universities, the government and industry, in order to develop technologies, improve practices and establish procedures and standards which support emergency responses in different levels such as individual, regional and national emergencies.

Some specific organizations in US are responsible for the emergency response planning. The EPA (Environmental Protection Agency) is responsible for developing plans and actions and for supporting industry in meeting the requirements of an efficient emergency framework response.

The companies in different industries are responsible for implementing their individual emergency plans. In addition, it´s necessary to have such individual emergency plan approved by the authorities.

The Coastguard is responsible for developing and supervising the emergency plan in the maritime zone as well as providing support in response to emergencies.

The EPA together with other organizations has developed specific tools to support emergency and contingency plan such as CAMEO, ALOHA and MARPLOT software.

The MARPLOT is an electronic map databank used to support risk assessment and verify the areas vulnerable to the accident impacts.

ALOHA software enables risk assessment for different accident scenarios based on chemical database features as well as the accident environment characteristics such as wind velocity, temperature, humidity, linkage configuration, type of vessel, *etc*. The Fig. **7** shows an example of consequence assessment in case of ammonia vessel linkage. It is also possible to check the toxic limits.

In Fig. **7**, the vulnerable area is plotted in wind direction. It can be seen that the red area is the most critical because of its impact, with it achieving a distance of 32 meters with 750 ppm of toxic ammonia concentration. This is tolerable for a maximum of 60 minutes. The orange area is less critical, although it can cause impacts on health for 400 meters with 150 ppm of ammonia concentration. The least critical area is the yellow one, around 800 meters with toxic concentrations of 2ppm ammonia. In fact, it´s also important to consider the community vulnerability as shows Fig. **8** based on MARPLOT software simulation

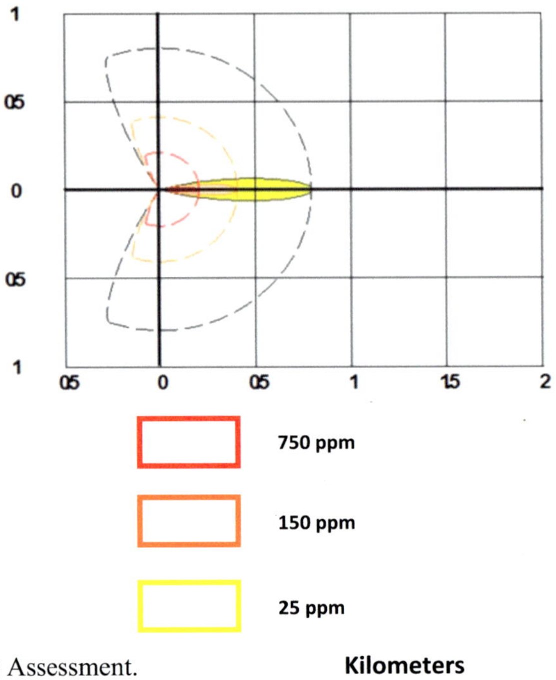

Figure 7: Consequence Assessment.

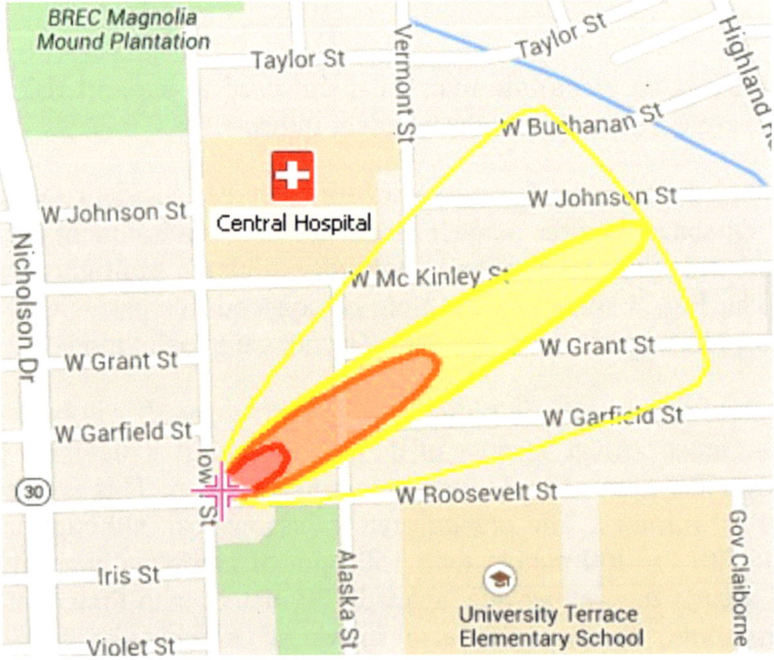

Figure 8: Marplot vulnerable accident area (Source: http://www2.epa.gov/cameo/marplot-software, 2015).

The Fig. **8** above shows different vulnerable area of accident consequence represented by different colors. Similar to previous ammonia example, the red color is more critical than the orange and the red one. The vulnerable areas are defined based on wind direction. In fact, depending on such direction, the central hospital may be affected by the accident consequence which would cause catastrophic effects for society.

The additional tool developed by EPA is the electronic sensitivity map which enables to plot in a regional map different natural resources which can be affected by an accident such as oil spill, toxic product spill or nuclear disaster.

The natural resources of ecosystems, is represented by colored lines to highlight the criticality of these resources as shown in Fig. **9**.

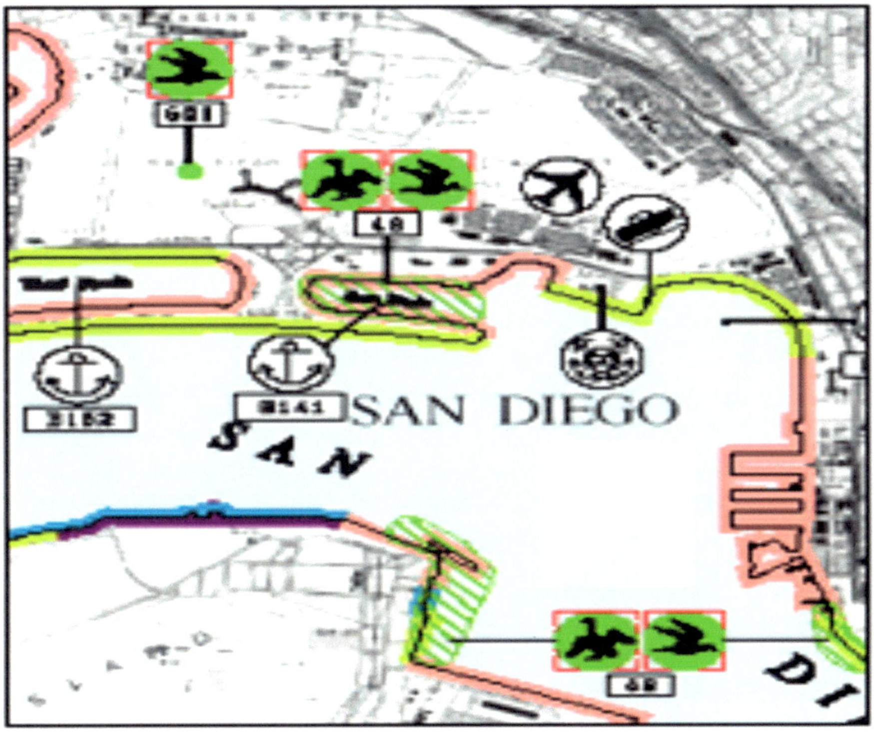

Figure 9: San Diego Sensitivity Map (Source: EPA website, 2008).

5.5.2. Canada Contingency Plan

The Emergency Response in Canada is also developed and supported by government. The Canada government structure the emergency response framework in different command levels supported by various staff levels and specialized groups. Such emergency response is also supported by different processes such as RMS (Response Management System) processes, Report processes and Incident Action Plan meeting (IAP).

The RMS process performs the emergency situation assessment to define the priorities during emergency response such as an area to be protected, response actions and resources. And the response level ensures an efficient response. The RMS (Response Management System) process flow is shown in Fig. **10**.

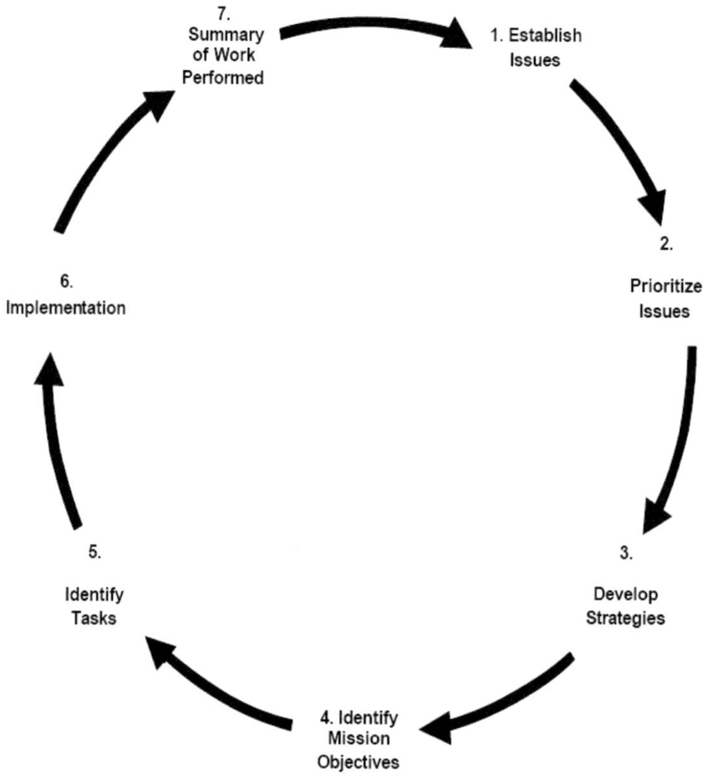

Figure 10: The RMS Process (Source: Response Management System user guide, 2006).

The reporting process provides support for emergency response analysis after the emergency response in order to enable improvement and learning in the whole process.

The Incident Action Plan meeting (IAP) is held during the emergency response in order to check actions and define new strategies if necessary whenever accident scenarios change. Basically, these tree process mentioned above complete the general emergency response as shows the Fig. **11**.

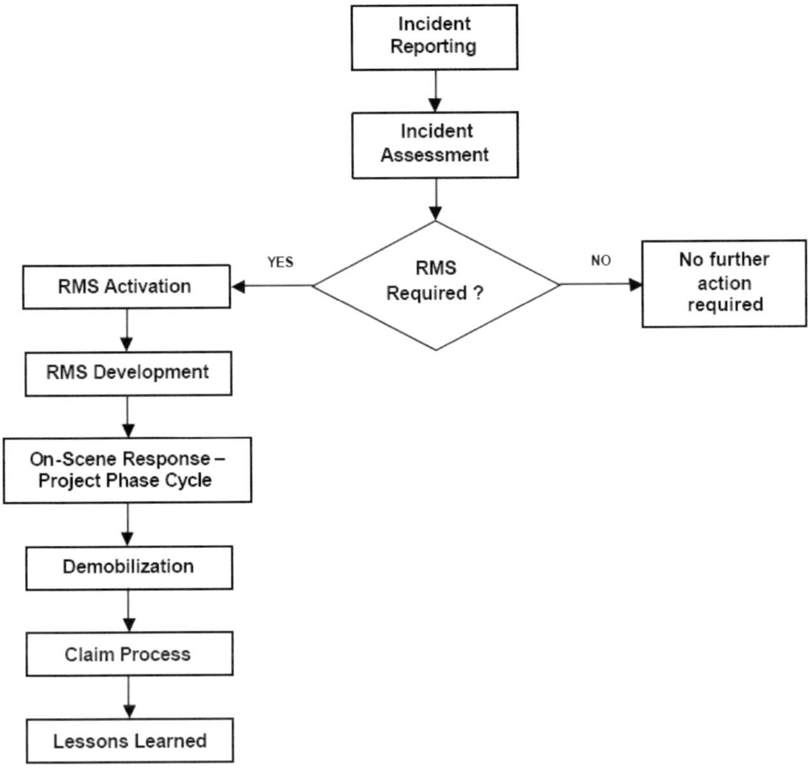

Figure 11: Emergency Management Process (Source: Response Management System user guide, 2006).

5.5.3. UK Contingency Plan

The UK emergency response plan is a very centralized emergency framework.

In case of marine emergencies, the MCA (Marine and Coastguard Agency) is responsible for developing emergency response plan baselines, approve and monitor private companies.

In the case of oil spill accidents the MCA defines three TIER levels to implement the emergency plan.

The TIER ONE, companies which processes or transports toxic products have to develop an individual emergency plan and submit it to the authorities for approval.

The TIER TWO, the cooperation between different private and government organization must to work together to contingency plan takes place.

Figure 12: Oil Spill Locations in the UK (2005) (Source: Annual Survey of reported discharge in United Kingdom, 2004).

The TIER THREE, national resources is necessary and also cooperation between private and government organization.

Depends on the type of emergency one of such tree TIER must take place. In fact, the profile of emergency around the UK territory helps to organize the organization emergency resources in order to implement the emergency responses. Therefore, it´s important to map the severity and frequency of emergencies as well as localization in order to provide the best resources to enable an efficient emergency response. The Fig. **12** shows the most critical areas and emergency resources located in the UK.

Such map is a very good practice to support decision about the emergency resource location and contingency plan's structure for each region. In fact, such map must be constantly updated in order to represent the actual emergence profile around some area or territory.

5.5.4. Brazil Contingency Plan

In Brazil the emergency plan must be applied in companies which manages hazard product. Particularly in Oil and Gas and Nuclear industries, the emergency and contingency plan has been well developed in the last two decades. The nuclear industry has a very good emergency plan at all levels, but the oil and gas industry is still developing the national contingency plan. The Law Brazilian 9966 defines three emergency plan levels, such as local, regional and national for all industries with managing hazard product.

The oil spill accident in 2001 had a huge environmental impact been considered a huge oil linkage (over 700 tons) and change the Brazilian oil and gas industry in terms of emergency response approach. After this accident, many improvements took place and the most significant was the contingency plan response organizational configuration as shown in Fig. **13**

For the last ten years, the Brazilian Oil and gas industry has been improving the regional contingency plans and nowadays starts the first steps to develop the national contingency plan. In fact, a huge effort has been made to achieve the highest level contingency plan and the regional contingency plan response exercises are a proof of such effort.

The Fig. **14** shows an example of regional contingency plan response exercise which involved oil and gas companies, suppliers, transportation company, mining companies, local authorities, environmental government authorities and other organizations. In this particular case, the emergency scenario was about a truck accident with toxic product spill in local river having an environment impact.

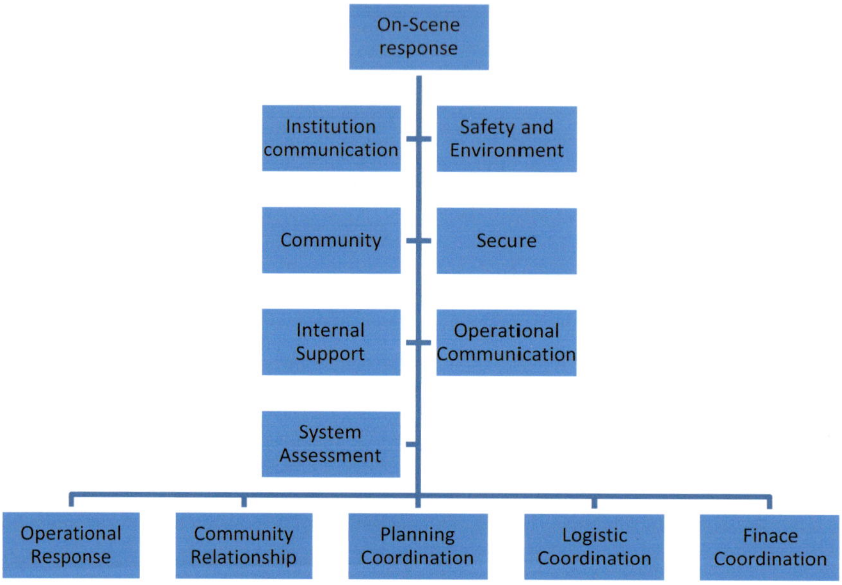

Figure 13: Emergency framework after accident (Source: Calixto E, 2011).

Figure 14: Location of Contingency Simulation Exercise (Source: Calixto E, 2011).

This case was one of the first initiatives to implement the regional contingency plan in Brazilian Oil and Gas industry and was very successful because enables different learning lessons to improve the region and also national contingency plan such as cooperation between different organizations, logistic limitations, the importance of communication and the complexity of contingency plan.

5.6. VULNERABILITY ANALYSIS

The vulnerability is defined as a lack of protection or fragile that one system has and can be exploited by external forces. Such lack of protection or fragile are related to external events like nature catastrophes, security information and terrorism attacks or internal events like sabotage.

In case of Systems' infrastructure, vulnerability describes how a system faces problems to carry out its intended function when exposed to materialize threats (Hofmann, 2012). The vulnerability of critical infrastructures as shown in Fig. **15** can be divided into several dimensions to form a general framework for analyzing vulnerability that is:

- Threat / hazard and unwanted event
- Exposure
- Susceptibility
- Coping capacity
- Criticality.

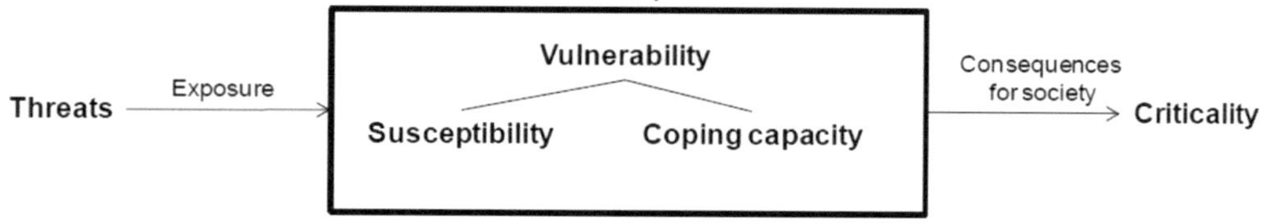

Figure 15: General Vulnerability Framework (Source: Hofmann, 2012).

Threat can be defined as any event with the potential to cause some damage to systems, society and environment. Threats can be categorized into nature/weather related threats, human threats and operational conditions threats. A threat may lead to an unwanted event, understood as a disruption of the system. The vulnerability regards susceptibility and coping capacity. The susceptibility describes if a threat leads to a disruption in the system and is depending on, for instance the technical components, the working force and the organization. On the system level other factors like institutional and social factors also have an influence on the susceptibility. A system is susceptible towards a threat if the threat leads to an unwanted event in the system. The coping capacity describes the ability of the system itself to cope with an unwanted event, limit negative effects, and restore the function of the system to a normal state.

Nature catastrophes are event triggered by nature forces like Tsunamis, Hurricanes, Tornados, volcanoes, Earthquakes, Thunderstorms and universe space threats (G. Woo *et al* 2006). Whenever such event occurs, industrial accident and public infrastructure rupture may take place which has extreme consequences for the whole society such as flooding area, transportation service disruption, environmental impact, health damages and death.

Throughout history, natural disasters have exacted a heavy toll of death and suffering and are increasing worldwide (Reyes, 2006). During the past 34 years they have claimed about four million lives worldwide, adversely affected the lives of at least a billion more people, and resulted in property damage exceeding $50 billion (Guha-Sapir and Lechat 1986b).

In order to face such vulnerabilities it is necessary to have Contingency Plans which consider such scenarios and effects. A very good approach to face such natural catastrophes is the SDMS (Systemic Disaster Management System) proposed by Dr. Reyes in 2006, that consider different level of response and required actions in response efficiently natural catastrophes. The SDMS (Systemic Disaster Management System) encompasses basically three elements such as accident environment, Total disaster operation Unit (TDO) and Total Disaster Management Unit (TDMU). The total disaster operation implements the emergency response in a disaster environment in order to mitigate the disaster effects.

The Total Disaster Management Unit (TDMU) has the main objective to drive targeted and objectives for the total disaster operation unit. Such function is achieved by different others systems such as system 2, 3, 4 and 5.

The system 2 is the communication channel between TDO and TDMU. Such system communicates the early warning to TDMU in order to start the emergency response.

The system 3 has basic function to define the safety policies and audit the total disaster operation (TDO) to check the compliance with such policies as well as check the risk level during emergency response.

The system 4 has the main objective to perform a disaster assessment in order to update the emergency response strategies and actions carried out by TDO.

The system 5 has the responsibility to take decision to drive the TDO actions based on information assessed by system 4 and policies defined by system 3.

The SDMS (Systemic Disaster Management System) can be represented graphically by Fig. **16** (Reyes, 2006).

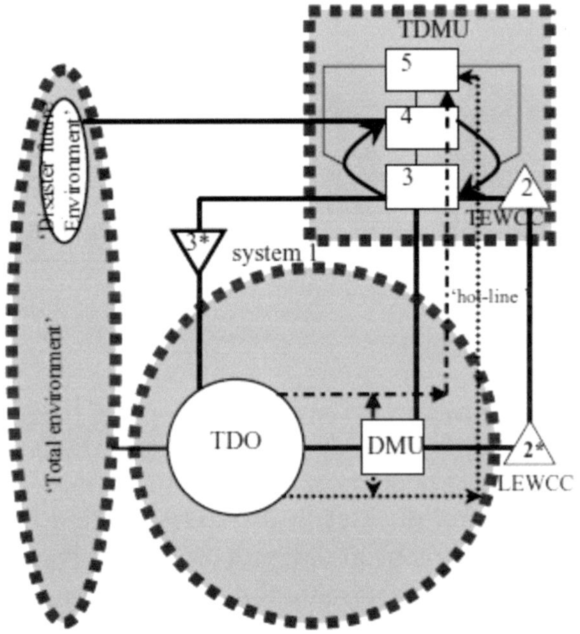

Figure 16: Systemic Disaster Management System (Source: Heyes, 2006).

In case of Natural catastrophes like a hurricane, earthquake or other such events may trigger a technological accident in different locations. Moreover, more than one different location may be affected by such event and requires Total Disasters Management Unit to support local emergency. By this way, the Total Disaster Operation can be divided per different Zones (ZA, ZB) or Regions (RA, RB) and be centralized in one singular Total Disaster Management Unit. Another option is to have two or more Systemic Disaster Management System as shown in Fig. **17**.

In this case, each system is independent from the other and attends different locations. Depends on decision intensity required as well as resources the better option might be to have two independent Systemic Disaster Management System with similar authority to response disaster effect in different locations.

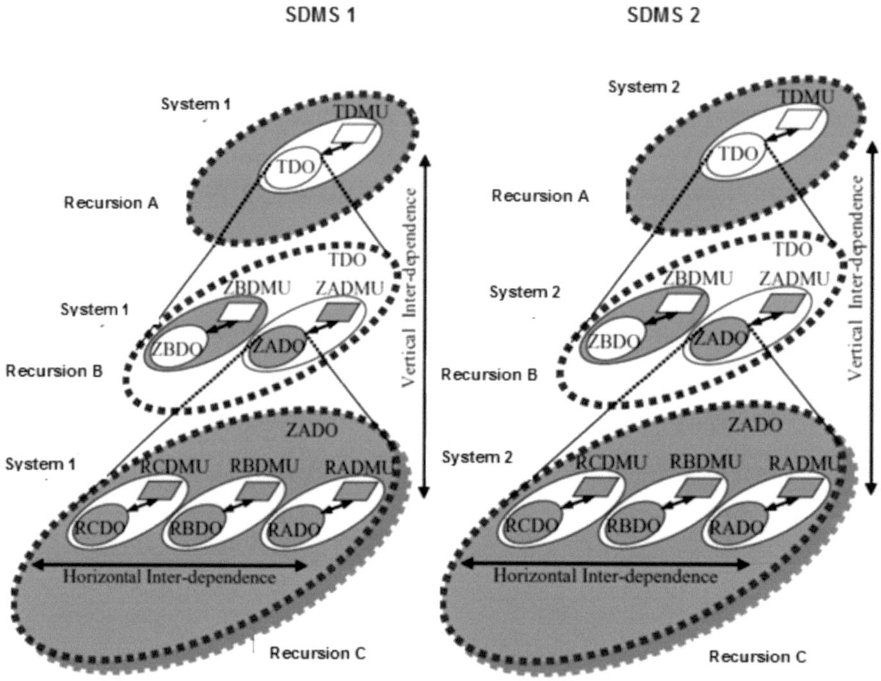

Figure 17: Recursive structure of a MULTI - SDMS Model (Adapted) (Source: Heyes, 2006).

In practical terms to have two or more Systemic Disaster Management System prepared for response Natural disaster in different location independently do not mean necessarily more investment in equipment, but for sure means more people prepared in all contingency levels of natural disaster response. In fact, to have two or more Systemic Disaster Management System is more complicated because requires more training, but by the other way round means more efficient way to response natural disaster in different locations. Maybe one natural disaster has punctual effect in one specific location or even in case that affect more than one location can be coordinated for one single Systemic Disaster Management System. The main assumption that support a decision to apply more than one Systemic Disaster Management System is that independent Systems have faster response when the decision is not centralized and such assumption becomes more true as much as complex the natural disaster. Delay in decision process can be crucial to save lives and that justifies the concept of Multi Systemic Disaster Management System.

In cases of sabotage and terrorism attack the delay in taking decision by centralizing command is a weakness that such treats, try to take advantages causing damage at different point in one system ore in several Systems (plants, Facilities, Government Units).

The terrorism attack is another vulnerability that many countries all over the world's face nowadays. A remarkable point is that even countries that theoretically are not targeted for terrorist attacks, they have plants and embassies in countries that are targeted and consequently they must take into account such vulnerability.

Being intelligent in the pursuit of their own self-defined objectives, terrorist strategies can be comprehended and modeled quantitatively using the resources of game theory, which is the mathematical theory of conflict (Woo, 2002). Terrorists can take advantage of the chaos, disorganization and disruption caused by a natural hazard event to breach the residual security at target sites.

One fundamental aspect to consider in modeling terrorism risk is the application of a multi-field approach (multi-disciplinary) at each step, *i.e.* study with various points of view as: economical and finances, justice (laws, regulations), politics et ideologies, sociology, ethnology and culture, artistic field et esthetics, philosophy, moral et ethics, theology et religions. The macro-steps of the methodology (Le Gallou *et al.* 1999) are the following:

- Determination of the objectives and finalities (fundamental objectives necessary for system existence) of the system, identification of the main constraints associated (due essentially to its environment);

- Determination of the system limits (borders), identification of the system environment (in space and other points of view), external links identification and associated constraints;

- Subsystems identification, internal links and associated constraints;

- Dynamical study of the system (temporal): diachronically (life cycle analysis) and synchronically (time analysis of the life phases) approach.

This methodology needs to be completed by a cross with study objectives to reduce the investigation field. The following step is to build adequate(s) model(s) adapted to the problem. In our case, it concerns risk analysis models. The general methodology includes also the use of model(s) and planning of the actions (solutions) in the real system (this part is not described in our present work).

The Ant Colony Optimization models real transportation problems taking into account the ant behavior when pursues to achieve a specific target such as transport resources from one point to another. The ACO was initially used for the resolution of traveling salesman problem TSP (Dorigo *et al*., 1991). Basically the TSP problem is described by the better route to go through a city considering the time and the cost of travel between different points.

In terror attack case, terrorist tries to find the safest way to get to his target. However a main difference arises between those two cases. In the attacker's situation, he doesn't move between a precise number of points (buildings, blocks…). At every stage he will have to choose between different points to move on so he can reach safely to his target.

The Fig. **18** describes graphically the TSP problem where it´s necessary to decide from the start point to the target point which is the best option considering possible options in each zone point.

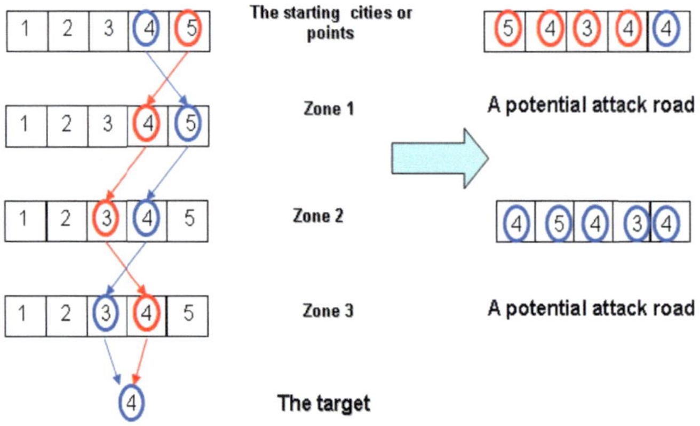

Figure 18: Example of solution construction (Source: Châtelet, 2007).

Other different models are also possible to describe terrorism actions and the effect on the system and test the efficiency of plan against such attack. By vulnerability point of view is important to know the threat probability as well as a probability to minimize their consequence or avoid them.

The third vulnerability that companies face nowadays is security information system attacked by hackers that can shut down plants or cause malfunction in systems and safety protection system that cause consequently an accident.

In general terms, in case of disaster events (Natural catastrophes, Terrorism attacks, sabotage) we need to consider the application tools and our entities of interest in order to define impact and the most appropriated response to mitigate such disastrous effect. The Fig. **19** below summaries, issues that must to be taken into account in respect to the vulnerability of the system.

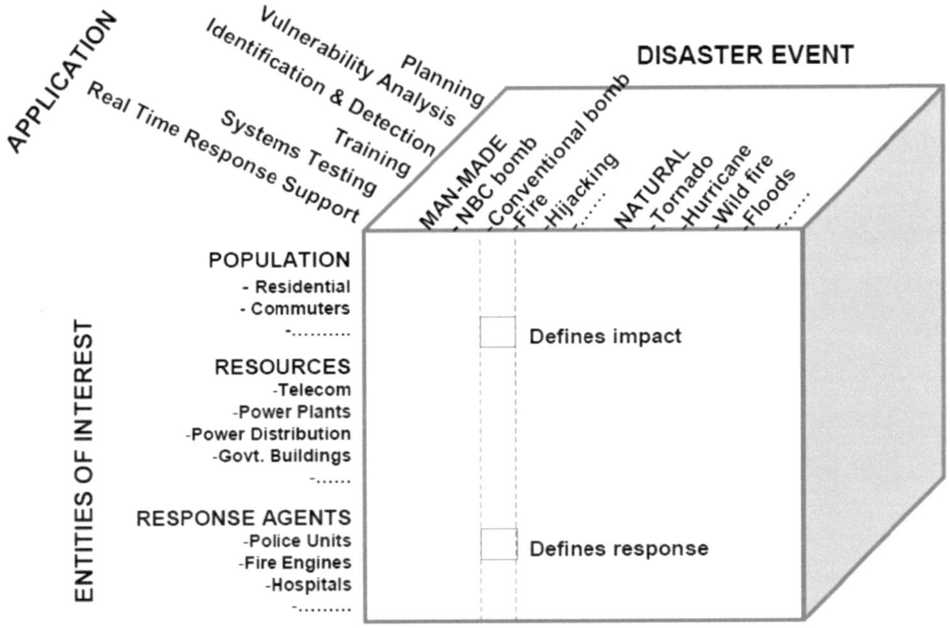

Figure 19: Integrated Emergency Response Framework (IERF) proposed by NIST (Source: Jain and Mclean, 2003).

Considering that such treats is really existing in the world it is necessary to have a measure of system vulnerabilities in order to monitor and mitigate the susceptibility of system and the whole society.

This book proposes a Bow Tie model to measure Systems and a Multi- Bow Tie Model to access a facility, city or country vulnerability which is exposed to multiples treats. As mentioned in chapter 2, The Bow Tie methodology is a risky method which considers on the left diagram side the probable cause of the incident, the incident in the middle and the consequences on the right size. Among the causes and incident is the control measures and between incidence and consequences are the recovery measures. In case of vulnerability analysis the causes are threats like natural disasters, terrorism attack and hacker's attacks. The control measures are protecting, check, monitoring and anticipate actions. The incident is the susceptibility of threats and recover measures a copying capacity to mitigate threats' effects. The Fig. **20** shows a Bow tie Model which describes the vulnerability of systems like industrial plants, trains, commercial building and aircrafts.

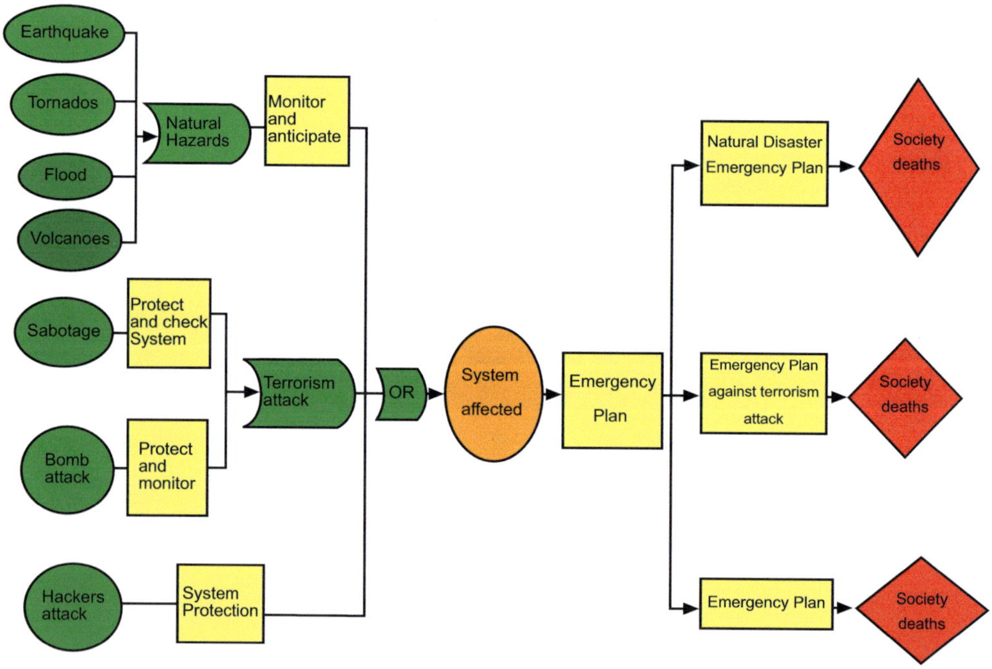

Legends:

Potential Causes (exposure)

Control Measures (Control Measures)

Loss of Control (suceptility)

Recovery Measures (coping capacity)

Consequences

Figure 20: Bow Tie Vulnerability Analysis.

The system vulnerability can be described mathematically by the frequency of the system is susceptible to all threats. Such frequency is a combination of treats frequency and control measures probabilities (copy probability). By this way the System Sulceptibility will be:

$$S = \sum_{i=0}^{n} F_{t_i} \times C_{M_i}$$

where:

$$S = System\ Sulceptibility$$

F_t = Treat frequency

C_M=Treat control measure failure probability (copy failure probability)

Therefore, the society's vulnerability will be:

$$V_{so} = \sum_{k=0}^{n} S \times Co_k$$

where:

$$V_{so} = Society\ Vulnerability$$

$$S = System\ Sulceptibility$$

k=the system or society affected by treats

Co_k= the coping probability of not mitigate the treats consequence

Depends on system accessed it´s easier to mitigate the vulnerability by reducing the susceptibility or by increasing the coping capacity. In case natural catastrophes, it´s hard to reduce the susceptibility by reducing the frequency of natural disaster or by avoiding their effect on systems, but it´s easier to mitigate the society's vulnerability by increasing the coy capacity efficiency like effective alarms, emergency response, safe places and efficient evacuation. Others threat like terrorism attack and hacker attach, the more effective is to reduce the susceptibility by monitoring the threats and reduce the frequency that such threats affect the system.

The treats events can have multi effects on different systems on the same location, in other words, city state or country. Because of that, is necessary to have a complete Vulnerability analysis considering all systems affected because is necessary for prior which location requires support and which kind of support. Therefore, a Multi Bow Tie is a more appropriate model and allows accessing all threats' effects on different systems with different consequences. The Fig. **21** shows the Multi Bow Tie model to have a complete Vulnerability analysis.

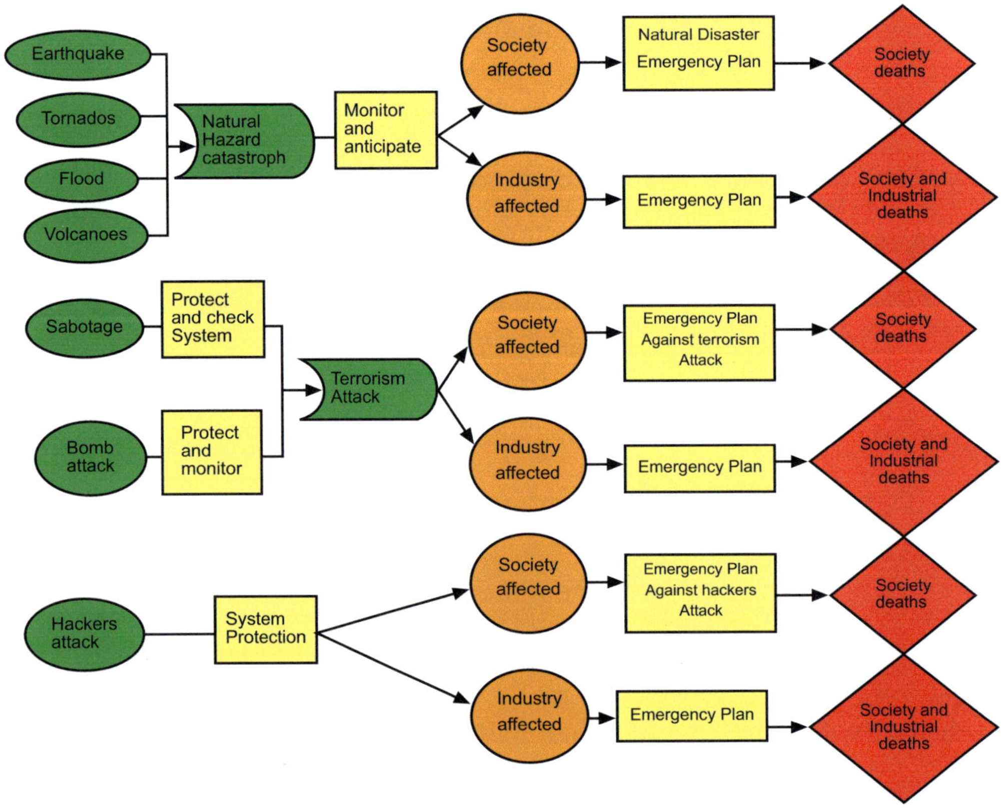

Figure 21: Multi Bow Tie Model for Vulnerability Analysis.

Considering that threats are able to affect system and society, it is necessary to consider different susceptibility for each threat group (Natural catastrophes, Terrorism Attack and Hackers Attacks). In addition, different emergency plans will be carried out depends on treat characteristics.

Considering that such traits can affect society or Industrial plants, there are other two possibilities that can describe in terms of probability to be multiplied by treating frequencies. The Multi Bow tie Model regards different vulnerabilities natures like Natural hazard catastrophes, terrorism attacks and hacker attacks. Consequently, the Total vulnerability is the sum of all vulnerabilities.

$$V_T = \sum_{k=0}^{n} F_{t_i} \times C_{M_i} \times Co_k$$

k=1,2,.n

where:

V_t=Total Final Vulnerability

F_t = Treat frequency

C_M=Treat control measure failure probability

k=the system or society affected by treats

Co_k= the coping probability of not mitigate the treats consequence

After defining the vulnerability in terms of frequency is also important to estimate the expected number of susceptible treats in order to help emergency response and security teams have a target and keep such number as low as possible. By this way is possible to define the expected number of susceptibility by.

$$E(N_i) = \int_o^T \rho(t)\,dt$$

The Crow Ansaa Model assumes that the intensity of the event is approximately Weibull event rate, thus intensity of event on time is:

$$\rho(t) = \frac{\beta}{\eta^\beta} T^{\beta-1}$$

Considering initial event rate as:

$$\lambda_i = \frac{1}{\eta^\beta}$$

If we consider the event as a threat, the cumulative threat rate is approximately threat intensity we have:

$$\lambda_c = \beta \lambda_i T^{\beta - 1}$$

when β=1,

$$E(N_s) = \lambda_c \mathrm{T}$$

where:

$E(N_s)$=Expected number of susceptible threats

λ_C = Threat frequency

T=Accumulated time

The equation above describes the threat intensity and depends on β value it´s increase, decrease on keep constant along time. Is very important to have in mind that β in Crow Anssa Model describes threat intensity behavior and have not relation with Weibull distribution shape parameter. In fact, β is a shape parameter of threat Intensity Function in Crow Anssa Model. Thus, in this model when β>1 means higher threat because threat intensity is increasing, in other words, the frequency of threats increases and control measures and copy measures actions are not reducing the vulnerability. When β<1, threat intensity is decreasing along time, in other words, threats frequency are reducing or control measures and copy measures actions are reducing the vulnerability. When β=1, the threat intensity is not getting higher or lower.

The expected number of catastrophic consequences in a cumulative time must be between 0 and 1. Therefore, we can consider low vulnerability for values between 0 and 0.4, medium vulnerability for values 0.4 and 0.7 and high vulnerability for values between 0.7 and 1. In case of low vulnerability the monitoring and data updated must be continuous but is not necessary improvements. In case of medium vulnerability is necessary to monitoring the threats, improve exist control measures or implement additional control measures as well as copy capacity improvement to achieve a low vulnerability level whenever is feasible. In case of high vulnerability is necessary to monitoring the threat and try to eliminate them whenever it´s possible, improve existing control measures or implement additional control measures as well as update and improve copy capacity. In addition, in order to mitigate the system and the society threat effect is recommended to shut down systems and isolate the possible society affected or move them to a safer location.

Finally, if copy capacities are not able to eliminate treats there will be consequences and society, industrial population or both will be affected. By this way is also important to estimate the number of deaths, causalities and cost caused by treats in order to have complete consequence analysis of vulnerability effect. Thus, the risk related to such treats is combination of vulnerability with the expected number of deaths, causalities or cost. The number of deaths is more effective to have a perception of whole societies and can be compared to other risk indicators (individual and societal risk). In fact, there's no an acceptance risk criterion to such threats events (natural catastrophes, terrorism attack and hacker attacks) nowadays and is a worldwide concept that as lower as possible better is to the whole society. The consequence can also be measured by cost and by this way it is possible to compare the investment necessary to improve the control measures as well as recovery measures.

REFERENCES

Apostolakis G., Michaud D. (2006). Screening vulnerabilities in water-supply networks. PSAM8, New Orleans. USA.

BBC, Tsunami disaster. 2005. URL:http://news.bbc.co.uk/ go/pr/fr/- 1/hi/world/asia-pacific/4136289.stm.

BBC,http://news.bbc.co.uk/go/pr/fr/-/1/hi/world/americas/4326084.stm (Published: 2005/10/10 11:51:05 GMT).

BBC,http://news.bbc.co.uk/1/hi/world/south_asia/4322624.stm (15/112005).

Booth, R.T., 2000. An iconoclastic critique of the Sacred Cows of Health & Safety. An analysis of the conventional wisdoms of health & safety management. Sampson Gamgee Memorial Lecture, Birmingham Medical Institute (unpublished).

Centers for Disease Control and Prevention (CDC). 1993a.

Centers for Disease Control and Prevention (CDC). 1993b.

Centers for Disease Control and Prevention (CDC). 1993c.

Chartoff, S.E., and Gren, J.M. (1997). Survey of Iowa emergency medical services on the effects of the floods.

Darby J. (2006). Evaluation of terrorist risk using belief and plausibility.PSAM8, 2006, New Orleans, USA.

D.C. Tanaka, K. (1996). The Kobe earthquake: The system response. A disaster report from Japan. Eur. J. Emerg. Med., 3, 263–269.

Deneubourg J. L.&Goss S. (1989). Collective patterns and decisionmaking, Ethology and Evolution. 1989; 1: 295–311.

Eriksson D. M. (1997). A principal exposition of Jean-Louis LeMoigne's Systemic theory. Cybernetics and Human Knowing, 1997, 4 (2–3).

Fontaine B., Debray B., Salvi O. (2006). Protection of hazardous installations: Complementary safety & security approaches. ESREL06, 2006, Estoril, Portugal.

Flood-related mortality. (1993). MMWR, 42,836–838. 1993, Missouri.

Freeman, S. (2005). A generation has been lost. Times Online(www. thetimes.co.uk), October 10, 2005.

Garrick B.J. *et al.* (2004). Confronting the risks of terrorism: making the right decisions. Reliability Engineering and System Safety, 2004, 86: 129–176.

Guha Sapir, D., and Lechat, M.F. (1986). Reducing the impact of natural disasters: Why aren't we better prepared? Health Policy and Plng., 1, 118.

Hagman, G. (1984). Prevention better than cure. Swedish Red Cross, Stockholm. International Decade for Natural Disaster Reduction (IDNDR) Secretariat.

Hessami A.G. (2006). Safety and security assurance – an integrated framework. ESREL 2006. Estoril, Portugal.

Holland J. H. (1975). Adaptation in Natural and Artificial Systems. University of Michigan Press, Ann Arbor, 1975.

Jain, Sanjay and McLean, Charles R. (2003). Modeling and Simulation for Emergency Response: Workshop Report, Standards and Tools. Modeling and Simulation for Emergency Response Workshop. NISTIR 7071, December 2003. http://www.mel.nist.gov/msidlibrary/doc/nistir7071.pdf

J.R. Santos-Reyes.(2006). Edge Hill railway accident: A systemic analysis. Safety and Reliability for Managing Risk – Guedes Soares & Zio (eds).© 2006 Taylor & Francis Group, London, ISBN 0-415-41620-5

Kunii, O., Akagi, M., and Kita, E. (1995). Health consequences and medical and public health response to the Great HanshinAwaji Earth-quake in Japan: A case study in disaster planning. Medicine and Global Survival, 2, 32–45.

Le Gallou F., Adjallah K. H., Châtelet E. (1999). Systemic approach and industrial ecology. Attempt of global methodological approach and finalization. International Conference on Industrial Ecology and Sustainability, 1999, Troyes, France.

Morbidity surveillance following the Midwest flood—Missouri, 1993. MMWR, 42, 797–798.

National Research Council (NRC). 1987. Confronting natural disasters: An international decade for natural disaster reduction, National Academy Press, Washington, D.C. Nations, New York.

Office of U.S. Foreign Disaster Assistance. (1999). Disaster history: Significant data on major disasters worldwide,1900–present. Agency for International Development, Washington,

Public health consequences of a flood disaster.(1993). MMWR, 42, 653–656. 1993, Iowa.

Rausand Marvin; Arnljot Høyland. (2004). System Reliability Theory: Models, Statistical Methods, and Applications, 2nd Edition. Wiley Series in Probability and Statistics 2004

Samrout M., Yalaoui F., Châtelet E. & Chebbo N. (2005). New methods to minimize the preventive maintenance cost of series-parallel systems using ant colony optimization.

Reliability Engineering and System Safety, 2005,

S.A. Soria, R. Josens, D.R. Parisi. (2013). Experimental evidence of the "Faster is Slower" effect in the evacuation of ants. Safety Science. Volume 50, Issue 7, August 2012, Pages 1584-1588.

StützleT. (2002). The Ant Colony Optimization Metaheuristic: Algorithms, Applications, and Advances. In F. Glover and G. Kochenberger, editors, Handbook of Metaheuristics, Kluwer Academic Publishers, Norwell, MA, 2002. 879

Tavares, Rodrigo Machado; Marshal, Steven. (2012). The development of a real performance based-solution through the use of People Movement Modelling Analysis (PeMMA) combined with fire modeling analysis. Safety Science. Volume 50, Issue 7, August 2012, Pages 1485–1489.

The international decade for natural disaster reduction: Action plan for 1998–1999

Geneva. United Nations Office for the Co-ordination of Humanitarian Assistance.

CHAPTER 6

Incident and Accident Analysis

Abstract: The accident is the worst event in the industry because it causes employee health damage. Depending on consequence effect the accident can be classified as a major accident or minor accidents. Much attention has been driven to avoid the major accident compared with the effort to avoid the minor accident. Despite the severity of a major accident, the minor accident also causes damage to the employee's health and has been occurring much more frequently in all industries. In order to avoid an accident, the main philosophy is to be preventive and pay attention to incident to avoid that accident happen. That requires full attention to unsafe asset conditions during all lifecycle phases. The incident and accident analysis is carried out by different methods like Ishikawa diagram, Why because diagram, event tree analysis, Fault tree analysis, Bow Tie Analysis. Such methods have different features. What define which is the best method to be applied is the type of accident that in some cases is complex due to different causes combinations and more than one consequence. This chapter aims to describe the incident and accident analysis methods with examples to clarify the drawbacks and advantages of each method.

Keywords: Incident, accident, Ishikawa Diagram, Why because diagram, event tree analysis, Fault tree analysis, Bow Tie Analysis.

6.1. INTRODUCTION

The incident and accident analysis has the main objective to perform an assessment about all causes, conditions, human factors which take influence on incident and accident in order to prevent such event based on workplace monitoring improvement processes.

The accident can be defined as "an undesired and unplanned event or sequence of events which cause health damage, assets disruption as well as environmental impact". Despite similar nature an incident has not a negative effect as a consequence.

On most of cases incident and accident are related to time because the incident, event happens previously without cause, any damage to the employee's health and facilities. That means before an accident, some incident with similar or same event and conditions happen. Thus the main question arises: why is so hard to avoid accidents if similar incident, event happens previously? In fact, there are some reasons which may explain that like the following:

– Missing information about the incident;

Eduardo Calixto

- Human factors are misunderstood during incident analysis;
- Recovery actions are not enough to avoid the incident turn into an accident;
- Recovery actions take too long to take place;
- Combination of failure on layers of protection;
- Operational and maintenance teams are not familiar with new technologies.

Missing information about incident does not allow taking place preventive measures to eliminate or mitigate the risk, in other words, to reduce the accident probability or reduce its consequence. One of the most common reasons to have such missing information is that many incidents are related to human factors, *e.g.*, human error. In this case, employees are afraid to report the incident and have some kind of punishment. That can be explained by one of the most common errors in incident and accident investigation and analyses is to look for someone to guilt. Many companies do not realize the real difference between human error and sabotage.

Sabotage is an intentional action to cause damage to a specific system and is not considered to be a human error. The point is that there is an interface between human error and sabotage, which is "not following the rules". In this case, although not following the procedures, employees do not have the intention to cause damage to the system most of the time and such behavior is due some stimulus like increasing production, not stopping production or even carrying on maintenance tasks as fast as possible. Even though, whenever a major accident happens and one of the root causes is human error, many companies try to find someone to blame for the accident and because of that many incident data are not reported by employees.

In addition, in many countries, it is not necessary to report incident events and because of that, many companies do not proceed with incident analysis to find out the real causes in order to mitigate the risk. Actually, such event happens in the ground area and in many cases the managers have no idea about such events.

Regarding human error, many factors influence in human performance and some of them are under company control like procedures, tools, technology, workplace conditions, workload, ergonomic. Such factors take influence on human performance and sometimes are the incident root causes. There are also other factors like psychology, sociologic and health that is not under company control but can be monitored by supervisors and leaders in order to detect when employees are not in their best conditions to carry on specific tasks which can lead to an incident.

Actually, human error is influenced by human performance factors and some of them can be avoided. If employees are well trained, there is a good workplace condition and workload is adequate such human error is reduced fairly. Nevertheless, there are many accident cases where such factors were not observed by the company, for example, even common issues like to allow employees without specific training or experienced enough to carry on complex and dangerous tasks. Thus, the first step that a company needs to carry out is to identify the critical human performance factors meanly in dangerous activities in order to prepare employees to avoid human error.

Even if incident analyses are carried out in many cases, such recovery actions are not enough to avoid accidents with similar causes. That interesting issue mostly happens because of the dynamic nature of the accident. Whenever recovery actions are proceeding like to install layers of protection, training employees, perform inspections, routine or even install more reliable and safer equipment, such actions must be monitored and the risk must be assessed constantly. Despite more reliable equipment and layers of protection are installed, such devices need to have a long term inspection and maintenance policy. The experienced and well trained personnel change from time to time to other management and new employees without too much experience take their places. Even for experienced professionals, it is necessary to update and training them constantly. Whenever some modifications in the process occur or even a new process condition take place, it is necessary to revise procedures and update employees. Thus, regarding all such recovery conditions, it is clear that is really hard to monitor and keep all of them under control for a long time. That explains why some accidents happen even though recovery action were implemented following a previous incident.

In some cases, the recovery actions are not the best ones, but the simplest or the cheapest. Because of that, whenever a similar condition or a little bit harder condition happens, the accident occurs. A good example of this is when it is necessary to carry out an inspection and preventive maintenance to maintain a layer of protection in a high availability level and due to reduce cost policy is decided to cut part or the total layer of protection preventive maintenance. In many cases when such devices fail, there is no plant shutdown. In other cases the number of trainees is cut off in order to reduce operational costs.

Actually, due the cost of recovery action and misperception of their advantages, such preventive actions are not implemented or are postponed even for more than one year later, because it was not considered in a budget and within this period of time the accident happens.

The preventive approach to try to avoid the accident basically is monitoring the process and human factors constantly and whenever an unsafe condition is detected, a preventive action is planned and implemented before the incident happens as shown in Fig. **1**. Once the preventive action is effective in avoiding the incident, it is necessary to standardize such action as well as the monitoring process and human factors and update procedures and trainee (?) employees as well as process modification or operation conditions happen.

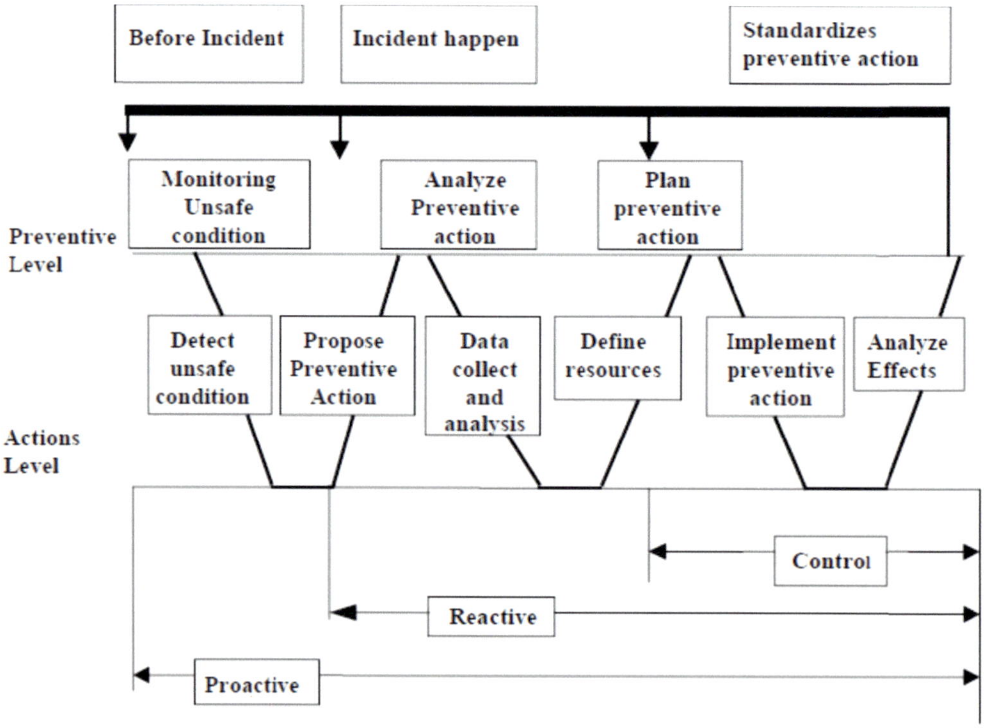

Figure 1: W V Model (Source: Shiba Shoji, 2002 adapted).

When all recovery actions take place on time and are monitored, it is also necessary to really understand the relation amongst layers of protection and to be aware that all events are probabilistic. Actually, it is necessary to have in mind that the hazard identified is based on knowledge of all events that exist until the present time. Sometimes an unknown process condition or even an environmental condition can trigger events that layers of protection are not able to prevent accidents. Even layers of protection have limits to operate under specific conditions. The layers of protection are designed to specific unsafe conditions and whenever such conditions are other than the technical limits, the accident can happen.

When a tsunami reached the Fukushima Nuclear Power Plant in 2011 (Japan), the cooling water pumps were not able to cool down the reactor because they were damaged by the tsunami waves, so in this case, the layers of protection were not available when required. Actually, the layers of protection were not designed to operate under such circumstances. The Plant was not designed to withstand a tsunami of such scale.

Finally, when maintenance and operations professionals are not familiar with new technologies, it is possible that incident and accident happen because they are not acquainted with new devices and are not able to use them to mitigate the incident. A common error when a new technology is implemented in a process it is do not train employees well enough to operate or carry on maintenance and in such case the device operates in an unsafe condition or even can cause an accident during a maintenance task. The appropriate steps to start up and shut down such devices must be clear and professionals must be trained enough to understand such steps. Another common mistake that happens in many cases, is related to the fact that operational or maintenance procedures and devices manuals are written in different languages. Even when there is some support from companies that supply such devices, in some cases, they speak in a foreign language that is not usual for employees who carry out maintenance tasks or operate such devices. In addition, even though all operational and maintenance procedures are well understood, there is some unknown failure that is not usual and when such events happen the employees do not know how to proceed.

All the situations discussed above explain why incident turn into an accident and when an accident happens the only thing that can be done it is to learn for not repeating in future. In order to understand accident circumstances, it is necessary to carry out the detailed and systematic analysis. Basically, there are three primary types of accident analyses:

- Assessment to define where accidents occur;

- Risk assessment to prioritize preventive actions to avoid such accident;
- Assessment to understand how accidents occurred, the direct and indirect causes;

To identify where accidents happen and which type of accident, that means, the level of consequences and damages to system and employees, it is necessary to have a precise report about the accident. Therefore, all discussion above about missing information can have influence on accident analysis as well as the recovery action's effectiveness.

Whenever an incident and accident historical data exist, it is possible to compare the type of accident, their causes and consequences with previous events. This enables to know if recovery actions are good enough to prevent and reduce the number of incidents and accidents. Therefore, it is possible to update the risk level and check if such occupational risk is under control or needs to be mitigated. The ideal situation is to detect the unsafe conditions and implement recovery actions after the incident or before it to prevent accidents to occur.

The preventive actions and monitoring unsafe conditions are the essence of safety management, but unfortunately they are hard to be implemented in practice for many companies in different industries. The monitoring unsafe condition requires constant inspections and attention to the process and workplace conditions. A very good example is what happening in the Nuclear industry, where all process unsafe conditions are identified. The Event Tree is developed and input on electronic data systems which enable the operator in the control room to know when a combination of unsafe conditions are triggered as shows Fig. **2**. Such conditions come out on the screen to operate as well as the procedure that must be followed to precede the preventive action and prevent accidents.

That is the best practice and state of art to monitoring the unsafe condition on the process, but even though some actions like inspections, corrective and preventive maintenance are carried out by employees and depends on their risk perception about unsafe conditions in such tasks to prevent accidents. In addition, all human performance factors which have an influence on human error can also lead to an accident to occur and because of that, the avoidance of incidents and accidents will always be a big challenge for all industries.

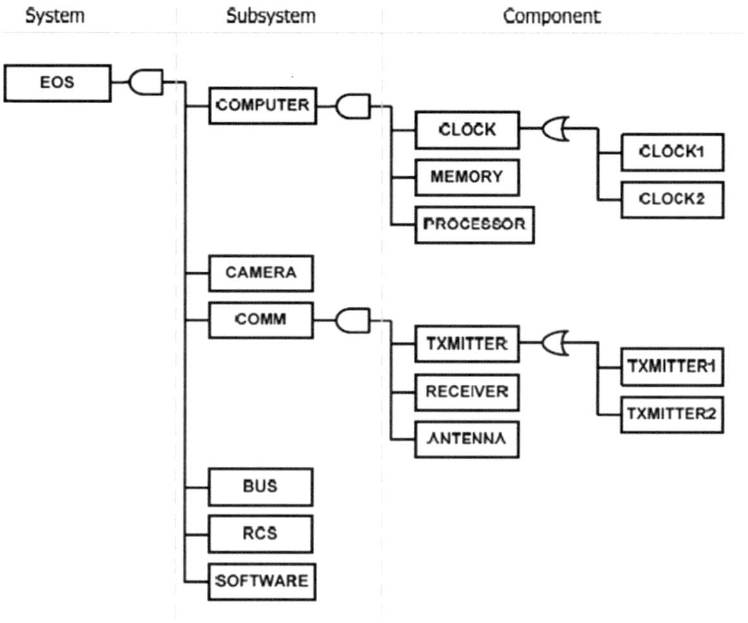

Figure 2: Nuclear Plant Fault tree (Source: Hamed, 2007).

In fact, when an accident happens, the only thing that must be done is to learn with such event in order to prevent that it will not happen in the future. Consequently, it is necessary to access the incident and accident causes and consequences and find out the real root causes. In some cases, this is really hard to be done because the accident area is destroyed or contaminated. Some accidents in Oil and Gas industry have as consequence fire or explosion, which causes a serious damage in the accident area, which turns it hard to assess the accident causes. Even though there are specialists who collect evidence in the accident area in order to get into some conclusions. That is similar, in the transportation sector, when catastrophic accidents happen with aircrafts, trains or ships. The nuclear industry has similar conditions and worse situation when a nuclear accident happens. In this case, a huge area inside or outside of a nuclear plant is contaminated, that makes harder to have access to the accident area. In such cases, an investigation takes longer time and in some cases can last years to find out the real root causes of accidents.

No matter how difficult the accident analysis is, the most important thing is to have a methodology to be carried out in order to assess the accident and find out the real causes.

There are different methodologies that must be applied in order to assess incidents and accidents.

In most of the cases the main objective is to identify the factors which took the influence on accident and the sequence of events, thereby traditional methods such as Why-because analysis, Sequence of events (domino effect), Fault tree analysis and others are applied. The next item will describe different approaches to assess incidents and accidents, their advantages and drawbacks as well as examples applied to different industries, which will make easier the reader's comprehension.

6.2. INCIDENT AND ACCIDENT ANALYSIS METHODS

Unfortunately, many incidents and accident happens because there is not a monitoring routine to process and human factors as well as assessment of critical new process or task condition. Because of that, incidents and accidents will always happen and under such circumstances, there is only an opportunity to learn with such events, in order to improve workplace conditions and to prevent future incidents and accidents. Therefore, it is necessary to have good methodologies to analyze events like the following ones:

- Ishikawa Diagram;
- Why Because Diagram;
- Sequence Time analysis;
- Event Tree Analysis;
- Fault Tree Analysis;
- Bow Tie Analysis;

This item will describe each one of such methods with real examples from different industry sectors and the next item will discuss two major accidents applying such methods.

6.3. ISHIKAWA DIAGRAM

Ishikawa diagrams were proposed by Kaoru Ishikawa in the 1960s, in order to solve quality management process problems in the Kawasaki shipyards.

The Ishikawa Diagram main objective is to perform problem assessment take into account different aspects such as people, methods, machines, materials, measurement and environments. The first step in the Ishikama Diagram method is to have a clear understanding about the problem which is intended to solve. The second step is to gather specialist to take part into a brainstorm to discuss about the problem cause and propose solutions. In fact, such methods have started to improve product and process quality by assessing the main cause of noncompliance. Despite an easy methodology, the most important thing is that participants have the same understanding about each subject mean. The usual subject means are:

- **People:** Human error causes.
- **Methods:** Methods and procedures causes.
- **Machines:** Equipment failures.
- **Materials:** Material specification and defect causes.
- **Measurements:** Data sources, index calculation or measurement mistakes.
- **Environment:** Operation condition causes.

Despite such subjects were established at the beginning of application, different other subjects may also be applied in order to understand a problem cause such as maintenance, operation, management, leadership, *etc*. The Fig. **3** show the traditional Ishikawa diagram with causes subjects. The Ishikawa diagram is known as a Fishbone diagram because it has a similar fish skeleton shape.

The Ishikawa Diagram can also be applied to incident and accident analysis and in this case the consequence will be an incident or accident and the causes will be related to human error performance factors, management, process condition or equipment. The causes assessed in incident and accident analysis can be predefined or even defined by specialist groups, when they assess the accident. In fact, on most of cases, it is better to predefine some usual causes like equipment failure, process conditions, management and human error. Thus, if necessary, any modification the specialist adapts other cause of an incident or accident analysis.

Consequently, it is possible to carry out an incident or accident analysis based on expert judgment about accident causes and then propose preventive actions, define resources and if feasible standardize such actions. Before applying incident and accident analysis some steps are always required like:

- The first step in incident and accident analysis is to analyze the incident or accident local.
- The second step is to collect evidences in an incident or accident local, take pictures and interview people.
- The third step is to define who are able to describe the event and help on incident or accident analysis.
- The fourth step is definitely a place to carry on incident or accident analysis and bring evidence, data and incident or accident historical data.

An example of an Ishikawa Diagram applied to an incident happens in a Plant where scaffold falls down on the floor before maintenance is carried out without causing any employee damage.

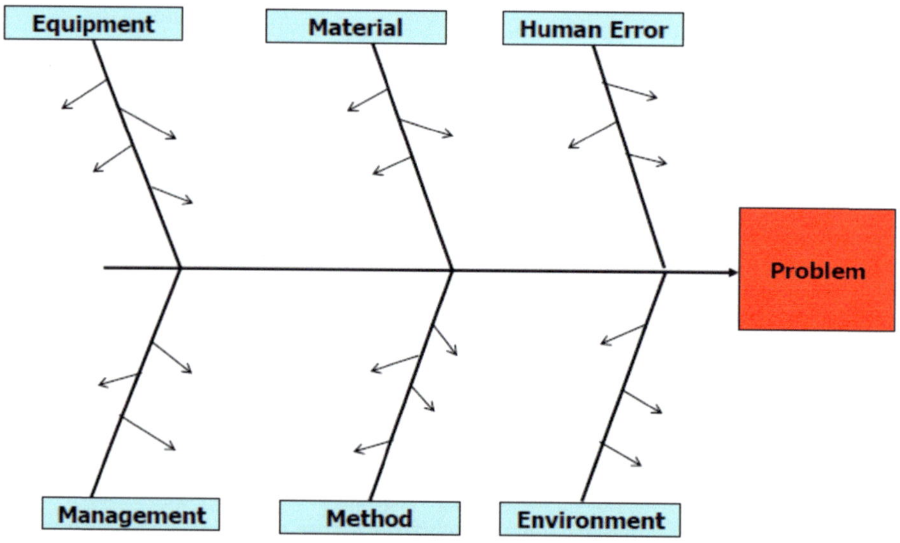

Figure 3: Ishikawa Diagram.

Such incident is usual to all industry whenever a work must be done in a high place but it is more usual in the Construction industry, that faces constantly such hazard. The scaffold is necessary whenever a high place must be accessed to perform a service that can be an equipment inspection or maintenance, or even a building repair, painting or roof replacement. Despite the fact that in this specific case was an incident, such event is serious because it is able to cause death to employees and an incident analysis was carried out to solve this problem and prevent accidents.

Thus, the first step is taking some pictures of the local of the incident and collect evidence. The Fig. **4** below shows a scaffold fall down and the local evidence collected shows that the scaffold was not well hold on the wall. The second evidence was that the preliminary hazard analysis was not conducted because the checklist document was not found out by the local of the incident.

Figure 4: Scaffold fall down incident (Source: http://scaffmag.com).

With such evidence and a meeting with safety specialist as well as operators involved in such task to build up scaffold was carried out a brainstorm to discuss the incident causes. In doing so applying the Ishikawa Diagram, based on brainstorm discussed we reached the conclusion that human error, management, method and environment took influence on the incident, as shown in Fig. **5**.

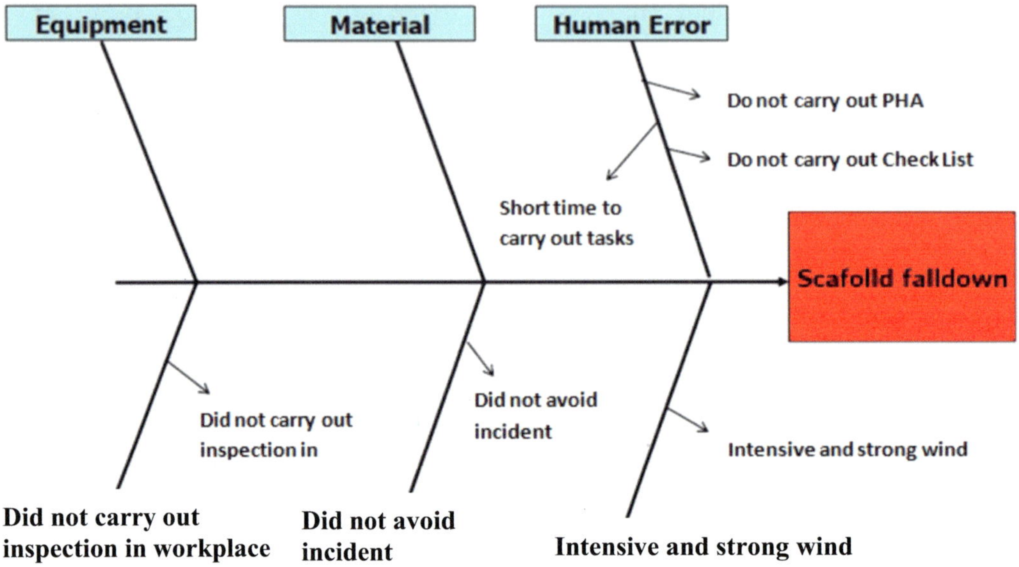

Figure 5: Ishikawa Scaffold fall down incident.

Considering management accident causes, during brainstorm, the group revealed that in this specific day there was not any inspection in workplaces, even knowing the hazardous serious activities that were being carried out on operational ground. The manager responsible for such safety inspection declared that the supervisor was contacted to carry out such inspection, but he mentioned that such inspection was not possible because he was involved in another preliminary hazard analysis at the same time that the scaffold was being assembled. Actually, such activity is a manager's responsibility and according to the company procedures he is not allowed to transfer such responsibility to others.

The human error was evident because the preliminary hazard analysis was not carried out as well as the checklist to work at a height activities was not used. The reason for that was that employees were required to assemble the scaffold as soon as possible and did not apply such safety procedures. In fact, because the employees were experienced in assembling scaffolds they believe that such scaffold was correctly assembled.

Regarding environmental issues, the wind started to blow more intensely after the scaffold was assembled and in this situation, it is defined in the company procedures that a new preliminary hazard analysis as well as a checklist must be carried out to guarantee that such scaffold is safety under this new environmental situation.

Finally, the incident analysis group also concluded that the method was not correct because the usual procedure was not able to prevent the incident. In addition, evidence of failure in equipment and material was not found out.

After incident analysis some actions like training employees in safety procedures have been conducted, which established the safety inspection as a basic condition to start such activities, with the purpose of preventing such incident to occur in the future. Thus, an action plan was performed as shown in the Table **1** below.

Table 1: Incident Action Plan

WHAT	WHO	WHEN	STATUS
1 – Spread out the incident analysis for all employees	Managers, Supervisors	One month from accident data	OK
2 – Trainee employees involved in high workplace activities	Safety engineer	Maximum one month for all employees	50%
3 – Modify safety procedures and establish that any high workplace service must be authorized for safety inspection.	Safety Engineer	One week after incident data	OK

The Ishikawa Diagram is a good method to be applied on incident and accident analysis and the main advantages of such methodology are:

- Easy to understand and be applied;
- Does not require a huge investment in training;
- Does not require a huge investment in software;
- It is flexible and can be applied to different types of incident and accident analysis;

The Ishikawa drawbacks are:

- Does not consider the events combination that trigger accident;

- Does not take into account the sequence of events combination that trigger accident;
- Does not consider the consequences of the accident;
- Does not consider causes of accident on time;
- Totally dependent of the group experience to describe the main accident causes.

Despite being a good incident and accident analysis tool, such drawbacks must be taking into account depending upon the situation and the type of result that is required for the analyses. It is well recommended for simple accident and incident cases where the causes of an incident or accident are evident.

6.4. WHY BECAUSE ANALYSIS

The Why because analysis (WBA) is a straight method to assess incidents and accidents and their main causes. This method has been applied in different industries all over the world. Basically, the method start with a question about the incident or accident and a logical tree is made up from the top to the bottom. The Why because analysis result is a schematic tree which has all causes and their sub causes. It is important to point out that such events do not have any kind of logic or dependency hierarchy and only describe events' causes and consequence relations. In order to clarify such methodology, a rail accident example is presented.

The Why-Because Analysis was applied to Glenbrook Rail Accident in order to define the accident causes. "The collision accident between two passenger trains travelling in the same direction occurs on 2 December 1999 at Glenbrook, in the Blue Mountains, west of Sydney, Australia".

An inter-urban train from the Blue Mountains to Sydney collided with the rear of an interstate train, the Indian Pacific, designated WL2, which had been waiting at Signal 40.8, which was showing "halt", and was starting to move off. The interurban train designated W534 had just passed signal 41.6, some 1.1 km before signal 40.8, after receiving clearance from the signaler to precede, for it was showing "halt". The interurban train driver accelerated to 50 kph in the block 1, and only saw the rear of the Indian Pacific a short distance before the collision (Ladkin, 2005)". Fig. **6** below shows the WBG.

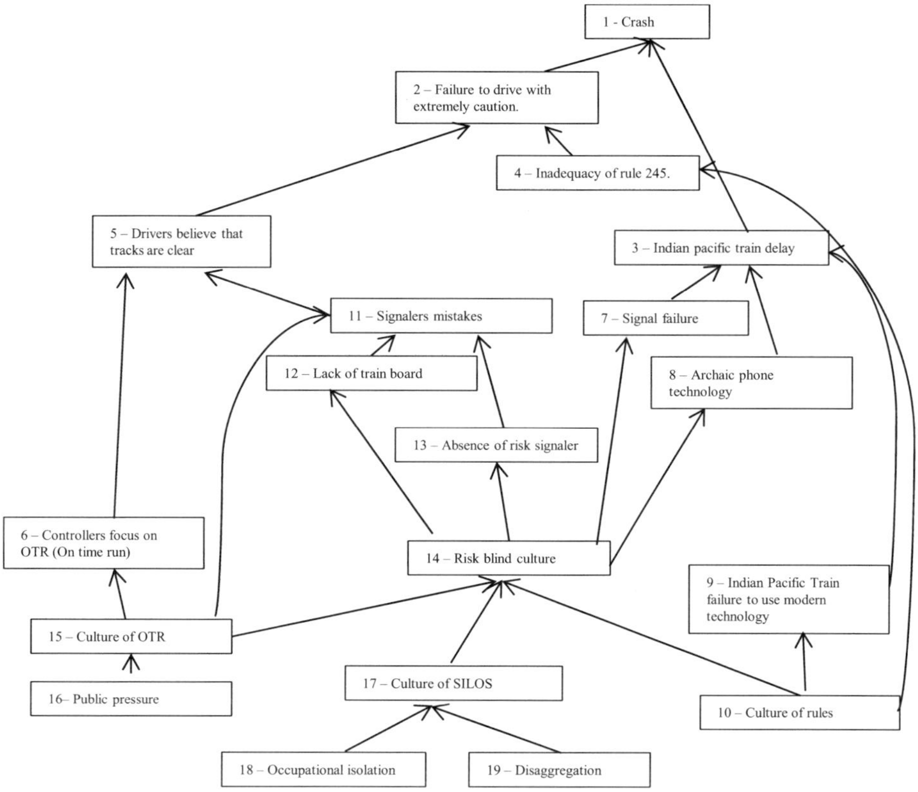

Figure 6: The Hopkins Accimap, rendered by the WB-Toolset (Source: Ladkin, 2005).

The Why because Graph at Fig. **6** above shows that there are technical and psychological causes related to a culture that contributed to the accident. Basically, the main causes of the accident were the failure of signalers, communication and imprudence of train drivers in driving faster. Actually, if we look into the causes of such mistakes we can observe that more causes related to procedures and unclear rules as well as culture contributed to such unsafe actions. Unclear rules and procedures are common causes of different types of incidents or accidents, but it is important to highlight the factors related to culture that can also be observed in many companies like "Silo Culture", "Culture of Rules" and "culture of on time run".

The "Silos Culture" occurs when a department or unit closes in on itself and starts acting for the benefit of itself rather than for the organization. There can be many reasons why silos culture develops, including management practices, target driven mentality and poor team leadership. In some cases, to remove silos' culture is difficult because it remains in some groups for years, but basically to remove silos culture is necessary to clarify the whole company objectives, processes and flow of products and information, try to operate in collaboration with other groups or departments as well as to choose good leaders and prepare them for leadership.

In this particular accident case the controllers were not aware of other departmental work and were blind about the influence of the other activities on their own activities. This is particularly clear when controllers allow a train run without a clear idea about the second train position.

The culture on time run is a presence in the transportation industry as well as many other industries. In fact, the culture based on time is present in many countries of the modern society nowadays and focus on being always on time and trying to reduce delays whenever it is possible. The point here is not to question such culture on time, but highlight that it is necessary to carry out always a tradeoff analysis, in order to decide about to be on time and the quality or risk of an accident that brings losses for the whole system and companies.

The culture of the following procedures is also present in the transportation industry as well in other industries. That is right regarding activities with catastrophic accident scenarios and in these cases it is clear that is very important to follow procedures. The main point here is to analyze such procedures in order to be made clear procedures for everyone and update them whenever necessary. In addition, it is also necessary to be cleared up that some actions are based on knowledge or skill. In these cases, procedures do not exert strong influence on the action performed.

The Risk Blind culture is related to not paying attention to the risk involved in some process due to other priorities, like the need to be on time or maximize production. Such risk Blind culture is also influenced by "on time run culture" as well as "silo culture". The Risk Blind culture is more general and has a different concept of the inattention blindness phenomenon. The Inattentional blindness, is a failure to notice a stimulus when there are other attentions demanding tasks that are being performed and it is not associated with any vision deficits. The Inattentional blindness happens because different stimulus overload the cognitive human capacity and consequently it is not possible to pay attention to all information and stimulus on same time.

Regarding the why because analysis methodology we can describe the following advantages:

- Easy to understand and to be applied;
- Does not require a huge investment in training;
- Does not require a huge investment in software;
- It is flexible and can be applied to different types of incident and accident analysis;

The why because analysis methodology drawbacks are:

- Does not consider the events combination that triggers an accident;
- Does not take into account the sequence of event combination that trigger an accident;
- Does not consider the consequences of the accident;
- Does not consider the causes of accident on time;
- It is totally dependent on the group experience to describe the main accident causes.

Despite being a good incident and accident analysis tool, such drawbacks must be taking into account, depending upon the situation and the type of result required. It is well recommended for accident and incident cases where the causes are evident and have not a time dependency or logic dependencies.

6.5. EVENT TREE ANALYSIS (ETA)

In incident or accident assessment the Event Tree Analysis has the main objective to assess an incident or accident that has one trigger event or a combination of events that result in a trigger event. In Most of the cases there are intermediate events between the trigger event (initial event) and final event (incident or accident). In such cases, one possibility is the accident will happen only if all intermediate events do not prevent or block the final event. This specific case is a LOPA (Layer of Protection Analysis) analysis, when intermediate events are layers of protection as explained in Chapter 3. When there are also events like manager decision, project mistakes and even layers of protection, the event tree is traditionally called Sequence Event Analysis, which will be discussed in the next item.

The other possibility is to have one trigger event or a combination of events that result in a trigger event and one or more than one consequence, depending upon what happened in the intermediate events. In Chapter 3, ETA examples like blow out accident or toxic release product accident were shown.

A good example of an ETA application of accident analysis is an explosion in a Chemical Plant. Fig. **7** below shows the accident area after the explosion.

Figure 7: Accident place after explosion (Source: CCPS Training course, 2011).

When the accident analysis group investigated the local of the accident, they realized that such area was confined to pipelines and probably to equipment like pumps. Such condition promotes an explosion, if a hazardous product cloud is in the presence of a heat source for a certain period of time. The second evidence of explosion condition was a confirmation that there was a pump in such a confined place where an explosion happened. The next step was the identification of a hazardous product leakage source, in this case a valve placed some meters away from the local of the explosion. There were corrosion signs in such valve, that enabled the hazardous product leakage and the wind direction drove the hazardous product cloud to a confined place, where the explosion occurred.

The accident investigation allowed understanding why explosion happens and not any other possible consequences, like jet fire or fireball. For example, a heat source close to the release source (valve) that would trigger a jet fire was not identified.

Fig. **8** shows a toxic explosion Event Tree that happens due to corrosion in a valve.

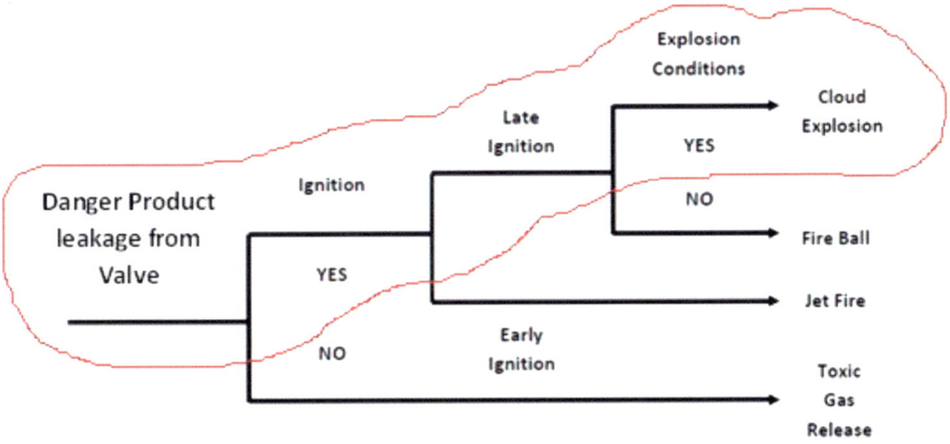

Figure 8: Event Tree of Explosion.

It is also possible to quantify the frequency of accidents as well as to estimate the risk of explosion like in Chapter 3 case, but in accident analysis the main objective is to assess the incident or accident causes.

Regarding the Event Tree Analysis methodology applied to an incident and accident analysis, we can describe the following advantages:

- Easy to understand and be applied;
- Does not require a huge investment in training;
- When qualitative, does not require a huge investment in software to be applied;
- It is flexible and can be applied to different types of incident and accident analysis;
- It considers logic dependences of event that trigger an accident;

The Event Tree Analysis methodology drawbacks are:

- Does not consider probabilistic dependencies of Event combination that trigger an accident;
- Does not take into account probabilistic dependencies of Event combination that trigger an accident;

Despite being a good incident and accident analysis tool, such drawbacks must be taking into account, depending upon the situation and the type of result required. It is well recommended for incident and accident cases where there is a sequence of events that leads to an incident or accident.

6.6. SEQUENCE ACCIDENT ANALYSIS

As stated above, in most of the cases there are intermediate events between the trigger event (initial event) and final event (incident or accident). In such cases, one possibility is that the accident will happen only if all intermediate events do not prevent or block the final event. This specific case is a LOPA (Layer of protection analysis) analysis when intermediate events are layers of protection, as explained in Chapter 3. When there are also events like manager decision, project mistakes and even layer of protection the event tree is traditional called Sequence Event Analysis. The sequence of event Analysis is traditionally represented by the Swiss cheese model, when all events that are able to prevent the accident fail as shown in Fig. **9**.

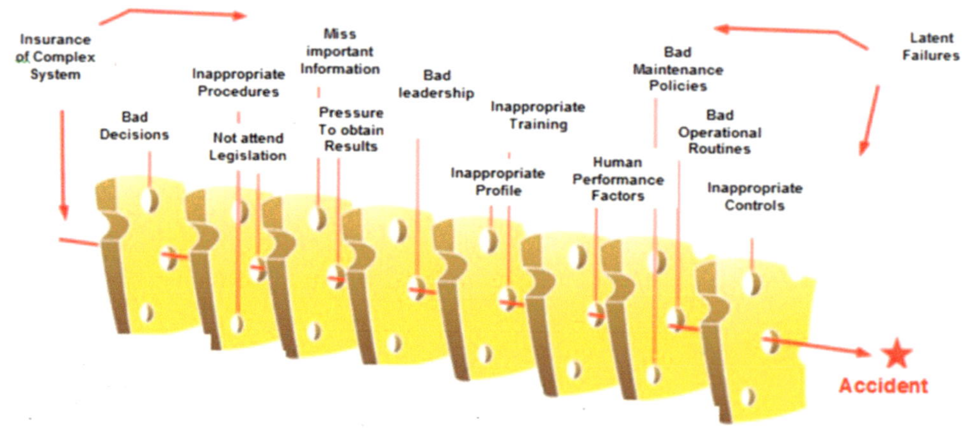

Figure 9: Swiss cheese model.

An example of a Sequence Accident Analysis can be applied in a huge oil spill in a refinery as shown in Fig. **10** due to sequences of event.

Figure 10: Oil spill accident (Source: http://static.guim.co.uk/sys-images/Guardian).

The first one was a bad decision to after installing a new oil tank, do not automatize the valve's action due to complex sequences of manual valve operation that, in case of failure, would lead in a huge oil spill. The second event was not updating all procedures. The third event was that the procedures did not regard the new accident scenarios based on risk analysis studies. The fourth mistake event was contracting a company to perform maintenance and operate tanks and process without a centralized leadership. In this case, the main company had no authority over contract company employees who operate and carry out maintenance on process tanks. Therefore, there was always a delay time to the contracted company fulfil the main company requirements.

The fifth mistake event was not training new employees well enough to operate the complex sequence of valves under highly demanding conditions.

The sixth mistake event was not paying attention to human performance factors, like poor workplace conditions, ergonomics, wrong work routine to maintenance and operation employees.

And the final mistake event was a loss of operational control in the operational routine that enabled one inexperienced operator mistake in a complex sequence of valves to trigger an oil spill to a local river. After one hour the oil spill was noticed and the emergency plan was triggered. Such task was observed previously by supervisors and all main process variables controlled and monitored. After the accident, a thousand gallons of oil were spilled into a local river. Fig. **11** summarizes the sequence of the accident analysis.

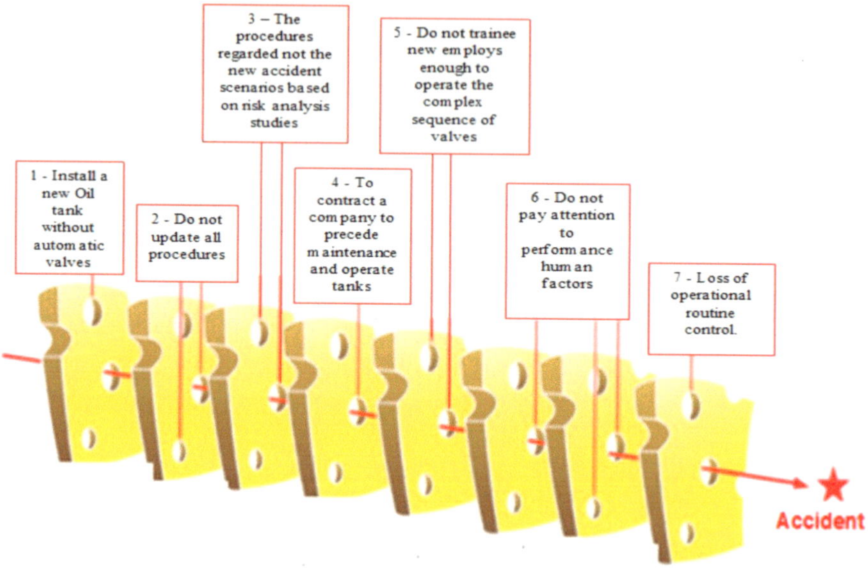

Figure 11: Sequence of the Accident Analysis.

Regarding the Sequence Accident Analysis methodology applied to an incident and accident assessment, we can describe the following advantages:

- Easy to understand and to be applied;
- Does not require a huge investment in training;
- Regards to decision and management mistakes;
- Is better applied to a sequence of events that triggers an accident;

The Sequence Accident Analysis methodology drawbacks are:

- Considers only a logic sequence of events that triggers the accident;
- Does not consider probabilistic dependencies of event combination that triggers the accident;
- Requires a lot of information that is related to bad decisions and managing a failure, which is not easy to obtain.

The Sequence Accident analysis regards bad decisions and management that in many cases create conditions to an accident when combined with latent failures. The methodology is appropriate to analyse a sequence of events that lead to an accident. In some cases, some of such events can be also analyzed by other methodologies like Fault Tree Analysis or Human Reliability Analysis. The next item will discuss the Fault Tree Analysis applied to incidents and accidents. The next chapter will discuss the human factors and human reliability analysis.

6.7. FAULT TREE ANALYSIS

As stated in Chapter 3, since 1961 the Fault Tree Analysis has been applied and the first application case was carried out by Watson to assess a missile control system. The FTA method provides a structured top down analysis by defining the top event and possible causes considering the combination of each event which is represented by logic gates.The logic Gates that basically are:

The Top Event triggered is necessary to satisfy the combination of all the events below the top event.

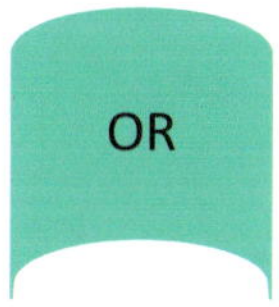

Logic Gate OR: One of the events below this gate must happen to trigger the logic gate OR.

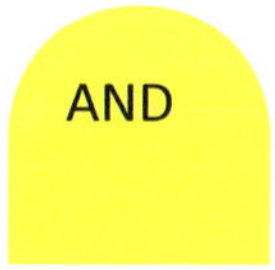

Logic Gate AND: All of the events below this gate must happen to trigger the logic gate AND.

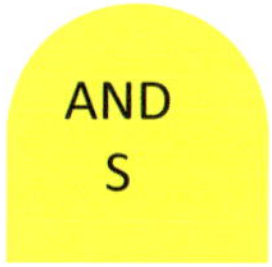

Stand By Event: All active and passive events must happen to trigger the fault in the standby event.

K/N Event: This gate is triggered when more than K of N events happen.

Basic Event: That is the last event on FTA.

Exclusive OR Gate: This gate is triggered when one of the events happen.

Priority AND Gate: This gate is triggered if all the events below happen in a defined sequence.

In order to understand the incident or accident causes the FTA can also be applied qualitatively and in this case the main objective is to understand the event combinations that trigger the top event.

An example of a Qualitative Fault Tree Analysis applied to assess incidents and accidents is a toxic product spill from a tank in a chemical plant. Such an accident happens after a shift changeover during the afternoon. The tank containing a toxic product operated normally, but with a wrong local level signal indication. The wrong indication was frozen at 70% and the usual procedure was maintaining the tank level on 90%. The level indicator problem was detected by the local operator that kept the control room operator updated about the tank level. In the control room panel the level indication was correct. When shift group changed and the problem with the local level indicator was not discussed with the shift team who started to work, so after 30 minutes, the new operator required to control room operator to fulfill the tank by more 20%. Despite the control room indicator was at 90%, the operator confirmed his request because the production was not in an ideal condition and he thought that was necessary to have more of such toxic product in the tank. Consequently, the level rose to 100% and the 10% of additional toxic product was released through the vent and reached the operational area. Two operators who were close to the tank were soaked by the toxic product, which caused some reaction in their skin. Fig. **12** shows the FTA represents the event combination that triggered the accident.

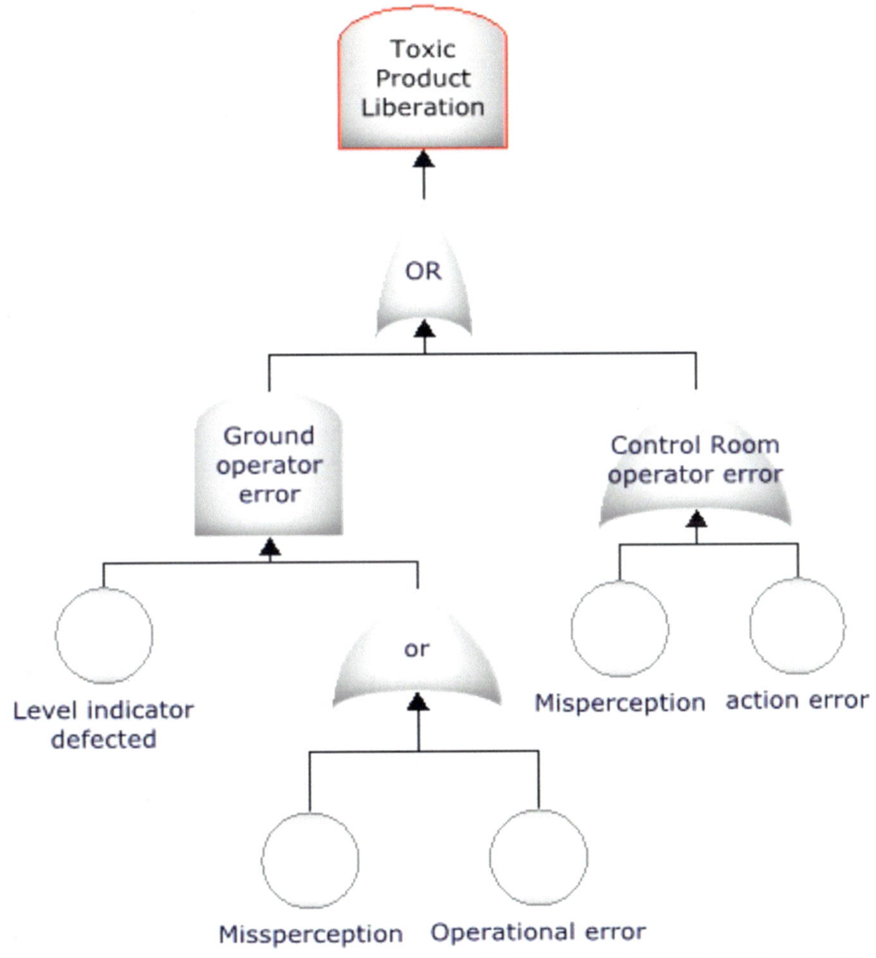

Figure 12: Toxic Product spill (FTA).

Regarding the Fault Tree Analysis methodology applied to an incident and accident assessment, we can describe the following advantages:

- Regards the logic combination of events that trigger a top event;
- Is a powerful tool to control the risk of accident by cut set events;
- Can be used to describe how event happens and be performed with other methodologies;

The Fault Tree Analysis methodology drawbacks are:

- Not easy to understand and to be applied;
- Requires investment in training;
- Considers only one consequence, which is the top event;

The Fault Tree analysis regards the combination of events which trigger a top event. It is also possible to include bad decisions and management errors and it is easier in the qualitative approach. The qualitative Fault tree Analysis has the main objective to understand the combination of events that triggers the top event. In some cases, in order to evaluate the risk with recommendations, will be necessary to calculate the Top event probability. In some cases, in order to clarify the consequences of the accident, it is necessary to go beyond the top event and in this case the Bow Tie will be the best approach. The Bow Tie Analysis will be discussed in the next item.

6.8. BOW TIE ANALYSIS

As described in Chapter 3, The Bow tie analysis is the newest quantitative risk analysis that has been in use since the 1970s, and has been incorporated into the Hazards Effects Management Plan methodology used by the Shell Oil Company in 90's.

The Bow Tie Analysis enables assess all combinations of events from incident causes to incident consequences regarding the layer of protections that prevent the accident and mitigate the consequences.

The Bow Tie Analysis can be also applied to incident and accident assessment. In this case, both the causes and consequences of the accident are clear in the Bow Tie graph. In addition, the control measures and recovery actions can also be implemented in Bow tie analysis, as incident and accident recommendations to reduce the chance of such event happens again.

An example of Bow Tie graph application is shown in Fig. **13**, that represents an accident due to a pump explosion. In a sequence of actions to shut down a Chemical Plant, in order to prevent backflow through the de-energized pump, employees manually closed the suction and discharge valves. When the power was restored, the pump started to run completely isolated. After some minutes the pump blew up and an operator who was at 10 m from the local of the accident was hit by parts of the equipment. The accident analysis conclusion was that the high pressure alarm did not work property as well as the relief valve and the accident could not be prevented by the control room operator because there was not any indication of high pressure of such pump in the control room panel.

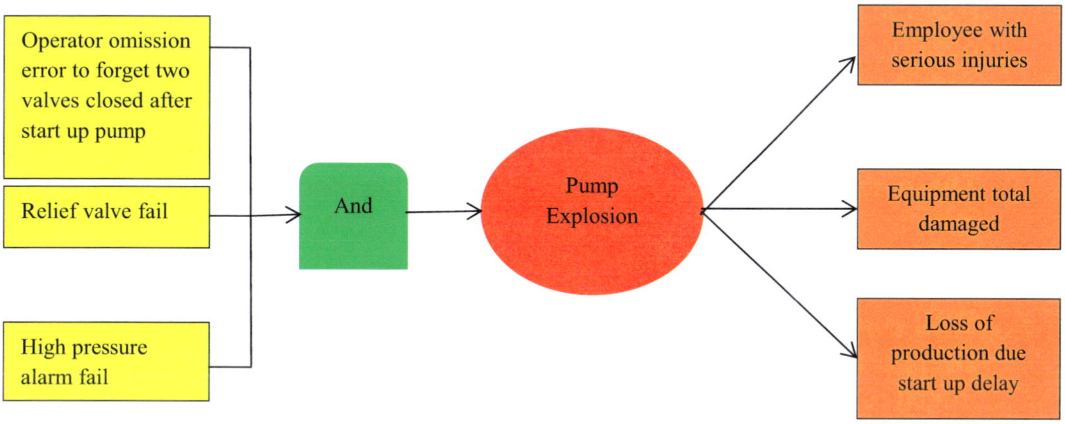

Figure 13: Pump Explosion (Bow Tie).

6.9. MAJOR ACCIDENT ANALYSIS - CASE STUDIES

6.9.1. The Deepwater Horizon Oil Rig Blowout

"On the evening of April 20th, 2010, a gas release and subsequent explosion occurred on the Deepwater Horizon oil rig working on the Macondo exploration well for BP in the Gulf of Mexico. Eleven people died as a result of the accident and others were injured. We deeply regret this loss of life and recognize the tremendous loss suffered by the families, friends and co-workers of those who died.

The fire burned for 36 hours before the rig sank, and hydrocarbons leaked into the Gulf of Mexico before the well was closed and sealed".

Based on the BP accident summary report (http://www.bp.com/), the eight key findings related to the causes of the accident are:

1. **The annulus cement barrier did not isolate the hydrocarbons**. The accident investigation team concluded that the migration of hydrocarbons happens because the cement was out of the design specification.

2. **The shoe track barriers did not isolate the hydrocarbons.** The accident investigation team concluded that the hydrocarbon ingress into the production casing happen because of functional failure in both shoe track cement and the float collar which allowed.

3. **The negative-pressure test was accepted.** During the negative-pressure test, the pressure readings and volume bled at the time of the negative-pressure test were indications of flow-path communication with the reservoir, showing that the integrity had not been achieved.

4. **Influx was not recognized until hydrocarbons were in the riser**. The hydrocarbon influx was not detected and the team took place preventive actions to control the well after hydrocarbons had passed through the BOP and into the riser.

5. **Well control response actions failed to regain control of the well.** Despite the team close the BOP and diverter, the fluids were not routed to the overboard diverter line.

6. **Diversion to the mud gas separator resulted in gas venting onto the rig.** The mud gas separator was diverted to to the MGS and increased the chance of ignition.

7. **The fire and gas system did not prevent hydrocarbon ignition**. Hydrocarbons migrated to electrically classified into areas with a high chance of ignition.

8. **The BOP emergency mode did not seal the well**. Three methods for operating the BOP in the emergency mode failed to seal the well.

After all barriers, the blowout accident took place on 31st December 2011 and demanded a huge contingency response which involved government and local residents. Such contingency response costed $14 billion, environmental impact as well as employee deaths.

Fig. **14** shows the Deepwater Horizon oil rig' component fail.

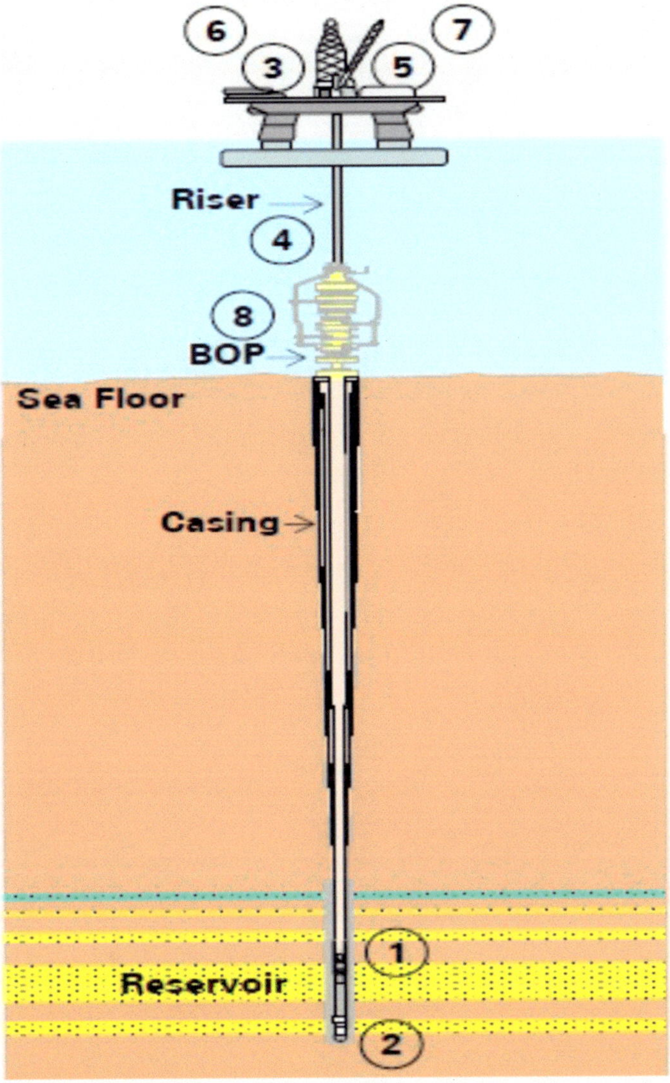

Figure 14: Events that triggered the accident (Source: www.bp.com.br).

The accident method which would be used to represent and understand the sequence of events that triggered the Blowout accident is The Sequence Accident Analysis as shows Fig. **15**. Although the events that lead to the accident is clear, it is also important to understand why such events happen, in order to prevent them in the future. By this way, each event can be considered as a top event and a combination of events that triggers such event can be assessed by an FTA methodology as defined on event 2 and 8 in Fig. **15**.

The Accident Sequence Analysis can successfully represent the events that lead to the Blowout as shows Fig. **15**.

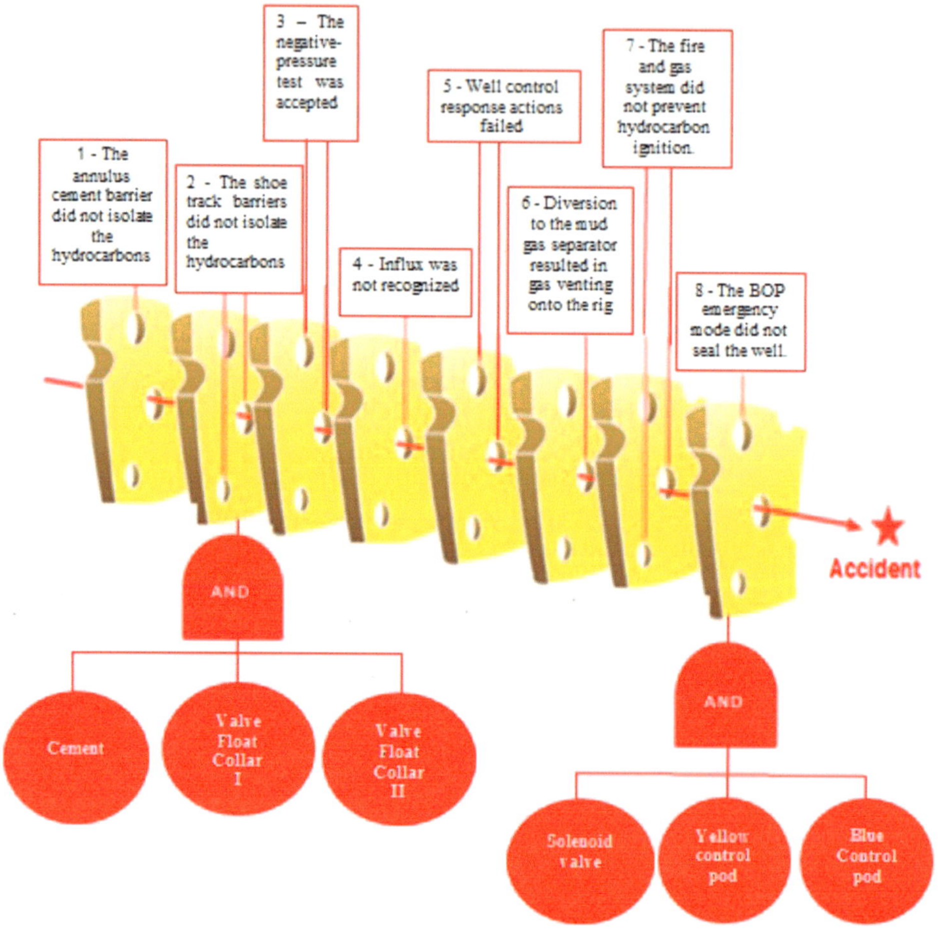

Figure 15: Sequence Accident Analysis + FTA.

We can reach the conclusion that even for a system that has a high level of layers of protection with an accident probability that is very low, it is always possible that such event will happen. More than having safety systems the big challenge is to keep such system at a high level of safety state and that depend upon design, reliable equipment, keeping lower the chance of human error as well as monitoring constantly the cut set event conditions. In the Nuclear industry, such cut set conditions are monitored constantly, but even under such circumstances, external events like natural catastrophes can change the safety level and trigger an accident like the one that happened in the Fukushima Nuclear Plant.

6.9.2. The Fukushima Nuclear Accident

"Fukushima prefecture provided a further report early in 2014 which said that the 'indirect' deaths in the prefecture were greater than the number (1607) killed in the quake and tsunami. Evacuees receive JPY 100,000 ($1,030) per month in psychological suffering compensation. In August 2012, Reconstruction Agency report also considered workers at the Fukushima power plant. By almost 1500 surveyed, many were stressed, due to evacuating their homes (70%), believing they had come close to death (53%), the loss of homes in the tsunami (32%), deaths of colleagues (20%) and of family members (6%) mostly in the tsunami. The death toll directly due to the nuclear accident or radiation exposure remained zero, but the stress and disruption due to the continuing evacuation remain high" (www.world-nuclear.org).

The sequence of accident event which triggered the nuclear accident was:

1. The Earthquake occurring on March 11th, 2011, at 14:46, shut down all reactors in operation in Units 1 to 3 at Fukushima Daiichi and Units 1 to 4 at Fukushima Daini Nuclear Power Plants including all off-site electric power supply energy. Despite the situation the emergency diesel generators maintained the reactor safety.

2. After Earthquake, The tsunami reached the nuclear power plants and the Fukushima Daiichi lost all cooling functions using AC power. In addition, the cooling seawater pumps were flooded and DC power shut down, resulting in lost of the reactor cooling function.

3. As a consequence of loss of cooling system, the radioactive materials in the fuel rods were released into the RPV and the chemical reaction between the fuel claddings (zirconium) and steam generate a substantial amount of hydrogen.

4. The pressure on PCV increase and the vents and relieves valve were used to relieve such pressure. Unfortunately, the probable hydrogen leakage in PCV causes an explosion in units 1 and 3 which destroyed the reactor structures.

The Fukushima Daiichi Units 1 to 3, the accident escalated into a chain of events and developed into a serious nuclear disaster. Fig. **16** shows the Fukushima plant equipment and Fig. **17** shows the Bow tie Analysis of this nuclear accident.

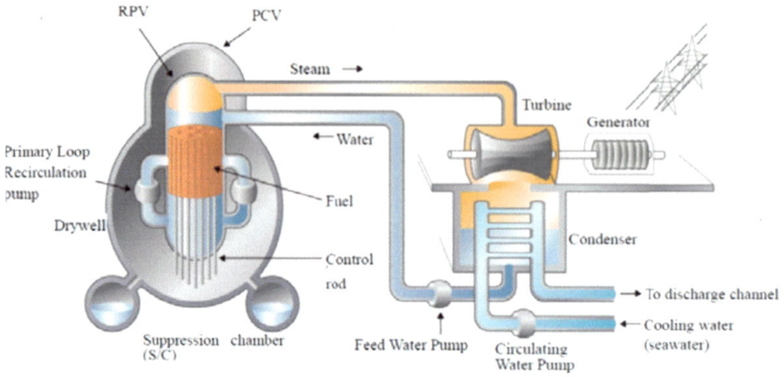

Figure 16: Nuclear Plant Equipment (Source: http://www.nirs.org/fukushima/naiic_report.pdf).

Figure 17: Bow Tie (Fukoshima Plant Accident).

We can reach to a conclusion that even in facilities with a high level of reliability in their safety system protection can also have an accident. In Fukushima Power

Plant Nuclear accident, external events like Earthquake and Tsunami were the main root causes of the accident.

In this case the discussion goes to vulnerability and such issues is also related to project team decided to put such dangerous plants in places vulnerable to natural catastrophes. However, in Japan case, we can understand because they have no other energy sources and not a huge territory to put such nuclear plants in a remote place. But there are many dangerous plants nowadays that are installed close to a dense population area or even in places vulnerable to natural catastrophes.

In the general case, one important issue that must be taken into account is the cost of prevention that, in many cases, is questioned in projects due to the rare possibility of catastrophic events. Secondly, in periods where the economy is not in good health, operational costs are cut off and many of such costs are related to preventive maintenance and inspections of safety protection systems.

Finally, the human factor is one of the key points in much accident analysis and human factor must always be considered as a critical factor to safety, in order to prevent an incident or an accident, as it will be discussed in the next chapter.

The incident and accident analysis methods can be applied to different types of analysis, but depend on cases and characteristic of incidents and accidents some methods can give better results than others. Table **2** shows the adequate incident and accident methods based on an incident or accident characteristic.

Table 2: Incident and Accident Analysis Methods Application

Method	The accident causes are clear	Logical combination of events	One consequence	More than one consequence
1 – Ishikawa Diagram	+++		+++	
2 – Why Because	++	+	+++	
3 – Event Tree Analysis	+	++	+++	+
4 – Sequence Analysis	++	++	+++	
5 – Fault Tree Analysis	+	+++	+++	
6 – Bow Tie Analysis	+	+++	+++	+++

+++ (The most adequate method to assess incident and accident faster, easily and consistently)
++ (Fair adequate method to assess incident and accident faster, easily and consistently)
+ (The least adequate method to access incident and accident faster, easily and consistently)
Remark: The blank that has no x means that the method is not suitable in such situation

The red marked blanks represents in which circumstances it is better to use such methods. For example, if the accident causes are clear and there is only one consequence, the most adequate method is Ishikawa Diagram. If the accident causes are clear, there is some logic event combination and there is only one consequence the most adequate method is Why because method.

In case of logic event combination the Why because method is not the most adequate one. In these cases other methods are more appropriate and the final decision is in the number of consequences. If there is more than one consequence the Bow Tie method is the most adequate one to be applied on incident and accident analysis.

There is also a possibility to combine two methods, for example, ETA or Sequence Accident Analysis and FTA like the "The Deepwater Horizon oil rig Blowout accident case study" but this should only be applied when any other possibility is not possible.

REFERENCES

Analysis of the results of the national campaign for setting of labour conditions in the chemical enterprises and units with high professional risk in compliance with the Health and Safety Conditions at Work. (2005). Retrieve from: http://git.mlsp.government.bg/Executive Agency General Labour Inspectorate (EAGLI). 2005a.

Analysis of the results of the national survey for provision of health and safety conditions at work in the main risk enterprises of the metallurgical industry. (2005). Retrieve from: http://git.mlsp.government.bg/ Executive Agency General Labour Inspectorate (EAGLI). 2005b.

Analysis of the results of the carried out national campaign: Control of implementations of obligations for provision of HSCW in the enterprises for production of thermal and electric energy. (2005). Retrieve from: http://git.mlsp.government.bg/

Blacket, C. (2005). Combining accident analysis techniques for organizational safety; PhD Thesis, School of Computer science and informatics National University of Ireland.

Brauchler, R & Landau. K. (1998). Task analysis. Part I –Guidelines for the practitioner. International Journal of Industrial Ergonomics 22 (1–2): 3–11

Brauchler, R. & Landau, K. (1998). Task analysis. Part II –The scientific basis (knowledge base for the guide).International Journal of Industrial Ergonomics 22 (1–3):87–99

Clemens, P.E. (2002). Human Factors and Operators Error. Retrieved from: www.ceet.niu.edu/tech/asse/humanfactors.pdf.

D. Doytchev & G. Szwillus. (2006). Combining Task Analysis and Fault Tree Analysis for accident and incident analysis: A case study from Bulgaria. Safety and Reliability for Managing Risk – Guedes Soares & Zio (eds). 2006 Taylor & Francis Group, London, ISBN 0-415-41620-5

Hollnagel, E. (1998). Cognitive Reliability and Error Analysis, Elsevier Science. ISBN 9780080428482

Hollnagel, E. (1999). Accident analysis and barrier functions, IFE (N); Version 1. Retrieve from: www.it.uu.se/research/project/train/papers/AccidentAnalysis.pdf

Ishikawa, Kaoru. (1968). Guide to Quality Control (Japanese): Gemba No QC Shuho by JUSE Press, Ltd., Tokyo.

Ishikawa, Kaoru. (1976). Guide to Quality Control, Asian Productivity Organization, UNIPUB, ISBN 92-833-1036-5

Johnson, C. (2003). Failure in Safety critical systems: A handbook of Incident and Accident Reporting Glasgow University Press.

Ladkin Peter B. (2005). Why-Because Analysis of the Glenbrook, NSW Rail Accident and Comparison with Hopkins's Accimap. Faculty of Technology, University of Bielefeld. Research Report RVS-RR-05-05 December, revised 19 December, 2005.

Non-profit Risk Management Centre. (2006). Retrieve from: http://nonprofitrisk.org/ws/c2/acc-inc-nm.htm.

Shiba Shoji and Walden David. (2002).Quality Process Improvement Tools and Techniques. Massachusetts Institute of Technology and Center for Quality of Management. revision 6.

http://www.bp.com. Accessed on 10-02-2013.

http://www.nirs.org/fukushima/naiic_report.pdf. Accessed on 10-02-2013.

http://www.world-nuclear.org/info/Safety-and-Security/Safety-of-Plants/Fukushima-Accident. Accessed on 29-12-2014.

CHAPTER 7

Human Factor

Abstract: All over the world, many incident and accident cases involve human error as one of the main causes. The Human factor must always be taken into account in dangerous activities and processes which involves risk with high severe consequences. The human reliability methods are addressed to assess the human performance factors as well as predict the human error probabilities. The methods such as THERP, OAT, ASEP, STAHR, SPAR-H, HEART, BBN support risk assessment as well as maintenance and operational activities in order to avoid human error. Despite the importance of human reliability analysis to mitigate the risk of human error, nowadays such methods are not too much applied in many asset projects or operational phases. This chapter aims to describe the human reliability concepts and methods with examples addressed to incident and accident cases as well as maintenance activities.

Keywords: Performance factor, human error, human reliability, THERP, OAT, ASEP, STAHR, SPAR-H, HEART, BBN.

7.1. INTRODUCTION

The human factor is related to safety from decision process when a specific technology is selected for execution of tasks in process, when operations and maintenance are carried out. Thus, even system with low level of human and machine interaction and interfaces has a human factors influence on project concepts, as well as in operation and maintenance tasks.

In many incident and accident analysis cases the root cause is human factors that influence human error. That makes the human reliability analysis a real critical issue to be taken into account in safety management. Unfortunately, the human reliability analysis methodology took place after the major accident happens in the last decades and was clear that a specific methodology to understand the influence of human factors on incident and accident would be developed. In fact, the effort to perform human reliability analysis started in 1960 ages when Williams suggests to consider Human Reliability in system reliability analysis and some studies related to equipment failures shown the human failures as root causes. In 1970 ages, the human reliability became more specific with IEEE human reliability report publication and a specific study about human factors related to atomic nuclear reactor carried out by Swain and Williams. From this point and on, different methodologies were developed to be qualitative and quantitative.

Eduardo Calixto

In fact, the human reliability methods can be grouped in three different generations. The first one, focus on methods to predict the human error probability such as a Technique for Human Error Rate Prediction (THERP), Operator Action Tree (OAT), Accident Sequence Evaluation program (ASEP), SLIM (Success Likehood Index methodology), SHARP (Systematic Human Action Reliability Procedure), STAH-R (Social technical Assessment of Human Reliability).

The second human reliability methods generation focuses on human performance factors such as Psychological, Physiologic, technological and social. In order to address human performance factor were developed methods such as Atheana and Cream.

The third generation methods have been developed to take into account the human performance factors influence in one each other such as Bayesian methods.

Then first step to apply Human reliability methods is understanding the main concepts related to Humana Reliability Analysis.

Human reliability is probability of human carried out specific tasks with satisfactory performance. Tasks may be related to repair, operation, safety action, analysis and all kinds of human action which take influence in system performance. Human error is contrary of Human reliability and basically the human error probability (P(HE)) is described as:

$$P(HE) = \frac{Number\ of\ Errors}{Number\ of\ error\ oportunities}$$

In general terms, Human Reliability methodologies regard that human error can be:

- "Omission Error" happens when one action is not performed due to lapse or misperception. As an instance, in preventive incident action, omission error is misperception of a valve open that must be closed. Considering maintenance activities, an omission error happens when due lapse, tasks are not performed which cause equipment failure in some period of time in the next future.

- "Commission Error" happens when the action is performed wrongly due to wrong quantity or quality of action or mistake in select or proceed sequentially. As an instance, in preventive incident action, a commission error happens when the valve that must be closed is kept open caused by operator mistake when doing a sequence of actions required in the procedure.

- "Intentional Error" happens when the action is deliberately carried out to cause system failures and damage. In fact, equipment degradation would occur when intentional sabotage takes place, as an instance, operator lets some tools into equipment intentionally to damage it. The worst case of intentional error is the sabotage to cause an accident. Such type of error must also be considered in order to reduce system vulnerability as discussed in Chapter 5.

- Despite the human error concept importance, the understanding of human performance factors will definitely help to minimize the human error probabilities.

Basically, the human performance factors can be divided in internal or external factors. The Internal Human Performances Factors are related to each individual characteristic and they are:

- Psychological: Stress, over psych workload, over cognitive workload, depression, demotivation, no concentration, sadness.

- Physiologic: health conditions, diseases.

The physiological factors are easier to be monitored rather than psychological. In fact, such factors indicate that employees are not in their best conditions and that is very important to be detected when a dangerous task will be carried out in order to avoid that human error lead to an incident or accident.

By the other way round, "The External Human Performances Factors" depends on company and society and they are:

- Technological: procedures, equipment, ergonomics.

- Social: bad social conditions, bad acceptance in group,

In fact, the organization can influence in internal social relations and technological factors, but has not control under external social issues.

In general terms, organizations must promote the best internal social relation between employees, provide the best procedures, ergonomics and work conditions as well as detect when employees are not in their better physiological and psychological conditions to minimize the chance of human error. It´s also necessary to be aware about factors that are not under their control such as external social conditions and psychological.

In fact, in a workplace which employees work motivated and happy the chance of human error tend to reduce. Therefore, it´s necessary that the organization creates a workplace which allows the employee's internal motivation lead their best performance. In addition, happiness in the workplace is also a factor which leads high performance and such high performance can also be understood as lack of human error. The Fig. **1** shows HRA factors which take Influence in human error.

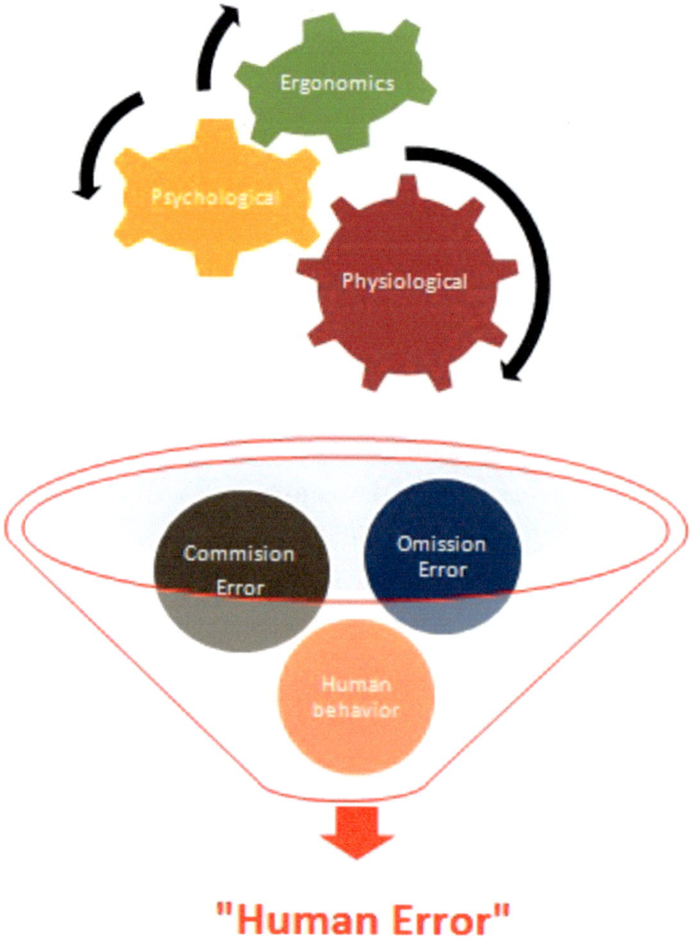

Figure 1: Human error.

In order to minimize human error different human reliability methods are applied and in many cases the general recommendation is to improve the current procedure or establish a specific one. Depends on the activity, the tasks required different type of employee behavior which is based on factors such as procedure, skill and knowledge. In fact, most of the tasks require a combination of such factors but have one of than predominant.

Whenever the task behavior is highly influenced by a procedure factor if it´s established a good procedure, it´s possible to minimize the human error probability. The emergency evacuation is an example of a task which is highly influenced by procedure factor and thereby the human error is minimized when the procedure is followed.

By the other hands, the same emergency requires from the control room operator to control and shut down the plant. Despite a set of procedures are followed, it´s also necessary to know the correct sequence as well as to decide the exact time to shut down each equipment. This case is an example of a task which is highly influenced by the procedure and knowledge factors.

Now we can imagine, for the same emergency case, one pump does not shut down by the control room command and it´s necessary to shut down on local. That´s requires from operational skills and knowledge. In other words, influenced by skill and knowledge factors.

The different tasks are based on different behavior which is influenced by different factors which requires a deep understanding. In fact, tasks which behavior are based on skill and knowledge are totally different in tasks which behavior is based on procedures. Therefore, in order to define human reliability analysis´s recommendations for different task, it´s necessary a deep understanding about the behavior as well as the human performance factor.

In general terms the human reliability analysis must answer at least five questions:

- How human error can influence on system performance?
- Which are the human error consequences?
- Which behavior factors influence on task?
- Which human performance factors take more influence in human error?

- What is necessary to improve human reliability to avoid or prevent human error?

Remarkably, applying human reliability analysis is to be prevented to avoid the incident and accident. That means whenever is identified in Risk analysis that a human error can lead an incident or accident is necessary to carry on such human reliability analysis to access the human performance and improve human performance factors in order to minimize the chance of human error. Unfortunately, on most of cases, such preventive applications of human reliability analysis are not applied preventively even when some incidents happen.

7.2. TECHNIQUE FOR HUMAN ERROR RATE PREDICTION (THERP)

The THERP was one of the first human reliability analysis methods developed by Swain in 1975 in order to analyze human error influence on nuclear reactor failures. The THERP methodology was published in 1983.

Such methodology has the main objective to predict the human error probability. Basically, in order to define the HEP, the different branch tree diagram is built up from the top to the bottom taking into account the tasks and the probability of omission and commission error as well as success for each task. In order to perform THERP analysis the sequence of steps are required such as:

1. Understand the system and human interface;
2. Identify human tasks;
3. Define the human error probability for each task;
4. Model the Human error tree diagram;
5. Estimate the Final Human Error Probability based on human error tree;
6. Propose recommendation to reduce the human error probability;
7. Estimate the recommendation effects on human error probability based on the human error tree diagram.

The first and second step is common for all human reliability analysis no matter the method applied. The third step is the most difficult and can be accomplished based on historical data or specialist opinion. Some organizations define omission and commission error probability foe different tasks in their procedure.

Whenever a specialist opinion takes place, it´s important to be on mind that it´s easier to estimate and agreed the frequency of occurrence of failure rather than probability because frequency is based on events a long time and probability are based on perception and classification of each one. However, in some cases the human error frequency or probability may not be constant. In cases such as risk analysis (FTA, ETA and BOW TIE), where human error is assessed together with equipment failures during a period of time it´s necessary to define human error probability density function. The simplest cases consider the constant failure rate and is represented by an exponential CDF. In fact, that´s a good human error representation considering such human error, mostly is a random event which fits in the equation:

$$F(t) = 1 - e^{-\lambda t}$$

where,

λ=expected number of human errors per time

T=time

F(t)=probability of human error occur until time t

The fourth step after estimate human error probability for each task is necessary to build up the human error tree. Such diagram represents human error failures by capital letters in the right side and success by lower case letter in the left side. The Fig. **2** shows an example of THERP analysis.

In order to exemplify the THERP application Aircraft accident analysis that happen in 2009 and in Brazil will be carried out.

The Embraer Legacy 600 was presented to the public in June 2001. Based on Embraer 135, exhibited several enhancements, including extra fuel tanks and winglets. Your avionics system (the set of electronic communication, navigation, flight tracking and weather indicators) included a sophisticated TCAS. The Traffic Collision Avoidance System, TCAS, allows the crew warnings about a possible air traffic in the opposite direction, so that the TCAS to work, you need both planes are equipped with the device and that obviously, are connected. It is also necessary that the transponders (the system that sends signals to the controllers on the ground, indicating flight data and aircraft identification) have been activated. No transponder, TCAS aircraft loses and becomes a threat hanging in the air.

After the usual negotiations, the air taxi company decided to buy EMB-135 BJ Legacy jets to incorporate them into their fleet. The delivery of the first unit was combined for the last week of September 2006, leaving Sao Jose dos Campos to Fort Lauderdale, Florida.

That same week, the pilots had flown together in Legacy. The route Sao Jose-Manaus-Fort Lauderdale would be the debut of the pilots flying on their own.

Embraer premises, plant engineers trained American pilots in the use of software Legacy That program was installed on a laptop to be used on the trip. The software included data on the route, including particularities of Manaus airport.

At 13:15 on 27 September 2006, the five passengers headed for the aircraft, which had been towed to the parking lot of the airport and stocked. The captain was on board, taking care of routine pre-flight. Meanwhile, the copilot remained in the delivery room, with a factory engineer, familiarizing himself with the weight and balance calculations Legacy, meeting software loaded on the laptop.

Besides the different directions of the first leg of the flight, San Jose-Manaus, the plan contained the corresponding cruising altitude.

As Brazilian aviation rules determine, from Brasilia, to take the path north in airway two-way, the plane would remain at levels where peers. As from north to south, the aircraft must fly at odd levels, this simple rule of thumb prevents a collision.

In the first call made to the N600XL airport tower, this will now display the runway in use and instructed on the taxiway to the head. Pilots also received the transponder code, 4574, of which would be identified on the radar screens of soil, accompanied by information on the altitude of the jet.

In possession of the code, Legacy, still standing in the yard, was released by the controller of Sao Jose dos Campos to fly at level 370 (37,000 feet) “until Eduardo Gomes International Airport in Manaus”. In fact, the Cindacta 1, in Brasilia, had authorized 370 (as had the flight plan for the first leg of the trip), but had not said that this altitude should be maintained until Manaus.

The pilot did not question the statement, which, however, did not follow the plan prepared for the trip. It merely acknowledges its receipt. Having spoken with the Legacy, the controller called Brasilia. He used colloquial language and acted as if the maintenance requests the same altitude throughout the route. The level 370 not only contradicted the flight plan as well as pre-established norms of altitude in the directions north and south-north-south.

After takeoff, when it reached 8000 feet, the Legacy was allowed to continue to rise. He also received instructions to contact the frequency of the Area Control Center in Brasilia, where an operator watched the movement of the plane, represented by the code 4574 on your radar screen. In the dark screen, a circle surrounded the block of data from Legacy, known in aviation as "target", composed of white letters and numerals. The existence of the circle was a sign that the aircraft signals emitted secondary radar. Therefore, your transponder functioned perfectly.

After passing the 200 level, and remain somewhat level 310, pending the resolution of traffic conflict with another aircraft, the jet was allowed to rise to 37,000 feet. In the cockpit, pilots, without familiarity with the rules of the Brazilian air traffic, understand, because of the initial instruction controller São José dos Campos, the level 370 should be observed to the destination.

Without noticing the mistake Brasilia, the plane flew in the opposite direction. However, this should not constitute serious problem, since the Legacy's transponder would soon send a signal to the radar screens of the Center for Integrated Air Defense and Air Traffic Control, the Cindacta 1, in Brasilia, showing the target Incorrect level. And even if none of these systems and regulations redundant work, the Legacy's TCAS takes care to indicate to the pilots, the evasive maneuvers required deflecting the jet of a possible oncoming traffic.

From the moment that leveled the X-Ray Lima at 37, 000 feet, the pilots began to work on a laptop. The pilots had doubts about the Legacy's performance during takeoff and landing (this the next day) at the airport in Manaus. Part of the airport runway was banned because of works.

In Brasilia, the sector controller did not notice that the 07 Legacy, under its responsibility, flew in the opposite direction. However, this was clearly displayed on the screen of your console. The data block plane showed: 370 (effective level) = 360 (set level).

At exactly 19h02 Zulu (Z = Zulu or Greenwich Mean Time, equivalent to 15.02 in Brasilia and Manaus at 14h02), the circle around the block of data from Legacy disappeared from screens Cindacta 1, meaning that the transponder was turned off.

Without realizing the loss of the signal secondary plane, called the Legacy Brasília not the radio. Meanwhile the pilots were still on the laptop, which alternated in the lap of one and another. Concentrates on the problem of track Manaus, the two did not see a small white message in each main panel screen: TCAS OFF (TCAS off) was what signaled the warning.

Always following the axis of the airway UZ6, the jet reached the position. There, according to the original flight plan, he should move from level 360, which had flown at any time, for 380 (38,000 feet).

Obeying the initial instruction of São José dos Campos, where they had taken off two hours before the pilots remained on odd level, FL370, exclusive of the north-south direction. A copy of the flight plan remained available in a bin between the two pilots. Also at their disposal were aeronautical charts, in which were specified in the correct levels of flying in both directions of the vertical two-way street.

Besides being on the wrong level, Legacy entered a transition zone, critical for radio transmissions, which many pilots call a black hole. With it, sometimes the planes cannot talk to the drivers, and *vice versa*. It was precisely what happened with the copilot, who was trying to call centers of the land.

There were 54 minutes, the transponder and therefore the TCAS collision avoidance device remained switched off without the air traffic controllers and pilots and double realized. So at 19h56m 54 Zulu, when the Legacy crossed the jungle, above the territory of the municipality of Mato Grosso Peixoto de Azevedo, something mysterious and terrifying happened.

Was heard in the cockpit of a dry sound impact, captured by microphones CAM (Cockpit Area Microphones) and registered in one of the black boxes, the CVR

(Cockpit Voice Recorder), voice recorder cockpit. The plane made a sharp turn to the left.

Applying the THERP method to access this accident the first step is establishing the sequence of events which lead to accident and then make up a THERP event tree. The sequence of events is defined along the diagram for uppercase letters are used to define events of failing and capital letters are defined to define success events. Furthermore values are defined probability of success and failure for each event type is represented by the letters S and F respectively. The sequences of human error are:

a = Lack of clarity in the information passed by CINDACTA 1 on the altitude of the flight.

b = Lack of awareness of the pilot on the difference in the information passed by CINDACTA 1 on the altitude and the flight altitude established in the flight plan.

c = Lack of awareness of Cindacta 1 that the aircraft was flying at the wrong altitude according to Brazilian standards.

d = Lack of awareness of Cindacta 1 that the transponder signal disappeared from the screen.

e = Lack of awareness of pilots warning "TCAS OFF" on the dashboard of the plane.

f = Lack of awareness Cindacta 4 that the plane's transponder signal did not appear on the screen of the air traffic controller.

The Fig. **2** shows the THERP diagram which represents all human error regarded on aircraft accident that are defined by the letter above.

By human reliability point of view, all related events are related to human failure once bad decision as events a, b and c are omission error due lack of risk perception of consequences. The probability of success is defined by equation:

$$P(success) = P(a) \times P(b) \times P(c) \times P(d) \times P(e) \times P(f)$$

As mentioned before, it´s possible to predict the human error based on historical data, specialist opinion and organizations procedures. There are also different source of human error probability in literature that can be used in case of lack of data as shows Table **1**.

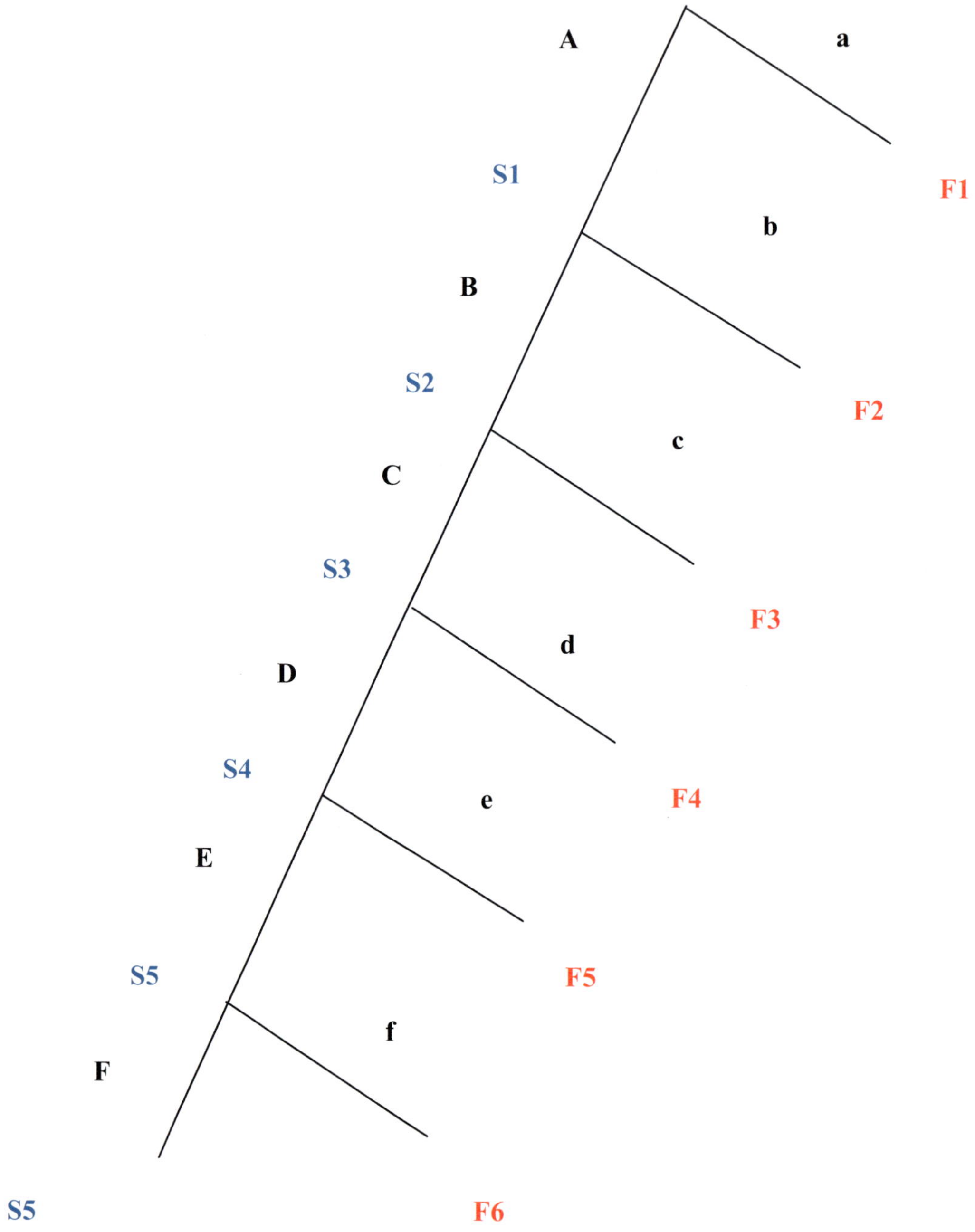

Figure 2: THERP tree.

Table 1: Human Error Probability (Source: Kumamoto, 1996)

Type of error	Time	Skillness	Procedure	Knowledge
Omission Error	Short	0,003	0,05	1
Omission Error	Long	0,0005	0,005	0,1
Comission Error	15 min	0,001	0,03	0,3
Comission Error	5 min	0,1	1	1

The Table **1** shows the human error probability which has different values depends on performance factor such as time as well as behavior factors such as skills, procedure and knowledge. The task duration is an interesting point which takes influence in human error probability. Thus, as short as task time, the higher is the human error probability. Despite a good reference of human error probability, such values must be confirmed by specialist.

In order to understand the THERP application quantitatively lets apply the failure probability on Table **3** in THERP equation. Thus we have:

P(accident) = P(a) x P(b) x P(c) x P(d) x P(e) x P(f)

Where:

P(a) = 0,3

P(b) = 1

P(c) = 0,1

P(d) = 0,1

P(e) = 1

P(f) = 0,1

P(accident) = (0,3) x (1) x (0,1) x (0,1) x P(1) x P(0,1) = 0,0003

The probability results by itself do not have a good idea about if the chance of an accident is high or low, but if using some risk criterion or even compare with the probability of an accident after improving performance human factors that reduce the human error probabilities. In first case if we considered that around 200 people deaths in such accident the risk is 6 x 10 E2, that is unacceptable.

As a conclusion, we can say that THERP analysis has the main advantages:

- THERP method is simple to be applied for simple cases to predict human error probability;
- The method has been widely applied in nuclear industry industry and has also been applied in other industry successfully;
- THERP method can be complex for human reliability analysis, which consider multiple tasks;
- In order to calculate human error probability is necessary to define the human error probability for each task that sometimes it may not be easy;
- Such methodology does not address any discussion about human performance factors which cause human error that can influence on recommendation.

7.3. OPERATOR ACTION TREE (OAT)

The OAT methodology was developed in 1982 in order to predict the human error probability based on tasks sequences. The first application was to predict the human error probability in the Susquehana Nuclear Plant. In fact, such method is similar to event tree analysis described in Chapter 3, however, focus on sequence of tasks which can fail due to human error.

In order to understand the OAT application shown on Fig. **3**, we will consider the same aircraft accident example used on item 7.2. Thus we identify the human error like:

Human Error 1 = Lack of clarity in the information passed by CINDACTA 1 on the altitude of the flight.

Human Error 2 = Lack of awareness of the pilot on the difference in the information passed by CINDACTA 1 on the altitude and the flight altitude established in the flight plan.

Human Error 3 = Lack of awareness of Cindacta 1 that the aircraft was flying at the wrong altitude according to Brazilian standards.

Human Error 4 = Lack of awareness of Cindacta 1 that the transponder signal disappeared from the screen.

Human Error 5 = Lack of awareness of pilots warning "TCAS OFF" on the dashboard of the plane.

Human Error 6 = Lack of awareness Cindacta 4 that the plane's transponder signal did not appear on the screen of the air traffic controller.

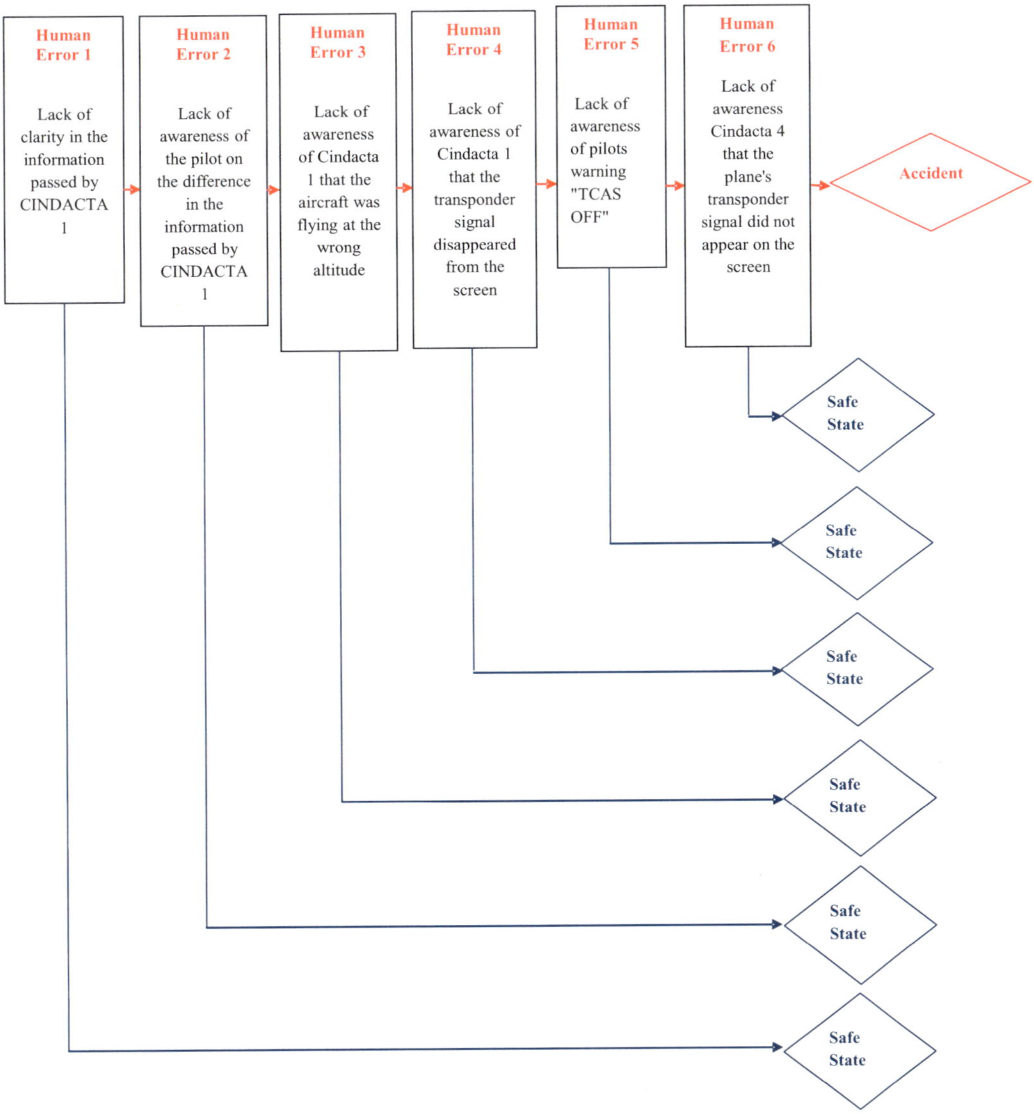

Figure 3: Aircraft accident OAT diagram.

If all human error represented in OAT diagram happen consequently the aircraft accident is the final result as happen in reality as shows Fig. **3**. If any of the uh errors were avoided by an effective action to put the airplane in a flight position safe, the accident would not happen. The next step of the methodology is to define the probability of each event of human error (1 to 6) and calculate the probability of various combinations of errors that cause the accident or if the correction takes flight to a safe state. The calculation of the probability of the accident is represented by the equation:

P(accident) = P(Human Error 1) x P(Human Error 2) x P(Human Error 3) x P(Human Error 4) x P(Human Error 5) x P(Human Error 6)

Considering the human error probability on Table **3** similar to THERP case we have:

where:

P(Humn Error 1) = 0,3;

P(Human Error 2) = 1

P(Human Error 3) = 0,1

P(Human Error 4) = 0,1

P(Human Error 5) = 1

P(Human Error 6) = 0,1

P(accident) = (0,3) x (1) x (0,1) x (0,1) x P(1) x P(0,1) = **0,0003**

Similar to THERP result, the probability results by itself do not have a good idea about if the chance of an accident is high or low, but if using some risk criterion or even compare with the probability of an accident after improving performance human factors that reduce the human error probabilities. In first case if we considered that around 200 people deaths in such accident the risk is 6 x 10 E2, that is unacceptable.

As a conclusion, we can say that OAT analysis has the main remarks:

- With OAT Tree is possible to access sequence of human tasks and identify the task with higher human error probability;

- The method is reliable because has been applied successfully in the nuclear industry and can also be applied to other industries to predict human error probability;
- It´s necessary to predict the human error probability for all tasks applied in OAT, which might be difficult without reliable data;
- Such methodology does not address any discussion about human performance factors which cause human error.

7.4. ACCIDENT SEQUENCE EVALUATION PROGRAM (ASEP)

The ASEP methodology objective is to predict the human error probability pre-accident and post-accident. Such methodology is well described in standard NUREG/CG-4772 published in 1987.

In general terms the pre-accident approach, focus on an event with influence on accident, such as a layer of protection failures caused by human error during installation, manufacturing, maintenance or operation. By the other hand, the post-accident approach focuses on human error in tasks which mitigate the accident consequence during emergency response.

7.4.1. Pre-Accident Analysis Methodology

The Pre-accident and Post Accident can consider the omission and commission error and also recovering actions.

The Basic Human Error Probability (BHEP) values is is 0,03 in the pre - accident analysis based on ASEP method. That includes the probability of omission, error (EOM) and commission error (ECOM). Based on ASEP procedure EOM and ECOM are:

P(EOM) = 0,02

P(ECOM) = 0,01

Thus:

$$F_T = 0,02 + \left(\left(1 - 0,02\right) \times 0,01\right) = 0,0298 \approx 0,03$$

$$F_T = 0,03$$

In fact, according with NUREG/CG-4772, the human error probabilities for pre-accident are defined based on basic and optimum condition combinations. The Table **2** summarizes such basic and optimum conditions possibilities.

Table 2: Basic and Optimum Conditions (Source: NUREG/CG-4772)

BASIC CONDITION	OPTIMUM CONDITION
"Basic condition 1 (BC1)" – No signal device to advice unsafe condition is available whenever such device is under maintenance conditions or other kind of intervention.	**"Optimum condition 1 (OC1)"**– Unavailable component status is indicated in the control room by some "compelling signal" such as an annunciation when the maintenance or calibration task or subsequent test is finished.
"Basic condition 2 (BC2)" – Component status is not verified by a Post-Maintenance (PM) or a Post-Calibration (PC) test.	**"Optimum condition 2 (OC2)"** – Component status is verifiable by a Post-Maintenance or Post-calibration test. If done correctly, full recovery is assumed.
"Basic condition 3 (BC3)" - There's no recovery factor to check unsafe condition.	**"Optimum condition 3" (OC3)"** – There is a requirement for a Recovery Factor (RF) involving a second person directly to verify component status after completion of a Post-Maintenance or Pos-Calibration task.
"Basic condition 4 (BC4)" - Check out of components status is not completely affect.	**"Optimum condition 4 (OC4)"** – There is a requirement for a current check on component status, using a written list.

Thus, considering basic and optimum conditions are suggested human error probabilities with error factor and upper bound which are comprised in ten cases that are (NUREG/CG-4772):

Case 1 – After a human error (omission or commission), neither Post – Maintenance (PM) nor Post-Calibration are not able to recover error as well as other recovers factors (RFs). Thus, all basic conditions are applied. The probability of human error is:

FT = 0,03 (EF=5 and UB=0,15).

Case 2 - After a human error (omission or commission), neither Post – Maintenance (PM) or Post-Calibration are not able to recover error as well as other recovers factors (RFs). Thus Basic Conditions 1 and 2 are applied as well as Optimum conditions 3 and 4. Therefore the probability of human error is: FT= 0,0003 (EF=16 and UB=0,05).

Case 3 - After a human error (omission or commission), neither Post – Maintenance (PM) nor Post-Calibration are not able to recover error as well as feedback signal but second person or other recover factor (RFs) is used. Thus Basic Conditions 1, 2 and 4 are applied as well as Optimum conditions 3. Therefore the probability of human error is: FT=0, 003 (EF=10 and UB=0,03).

Case 4 - After a human error (omission or commission), neither Post – Maintenance (PM) nor Post-Calibration are not able to recover error as well as feedback signals but the periodic check is performed. Thus Basic Conditions 1,2 and 3 are applied as well as Optimum conditions 4. Therefore, the probability of human error is: FT = 0,003 (EF=10 and UB=0,03).

Case 5 - After a human error (omission or commission), Post – Maintenance (PM) or Post-Calibration is able to recover error and at least Optimum conditions 1 is applied. Therefore the probability of human error is: FT=negligible (UB= 0, 00001).

Case 6 - After a human error (omission or commission), Post – Maintenance (PM) or Post-Calibration is able to recover error. Thus Basic Conditions 1, 3 and 4 are applied as well as Optimum conditions 4. Therefore the probability of human error is: FT = 0,0003 (EF=10 and UB=0,003).

Case 7 – After a human error (omission or commission), Post – Maintenance (PM) or Post-Calibration is able to recover error. Thus Basic Conditions 1 is applied as well as Optimum conditions 2, 3 and 4. Therefore the probability of human error is: FT = 0,00003 (EF=16 and UB=0,0005).

Case 8 - After a human error (omission or commission), Post – Maintenance (PM) or Post-Calibration is able to recover error. In addition, second person to recover error is used. Thus Basic Conditions 1 and 4 are applied as well as Optimum conditions 2, 3 and 4. Therefore the probability of human error is: FT = 0,0003 (EF=10 and UB=0,003).

Case 9 - After a human error (omission or commission), Post – Maintenance (PM) or Post-Calibration is able to recover error. In addition, periodic test is performed. Thus Basic Conditions 1 and 3 are applied as well as Optimum conditions 2 and 4. Therefore the probability of human error is: FT = 0,00003 (EF=16 and UB=0,0005).

In order to understand better ASEP methodology applied to Pre-Accident HRA, a group of LPG Spheres storage accident will be exemplified and assessed by ASEP methodology.

At approximately 05:35 hours on 19 November 1984 a major fire and a series of catastrophic explosions occurred in PEMEX LPG Terminal at San Juan Ixhuatepec, Mexico City. As a consequence of these events some 500 individuals were killed and the terminal destroyed as shows Fig. **4**.

Three refineries supplied the facility with LPG on a daily basis. The plant was being filled from a refinery 400 km away, as on the previous day it had become almost empty. Two large spheres and 48 cylindrical vessels were filled to 90% and 4 smaller spheres to 50% full.

A drop in pressure was noticed in the control room and also at a pipeline pumping station. An 8-inch pipe between a sphere and a series of cylinders had ruptured. Unfortunately the operators could not identify the cause of the pressure drop. The release of LPG continued for about 5-10 minutes when the gas cloud, estimated at 200 m x 150 m x 2 m high, drifted to a flare stack. It ignited, causing violent ground shock. A number of ground fires occurred. Workers on the plant now tried to deal with the escape taking various actions. At a late stage somebody pressed the emergency shutdown button (www.hse.gov.uk).

Figure 4: BLEVE in LPG Sphere Mexico - 1984 (LPG Sphere Pre-Accident).

In order to avoid this type of accident tree tasks steps to check LPG spheres and pipelines must be carried out constantly such as:

I. Check out if there is vehicles in operational area around the LPG Sphere (Success: a – fail: A);

II. Check out if there is maintenance or other service with ignition sources being performed around the LPG Sphere area (Success: b – fail: B);

III. Check out if there is leakage on pipelines and spheres (Success: c – fail: C);

The tasks can be represented by the Human Tree Diagram in Fig. **5**. Remarkably, is to remember that the first three tasks do not trigger an accident by itself, but when violated is a human error in terms of the procedure due to unsafe conditions in case of pipeline or spheres leakage.

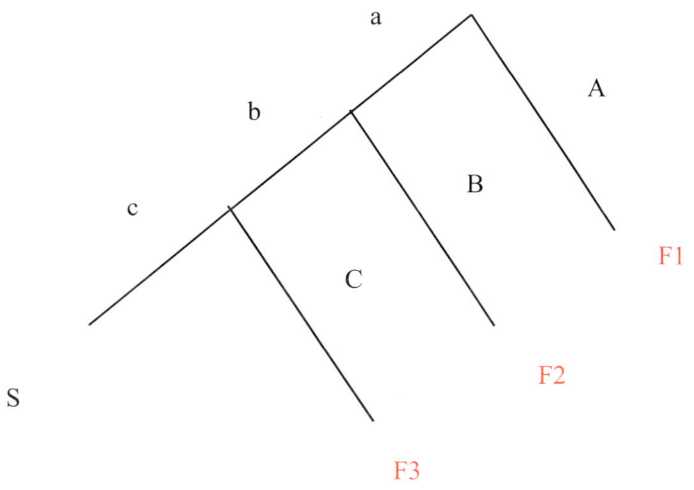

Figure 5: ASEP Human Tree Diagram (LPG Sphere Pre-Accident).

The omission human error occurred on task three, because a leakage on pipelines to feed spheres last for 10 minutes without be detected by operators despite the drop down pressure on the pipeline.

Unfortunately, there was not recover factors. Consequently, based on ASEP pre-accident approach cases, the probability for each task is 0,03 ("Case 1– page 22"). Therefore, the probability to occur leakage (Human Error Probability – HEP) is:

$$HEP = 1\text{-}P(s) = 1\text{-}(0{,}97x0{,}97x0{,}97) = 7{,}73\%$$

Whether the recover action or a second person to check out the steps 1, 2 and 3 took place, the probability of failure in such tasks, would have reduced to 0,0003 ("Case 4 – page 23") and the new Probability of leakage in LPG Sphere would be:

$$HEP = 1\text{-}P(s) = 1\text{-}(0{,}9997x0{,}9997x0{,}9997) = 0{,}08\%$$

The ASEP methodology also considering dependencies in predicting the human error probability. In fact, there are two types of dependence: between person dependence and within-person's dependence. In the first case, the between-person's dependence is happening when another person checks first person task. In the second case, the within-person`s dependence will happen when a set of operation take place for only one person.

7.4.2. Post-Accident Analysis Methodology

The ASEP methodology discussed until now is to access Pre-Accident condition, but NUREG/CR 4772 has also proposed methodology to access Human Error Probability for POs-Accident analysis. In this case, the time to detect accident is very important and takes high influence on Human Error Probability in Post-Accident actions.

In fact, detect accident on time, perform correct diagnose and take correct decision is essential to perform corrective action to control accident scenario. If diagnose and decision take longer than necessary and consequently there´s not time enough to perform corrective action the accident will not be under control as well as if misdiagnose or wrong decision take place. In addition, even though correct diagnose and decision on property time occur, if corrective action is not correct or not performed in record time the accident will not be under control. A remarkable point is that in such model is being regarded that all resources to control accident situation are available that in real case is not always like that. Therefore, the total time to perform corrective action is the sum of diagnose time and action time as shown in Fig. **6**.

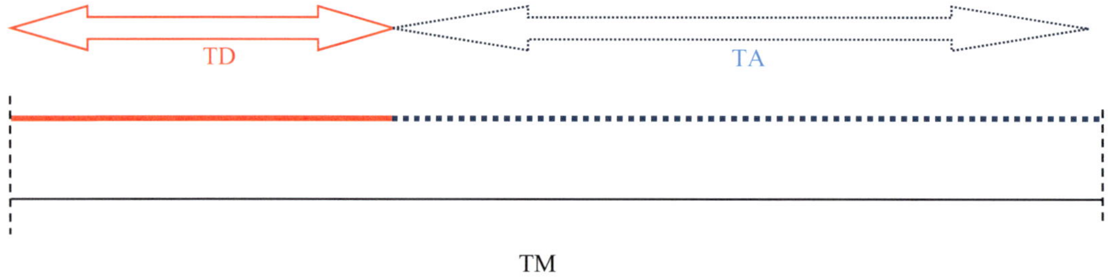

Figure 6: Time to response correction action in accident.

TM=TD+TA

where:

TM = Maximum time to detect an accident, diagnose, take a decision and perform a post- diagnose action to control accident.

TD = Time to detect an accident, diagnose, take a decision to define actions to control accident.

TA = time to perform a post-diagnose action to control accident.

The probability of success or failure in Post-Accident analysis is dependent on time. Thus, as shorter as time to diagnose or perform corrective action higher is probability of human error in recovering accident situation. Basically detection and diagnosis involves knowledge-based behavior and Post-diagnosis actions involve rule-based-behavior or skill-based behavior.

The ASEP methodology to access Post-Accident proposes the graph shows in Fig. **7** which comprise time and Human Error Probability (HEP) for diagnose action. Thus, is possible to estimate human probability error based on time having upper and lower limits to be regarded depends on how conservative is the analysis. In fact, the nominal model has more conservative values than Screening Model also proposed to analyze Human Error Probability in the diagnostic task within time.

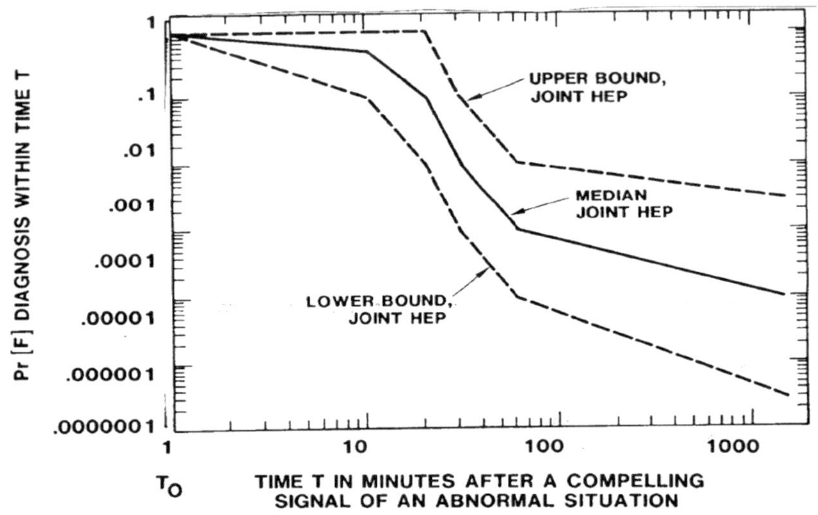

Figure 7: Nominal Diagnose Model (Source: NUREG/CR 4772).

After estimate human probability error in diagnosis is necessary to estimate Human Error Probability (HEP) to Post-Accident diagnose action. Based on ASEP methodology proposed in NUREG/CR 4772 such probability is related to particular conditions as shown in Table **3**.

Table 3: Post-Accident diagnose Action Human Error Probability (Nominal Diagnose Model) (Source: NUREG/CR 4772)

HEP	EF	ASSUMPTIONS
100%		Action outside control room is required.
100%		Is necessary to perform skill-based behavior action or rule-based behavior action when no procedure is available.
5%	5	Perform a critical procedural action correctly under moderately high stress.
25%	5	Perform a critical procedural action correctly under extremely high stress.
1%	5	Perform a post-diagnosis action can be classified as skill-based actions and there is a backup written procedure.

HEP=Human Error Probability
EF=Error Factor

Considering similar BLEVE accident described on previous item which the time to avoid accident requires 10 minutes to diagnose and at most 50 minutes to post-diagnose action. Based on Fig. **7**, the Human Error probability to diagnose is 90% and in this case due to have a not clear accident scenario, such situations are not detected with not time enough to avoid the accident. That means that diagnose time is not enough to define a recovery action to prevent the accident. The Human Error Probability for post-diagnose action considering that action outside control room is required is 100%. Therefore, the total Error probability is calculated by the human event tree as shows Fig. **8**.

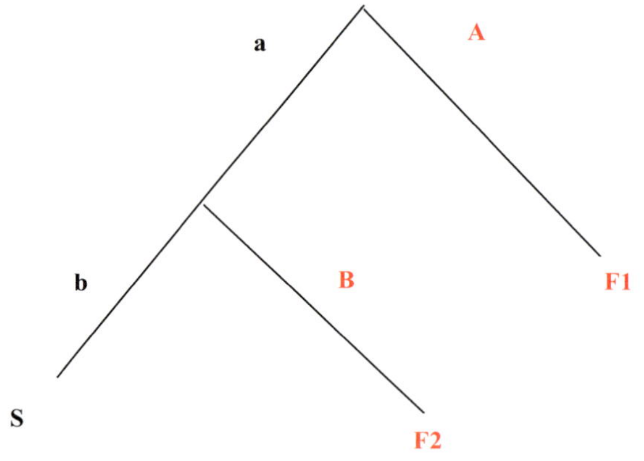

Figure 8: Human Event Tree: Post-Accident Analysis (LPG Sphere fire).

A=Human error diagnose

B=Human error Pos–diagnose action

P(Bleve) =1–P(S) =1–(P(a) x P(b))

If,

P(a) =1–P(A) =1–0,9=0,1

P(b) =1–P(B) =1–1=0

Thus,

$$P(Bleve) = 1 - P(S) = 1 - (P(a) \times P(b)) = 1 - (0,1 \times 0) = 100\%$$

The probability of BLEVE in Sphere was 100% based on ASEP procedure. That was a real accident case result due to lack of perception on pipeline leakage.

The purposes of this chapter were to introduce the ASEP concepts to be applied in different industries for different situations. In general terms the remarks about ASEP methodology are:

- The approach to be applied to pre-accident and post-accident are very clear and feasible to be applied in real cases;
- The performance factors are not taken into account which in some cases will not help to minimize the human error;
- Complex accident analysis may be difficult to be assessed based on THERP human tree.

7.5. SOCIAL TECHNICAL ANALYSIS OF HUMAN RELIABILITY (STAHR)

The methodology was developed by Philip (1982) and regards specialist opinion about human performance factors which take influence in human error. Furthermore, is necessary to make up Human Reliability Tree that comprises a human error and performance factor associated in order to represent the human error situation under assessment. The Such Human Reliability Tree is only a schematic representation about the human performance factors which influence on human error and have no relation to human reliability tree from THERP method.

Furthermore, such Human Reliability Tree representations has the human error probabilities calculated based on specialist opinion which take into account specific weight for each human performance factors influence on human error in order to calculate the final human error probability.

In order to understand, the better such methodology an example for STAHR method will be carried out. Thus, such method will be applied to an example of human error in valve maintenance.

A hydrogen sulfide gas leakage accident occurred at Plant, when off-site piping repair work was carried out as a part of shutdown maintenance. A block valve at the outlet of a pressure control valve was removed. At the same time, repair to an air supply line for the pressure control valve was also underway. The air supply to the pressure control valve was stopped, and the control valve was fully opened. As the inlet block valve of the pressure control valve was also open, hydrogen sulfide gas leaked upon removing the outlet block valve. A blind plate was inserted, but the position was wrong, so it also came off when the outlet block valve was removed. A blind plate was inserted at the upstream-side flange of an outlet block valve of a pressure control valve.

Based on accident analysis, the root cause of the accident was two performance factors that are procedure and communication. Both human performance factors must have their root causes analyzed and discussed. The Human Reliability Tree was building up and is represented at shows Fig. **9**.

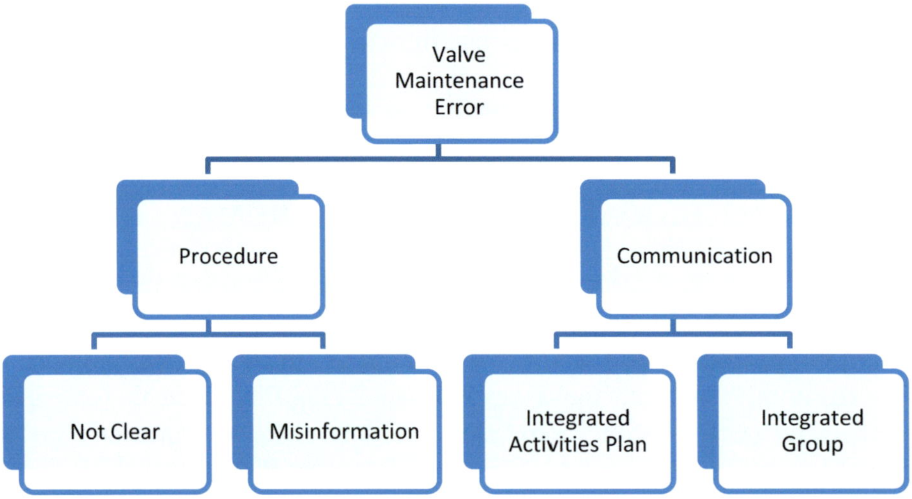

Figure 9: Human Reliability Tree (STAHR).

Based on STAHR procedure, such human performance factors, root causes must have a weight that reveals its importance to each human performance factors. Thus, in procedure case, the main root causes are "Not clear information" and "Miss Information". Based on specialist opinion the "Not clear information" have a high influence on procedure quality (Much - 0.7 and Low – 0.3) as well as "Miss Information" (Much - 0.8 and Low – 0.2).The Table **4** show the final probability of procedure quality that influence in the end on valve maintenance activity error. Once analyze procedure quality, the combination of root causes ("Not clear information" and "Miss Information") were taken into account.

Table 4: Procedure (STAHR)

Procedure							
If	and	So	Probability to have procedure quality		Final weight (Not clear and miss information)		
Not Clear	Miss information		Good	Bad	Not clear	Miss information	Rsult
Much	Much		0.1	0.9	0.7	0.8	0.56
Much	Low		0.2	0.8	0.7	0.2	0.14
Low	Much		0	1	0.3	0.8	0.24
Low	Low		0.8	0.2	0.3	0.2	0.06
total			0.132	0.868			

The next step is to precede the same calculation to analyze and define the probability of quality of communication that take also influence on valve maintenance procedure. In communication case, the root causes are "Integrated group activities" and "Integrated group". That means respectively all activities are integrated as well as group member. Based on specialist opinion, the "Integrated group activities" have a high influence on procedure quality (Much - 0.8 and Low – 0.2) as well as "Integrated group" (Much - 0.9 and Low – 0.1).The Table **5** shows the final probability of communication quality. Once analyze communication quality, the combination of root causes "Integrated group activities" and "Integrated group" were also taken into account.

The final and next step is combining both human performance factors (Procedure and communication) to define the human error probability on Valve Maintenance. The Table **6** below shows the valve maintenance human error probability. The probabilities of having a good and bad procedure and communication quality defined on Tables **4** and **5** are applied on Table **6** like weights in order to calculate the final human error probability.

Table 5: Communication (STAHR)

Communication							
If	and	So	Probability that communication quality is :		Final weights (activities plan and group integration)		
Integrated activities plan	Integrated group		Good	Bad	Integrated activities plan	Integrated group	Results
Much	Much		0.9	0.1	0.8	0.9	0.72
Much	Low		0.2	0.8	0.8	0.1	0.08
Low	Much		0.1	0.9	0.2	0.9	0.18
Low	Low		0	1	0.2	0.1	0.02
total			0.682	0.318			

Table 6: Communication (STAHR)

7 - Valve maintenance error							
If	and	So	Probability to have valve maintenance		Final weight (Procedure and Communication)		
Procedure	Communication		Success	Failure	Procedure	Communication	Result
Good	Bad		0.6	0.4	0.132	0.318	0.041976
Bad	Good		0.6	0.4	0.868	0.682	0.591976
Good	Good		0.9	0.1	0.132	0.682	0.090024
Bad	Bad		0.1	0.9	0.868	0.318	0.276024
total			0.488995	0.5110048			

As conclusion of STAHR application, we have a high probability (51%) to have failed when the valve is under maintenance because high influence of procedure and communication. The procedure is explained by to have not clear and miss information and communication is explained by to have not integrated on group activities as well as group member integration.

Basically, in this case there were two human performance factors, but it is possible to have more as much as necessary to explain such human performance influence on human error. The important remarks as much as performance factors are included in such analysis more complicated to specialist access, the combination of such human performance factors together in human error and such issue can take influence on final human error calculation. Consequently, whenever is possible is advisable to consider the human performance factors that have a relevant influence on human error.

Thus we conclude that STAH-R method coins are:

- Simple to be applied;
- Required experience specialists estimate, weigh and probabilities;
- Regards human performance factor in human reliability analysis;
- Allows fast HEP calculation;

The STAH-R drawbacks are:

- Depends too much on specialist point of view;
- Requires more applications to be validated.

7.6. STANDARDIZED PLANT ANALYSIS, RISK HUMAN RELIABILITY (SPAR-H)

In order to support the Accident Sequence Precursor Program (ASP), the U.S. Nuclear Regulatory Commission (NRC), in conjunction with the Idaho National Laboratory (INL), in 1994 developed the Accident Sequence Precursor Standardized Plant Analysis, Risk Model (ASP/SPAR) human reliability analysis (HRA) method, which was used in the development of nuclear power plant (NPP) models. Based on experience gained in field testing, this method was updated in 1999 and renamed SPAR-H, for Standardized Plant Analysis, Risk-Human Reliability Analysis method (NUREG, /CR-6883).

The main objective is to define Human Error Probability considering human performance factors influence. Such methodology requires a specialist opinion in order to define values for standardized human performance. The PSF (performance factors) is used to define the HEP (Human Error Probability) in equation 1 as shown below.

Equation 1

$$HEP = \frac{NHEP \cdot PSF_{composite}}{NHEP \cdot \left(PSF_{composite} - 1\right) + 1}$$

Such method establishes value of HEP to omission, error (0,01) and commission error (0,001). In addition, the SPAR-H method is based on eight performances shaping factors (Gertman, 2005) that encapsulate the majority of the contributors to human error. These eight performances shaping factors are as follows: available time to complete task, stress and stressors, experience and training, task complexity, ergonomics, the quality of any procedures in use, fitness for duty, and work processes. Each performance shaping factor features are stated on list with different levels and associated multipliers. For example, the presence of extremely high stress would receive a higher multiplier (5) than nominal stress (1). Table **7** below shows PSF values to define PSF composite.

Table 7: PSF values (Source – NUREG, CR-6883)

PSFs	PSF Level	Multiplier for Action
Available time	Inadequate Time	P(f)=1
	Time Available ≈ Time required	10
	Nominal time	1
	Time Available ≥ 5x Time required	0.1
	Time Available ≥ 50x Time required	0.01
	Insufficient information	1
Stress	Extreme	5
	High	2
	Nominal	1
	Insufficient information	1
Complexity	Highly complex	5
	Moderately complex	2
	Nominal	1
	Insufficient information	1
Experience/ Training	Low	3
	Nominal	1
	High	0.5
	Insufficient information	1
Procedures	Not Available	50
	Incomplete	20
	Available, but poor	5
	Nominal	1
	Insufficient information	1
Ergonomics	Missing /Misleading	50
	Poor	10
	Nominal	1
	Good	0.5
	Insufficient information	1
Fitness for duty	Unfit	P(f)=1
	Degrate fitness	5
	Nominal	1
	Insufficient information	1
Work process	Poor	5
	Nominal	1
	Good	0.5
	Insufficient information	1

The SPAR-H method is a human reliability method based on specialist opinion about pre-defined values as well as basic calculation which is available in the behavioral sciences literature (NUREG, /CR-6883).

The main question about human factors in SPAR-H is the relation between such human factors and how does it take influence in human reliability. The relation between performance factors can be represented by Fig. **10**.

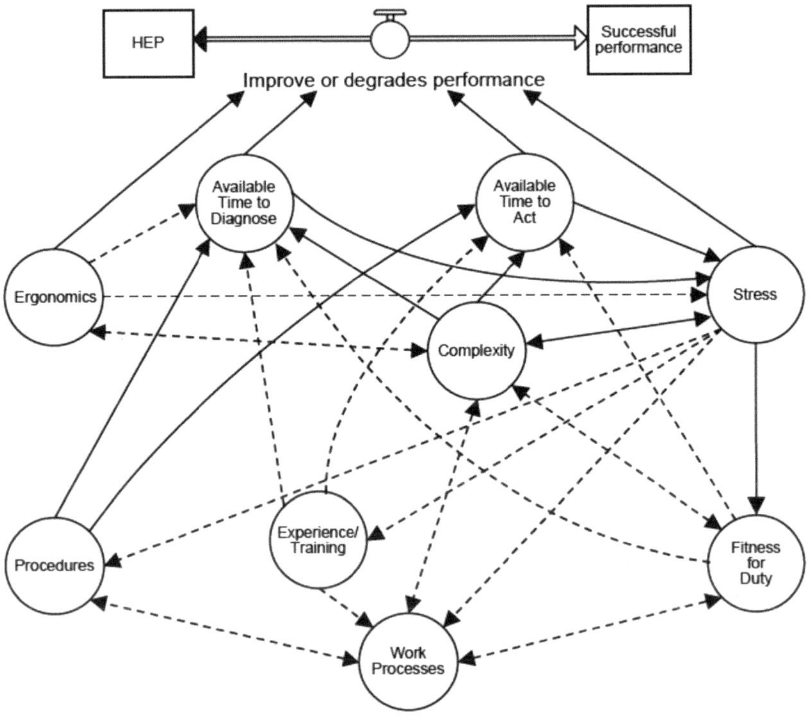

Figure 10: Path diagram showing relationships among PSFs (Source: NUREG, CR-6883)

In order to exemplify SPAR-H methodology an example of human error in furnace operation that lead loss of production due high temperature set up by the operator that damage the furnace tubes during a start up after a programed maintenance. Based on such incident analysis, such human error was influenced mainly by the high stress level to achieve production target, short time to precede, high complexity and low experience. Such incident on most of cases lead to an unsafe condition and must be always avoided. Based on such information the human performance factor analyzed by specialized based on SPAH-R methodology is shown and highlighted in red on Table **8** shows classifications for human performance factors highlighted in red.

Table 8: PSF values (Furnace Incident) (Source – NUREG, CR-6883)

PSFs	PSF Level	Multiplier for Action
Available time	Inadequate Time	P(f)=1
	Time Available » Time required	10
	Nominal time	1
	Time Available ^{3} 5x Time required	0.1
	Time Available ^{3} 50x Time required	0.01
	Insufficient information	1
Stress	Extreme	5
	High	2
	Nominal	1
	Insufficient information	1
Complexity	Highly complex	5
	Moderately complex	2
	Nominal	1
	Insufficient information	1
Experience/ Training	Low	3
	Nominal	1
	High	0.5
	Insufficient information	1
Procedures	Not Available	50
	Incomplete	20
	Available, but poor	5
	Nominal	1
	Insufficient information	1
Ergonomics	Missing /Misleading	50
	Poor	10
	Nominal	1
	Good	0.5
	Insufficient information	1
Fitness for duty	Unfit	P(f)=1
	Degrate fitness	5
	Nominal	1
	Insufficient information	1
Work process	Poor	5
	Nominal	1
	Good	0.5
	Insufficient information	1

The next step is to calculate the human performance composite that is calculated from the values of each PFS value presented in Table **11** as shows the equation below.

$$PSF_{composite} = \left(\left(Availabity\ time_{Multiplier}\right) \times \left(Stress_{Multiplier}\right) \times \ldots \times \left(Work\ process_{Multiplier}\right)\right)$$

$$PSF_{composite} = \left((10) \times (5) \times \ldots \times (1)\right) = 750$$

Thus, once defined the PFS composite and considering the commission error as 0.1, the next and final step is to apply such values in equation below and consequently we have:

$$HEP = \frac{NHEP \cdot PSF_{composite}}{NHEP \cdot \left(PSF_{composite} - 1\right) + 1}$$

$$HEP = \frac{0.1 \times 750}{0.1 \times (750 - 1) + 1} = 98{,}8\%$$

We can realize that despite a very good procedure to start up the furnace, such task requires high experience because operator need to identify when is the best point to go to the next range of temperature. In addition is important to state that the omission and commission error probability may have different values of 0.01 and 0.1 based on specialist opinion. In general aspects, the SPAR-H has as advantages:

⇒ Simple to be applied;

⇒ Has defined values to commission and omission error;

⇒ Allows fast HEP calculation;

The SPAR-H drawbacks are:

⇒ Do not consider the direct effect among PSFs;

⇒ Depends on the case, is necessary to consider other PFS (that is not on Table **10**);

7.7. HUMAN ERROR ASSESSMENT REDUCTION TECHNIQUE (HEART)

In 1985, the HEART technique was presented for Williams and after three years described in detail. Thus, basically this methodology is applied to analyze Human tasks with defined values for human error probability (Nominal Human Reliability) related to activities and for a context which each activity is being involved. Based on such values, calculates the Final Human Error probability formula considering the activities and Error Producing Conditions. The general application steps are basically as follows:

1. Define Activity;
2. Define correspondent Generic task and define Nominal Human Unreliability;
3. Define Error producing Condition related to the activity;
4. Assessing rating of Error Producing Condition;
5. Calculate the final HEP;

The final HEP is calculated by the equation:

$$Final\ HEP = GEP \times \prod R(i) \times (Wi - 1) + 1$$

where:

GEP = Generic Error probability (is defined in generic task table)

R (I) =Value of context task (is defined based on generic context task table values)

W (I) =Weigh for each context task defined for specialist opinion.

In order to define final HEP the first step has defined the task that is better defined in Table **9**. Thus Nominal Human Unreliability is choosing from proposal range values.

Thus the main idea is to check which generic task (from A to H) feet better on the task under reliability analysis and further define which is the human error probability value (Nominal Human Unreliability) based on Table **9**. The Nominal Human Unreliability value in Table **9** must be defined for specialist opinion or if there will be any data bank available to compare with such value.

Table 9: Generic Tasks and Nominal Human Unreliability (Source: Willians, 1988)

	Generic Tasks	Nominal Human Unreliability	
A	Totally unfamiliar, performed at speed with no real idea of likely consequences.	0,55	(0,35-0,97)
B	Shift or restore system to a new or original state on a single attempt without supervision or procedures.	0,26	(0,14-0,42)
C	Complex task requiring high level of comprehention and skill.	0,16	(0,12-0,28)
D	Fairly simple task performed rapidly or given scant attention.	0,09	(0,06-0,13)
E	Routine, highly practised, rapid task involving relatively low level of skill	0,02	(0,07-0,045)
F	Restore or shift a system to original or new state following procedures with some checking.	0,003	(0,0008-0,007)
G	Completely familiar, well-designed, highly practised, routine task ocurring several times per day, performed to highest possibe standards by highly motivated, highly-trained and experienced personnel, with time to correct potential error, but without the benefit of significant job aid.	0,0004	(0,00008-0,009)
H	Respond correctly to system command even when there is an augments or automated supervisory system providing accurate interpratation of system state	0,00002	(0,000006-0,009) 5th-95th percentible bound

The following step is defined which human performance factors that are in HEART methodology is called Error Producing Condition (EPC) is related to tasks. Each Error Producing Condition has a specific weight as shown in Table **10**. In this case, more than one Error Producing Condition (EPC) item can be chosen for different tasks and further will be applied the formula to calculate Final Human Error Probability.

In order to understand the HEART method a case study about a human error on start up furnace will be carried out. In this case, the human error caused an environment impact due high level of toxic gases spilled. In order to simplify the analysis and reader understand the whole start up furnace will be defined by two generic tasks which comprise several other tasks. Those tasks start up the furnace by different range of temperature and the second one is star up the electrostatic

Table 10: Error Producing Condition (Source: Willians, 1988)

	Error Producing Condition	Weight
1	Unfamiliarity with a situation which is potentially important but which only occurs infrequently or which is novel	17
2	A shortage of time available for error detection and correction	11
3	A low signal-noise ratio	10
4	A means of suppressing or over-riding information or features which is too easily accessible	9
5	No means of coveying spatial and functional information to operators in a form which they can readily assimilate	8
6	A mismatch between an operator's model of the words and that imagine bt designer	8
7	No obvious means of revesing an unintended action	8
8	A channel capacity overload, particularly one casue by simutaneous presentation of non-redundant information	6
9	A need to unleam a technique and apply one which requires the application of an opposing philosophy	6
10	The need to utransffer specific knowledge from task to task wiithout loss	5.5
11	Ambiquity in the required peformance standards	5
12	A means of su[[ressing or over-riding information of features which is too easily accessible	4
13	A mismatch between perceived and real risk	4
14	No clear, direct and timely confirmation of an intended action from the portion of the system over which control is exerced	4
15	Operator inexperince (e.g, a newly qualified tradesman but not an expert)	3
16	An impoverished quality of information conveyed by procedures and person-person interaction	3
17	Little or no independent checking or testing or output	3
18	A conflict between immediate and long term objective	2.5
19	Ambiguity in the required performance standard	2.5
20	A mismatch between the educational achievement level of an individual	2
21	An incentive to use other more dangerous procedures	2
22	Little oportunities to exercise mind and body outside the immediate confines of a job	1.8
23	Unreliable instrumentation (enough that is is noticed)	1.6
24	A need for absolute judgements which are beyond the capabilities or experience of an operator	1.6
25	Unclear allocation of function and responsibility	1.6
26	No obvious way to keep track of progress during activitty	1.4
27	A danger that finite physical capabilities will be exceeded	1.4
28	Little or intrinsic meaning in a task	1.4
29	High level emotional stress	1.3
30	Evidence of ill-health amongst operatives especially fever	1.2
31	Low workforce morale	1.2
32	Inconsistency of meaning of displays and procedures	1.2
33	A ppor or hostile environment	1.15
34	Prolonged inactivity or highly repetitious cycling of low mental workload tasks (1 st half hour)	1.1
34	(thereafter)	1.05
35	Disruption of normal work sleep cycles	1.1
36	Task pacing caused by the intervention of others	1.06
37	Additional team members over and above those necessary to perform task normally and satisfactorily. (per additional team member)	1.03
38	Age of personel performing percentual tasks	1.02

precipitator (EP). The EP is a particulate collection device that removes particles from a flowing gas (from the furnace) using the force of an induced electrostatic charge in order to keep the acceptable level of toxic gas emission. Both equipment, furnace and EP must operate together otherwise the level of toxic gas emission go higher than specified and cause an environment impact. Consequently, after a preventive maintenance in a furnace, both equipment were started up, but in order to start up fast the production the start up procedure was not taken into account and both equipment was not synchronized. Thus consequently, high levels of toxic emission cause an environment impact. Based on HEART procedure, the first step is to choose a generic task which fit better in case study and in this case we chose the generic task "B" (Shift or restore system to a new or original state on a single attempt without supervision or procedure). The second step is choosing the nominal human unreliability that in this case will be "0.14" based on specialist opinion. The third step is to define the error producing condition and in this case is "Operator inexperience" that is number 15 in Table **10**. The value applied for this error producing condition is 3. The next step is defining a weigh that such error producing condition and that reflect how much such error producing condition influence on human error. Such value must vary from 0 to 1 (0% to 100%). The final step is to apply the formula below for each task. The Final human error probability is presented in Table **11**.

$$Final\ HEP = GEP \times \prod R(i) \times (Wi - 1) + 1$$

On column seven of Table **11** the partial formula applied is:

$$Partil\ HEP = \prod R(i) \times (Wi - 1) + 1$$

In this case, as we have only one error producing condition, the formula is simplified by

$$Partil\ HEP = R(i) \times (Wi - 1) + 1$$

where

$$R(i) = weigh$$

$$W(i) = value \text{ (Column 5)}$$

Table 11: Human error probability (priori)

Task	Task classification	Nominal Human unreliability	Error producing condition	Value	Weigh	Partial Human error probability	Human error probability
Start up furnace	Action	14.0%	Operador inexperience	3	90%	2.8	39%
Start up Eletrostatic precipitator	Action	14.0%	Operador inexperience	3	90%	2.8	39%
Human error (Start up furnace)							78%

Once the furnace start up is proceeding following procedures under an experience supervisor the Nominal Human Error unreliability reduce from 14% to 0.3% based on Table **10** classification. That means a change from generic task "B" to the generic task "F". The Table **12** below shows the reduction of human error probability in startup furnace that reduce from 78% to 2%.

Table 12: Human error probability (posteriori)

Task	Task classification	Nominal Human unreliability	Error producing condition	Value	Weigh	Partial Human error probability	Human error probability
Start up furnace	Action	0.3%	Operador inexperience	3	90%	2.8	1%
Start up Eletrostatic	Action	0.3%	Operador inexperience	3	90%	2.8	1%
Human error (Start up furnace)							2%

In general aspects, the HEART has as advantages:

⇒ Simple to be applied;

⇒ Has defined values to Nominal human unreliability based on generic task;

⇒ Has defined values to error producing condition;

⇒ Allows fast HEP calculation;

The HEART drawbacks are:

⇒ Depends on the case, is necessary to consider other generic task and define new nominal human unreliability;

⇒ Depends on the case, is necessary to consider other error producing condition and define new weight;

7.8. BAYESIAN BELIEF NETWORK METHOD (BBN)

The Network Bayesian methodology was developed in 80 ages to make easier prediction an Artificial Intelligence analysis (Pearl, 2000). It can be defined as graphic frameworks which represents arguments in uncertain domains (Korb & Nicholson, 2003). Such framework is unicycle Graphs and because that cannot make up closed cycles and have only one direction. The node represents random variables and arcs represent direct dependency between variable relations. The arc direction represents cause effect relation between variables (Menezes, 2005). In Fig. **11**, the Bayesian Network is represented being node C, consequence from cause A and B.

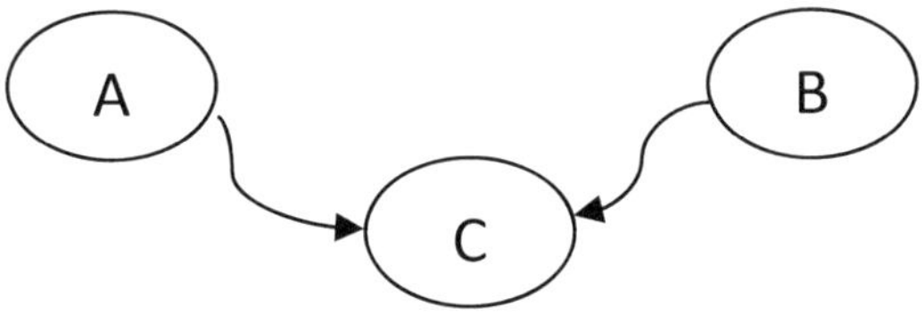

Figure 11: Bayesian Network.

In Fig. **11**, node A and B are the fathers of C and node C is called son of A and B. In such representation in each node there´s conditional probabilities which represent variable values of such event. The Human Performance factor represented by nodes A and B can be internal (Stress, over workload, depression, demotivation, health conditions, deceases) or external (procedures, equipment, work conditions, bad social conditions, bad acceptance in the group). Into each node there´s a conditional PDF (probability density function) which represent variable values over a period time when the random variable represents such event or we can also have constant conditional probability value. The equation 3 below shows Bayes equation:

$$P(A|B) = \frac{P(B|A) \times P(A)}{P(B)}$$

where:

$P(A|B)$ = posteriori probability of A when B is known

$P(B|A)$ = Maxi likelihood related to B with A evidence occurrence.

$P(A)$ = priori probability of event A.

The equation above is a Bayesian representation for two conditional events, but in some cases more events comprise Bayesian Network getting harder to calculate. In fact, as bigger as Bayesian networks, more complex to calculate and it is advisable to perform such calculation by software whenever it´s possible. In addition, as bigger as Bayesian Networks, performance factors associated with human error are harder to obtain precise and predict such conditional probability is also harder. In general terms, the Bayesian Network probability can be represented by the equation:

$$P(U) = P(X_1, X_2, X_3 .... X_n) = \prod_{i=1}^{n} P(X_i / Pf(X_i))$$

where:

$P(U)$ = Probability;

$P(X_i / Pf(X_i))$

The = Conditional probability of X related to their network father.

The methodology Bayesian belief networks (BBN) provides a greater flexibility as not it only allows for a more realistic representation of the dynamic nature of man-system, but also allows for representation of the relationship of dependence among the events and performance shaping factors (Drouguett,2007).

In order to clarify such methodology, an example of Bayesian Network is applied to assess a human error in pipeline repair task applied to refinery process plant vessel. Such task requires to isolate pipeline, check the pressure before performing repair task. Human error in such maintenance task may cause an accident with fatalities. Thus, is essential to take into account training, procedure and stress as human performance factors. The Bayesian Network represents the Process pipeline repair task on Fig. **12**.

Let T1 be the variable related to the level of repair training in such a way that T1=0 implies an adequate training and T1=1 represents an inadequate training. In the same way, let P2 and S3 be the variables associated with the adequateness of execution of the available procedure and the level of stress on repair task respectively. Finally, let C be the human performance, where C=0 implies in human adequate performance and C=1 in human error.

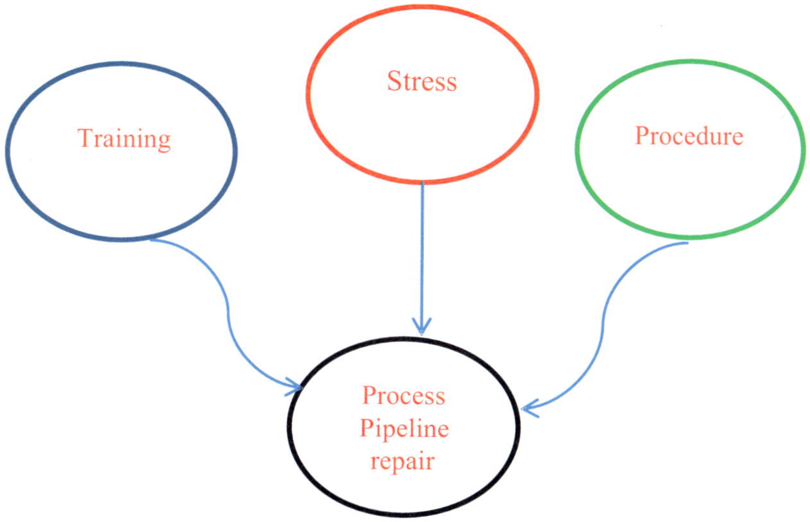

Figure 12: Bayesian Network (Process pipeline process).

Thus, HEP = P(C=1). In general, P (C=1) is represented by the equation:

$$HEP = \sum_{i=0}^{1}\sum_{j=0}^{1}\sum_{k=0}^{1} P(T_1 = i) \times P(P_2 = j) \times P(S_3 = k) \times P(C = 1 \mid T_1 = i, P_2 = j, S_3 = k)$$

where:

$\bar{C}$=Human error in repair task=(C=1)

T=Good training ($T_i = 0$)

$\bar{T}$=Bad training($T_i = 1$)

P=Good procedure($P_i = 0$)

$\bar{P}$=Bad procedure ($P_i = 1$)

S=Good stress level($S_i = 0$)

$\bar{S}$=Bad stress level($S_i = 1$)

Thus,

$$HEP = P(\bar{C}|\bar{T},\bar{P},\bar{S}) \times P(\bar{T}) \times P(\bar{P}) \times P(\bar{S}) + P(\bar{C}|T,\bar{P},\bar{S}) \times P(T) \times P(\bar{P}) \times P(\bar{S}) + P(\bar{C}|\bar{T},\bar{P},S) \times P(\bar{T}) \times P(\bar{P}) \times P(S) + P(\bar{C}|T,\bar{P},S) \times P(T) \times P(\bar{P}) \times P(S) + P(\bar{C}|\bar{T},P,\bar{S}) \times P(\bar{T}) \times P(P) \times P(\bar{S}) + P(\bar{C}|T,P,\bar{S}) \times P(T) \times P(P) \times P(\bar{S}) + P(\bar{C}|\bar{T},P,S) \times P(\bar{T}) \times P(P) \times P(S) + P(\bar{C}|T,P,S) \times P(T) \times P(P) \times P(S)$$

Such probability values are estimated by a specialist opinion by questionnaire below:

Specialist Opinion Questionnaire:

1. What is the probability to failure in **"repair task"** if training is not good? (Optimist = 30% and Pessimist = 70%)

2. What is the probability to fail in **"repair task"** if the procedure is not good? (Optimist = 50% and Pessimist = 80%)

3. What is the probability to fail in **"repair task"** if stress is not good? (Optimist = 10% and Pessimist = 60%)

4. What is the probability to fail in **"repair task"** if training, procedure and stress are not good? (Optimist = 90% and Pessimist = 100%)

5. What is the probability to fail in **"repair task"** if procedure and stress are not good and training is good? (Optimist = 80% and Pessimist = 90%)

6. What is the probability to fail in **"repair task"** if procedure and training are not good and stress is good? (Optimist = 80% and Pessimist = 90%)

7. What is the probability to fail in **"repair task"** if the procedure is not good and stress and training is good? (Optimist = 60% and Pessimist = 70%)

8. What is the probability to fail in **"repair task"** if stress and training are not good and procedure is good? (Optimist = 20% and Pessimist = 30%)

9. What is the probability to fail in **"repair task"** if stress is not good and procedure and training is good? (Optimist = 10% and Pessimist = 20%)

10. What is the probability to fail in **"repair task"** if training is not good and procedure and stress is good? (Optimist = 20% and Pessimist = 40%)

11. What is the probability to fail in **"repair task"** if stress, procedure and training are good? (Optimist = 1% and Pessimist = 2%).

Therefore, substituting probability values in equation 1 we have:

$$
\begin{aligned}
HEP = \ & [P(\bar{C}|\bar{T},\bar{P},\bar{S}) \times P(\bar{T}) \times P(\bar{P}) \times P(\bar{S})] \\
& + [P(\bar{C}|T,\bar{P},\bar{S}) \times P(T) \times P(\bar{P}) \times P(\bar{S})] \\
& + [P(\bar{C}|\bar{T},\bar{P},S) \times P(\bar{T}) \times P(\bar{P}) \times P(S)] \\
& + [P(\bar{C}|T,\bar{P},S) \times P(T) \times P(\bar{P}) \times P(S)] \\
& + [P(\bar{C}|\bar{T},P,\bar{S}) \times P(\bar{T}) \times P(P) \times P(\bar{S})] \\
& + [P(\bar{C}|T,P,\bar{S}) \times P(T) \times P(P) \times P(\bar{S})] \\
& + [P(\bar{C}|\bar{T},P,S) \times P(\bar{T}) \times P(P) \times P(S)] \\
& + [P(\bar{C}|T,P,S) \times P(T) \times P(P) \times P(S)]
\end{aligned}
$$

$$
\begin{aligned}
HEP = \ & [1 \times 0{,}7 \times 0{,}8 \times 0{,}6] + [0{,}9 \times 0{,}3 \times 0{,}8 \times 0{,}6] \\
& + [0{,}9 \times 0{,}7 \times 0{,}8 \times 0{,}4] + [0{,}7 \times 0{,}3 \times 0{,}8 \times 0{,}4] \\
& + [0{,}3 \times 0{,}7 \times 0{,}2 \times 0{,}6] + [0{,}2 \times 0{,}3 \times 0{,}2 \times 0{,}6] \\
& + [0{,}4 \times 0{,}7 \times 0{,}2 \times 0{,}4] + [0{,}02 \times 0{,}3 \times 0{,}2 \times 0{,}4] = 78{,}96\%
\end{aligned}
$$

Thus, substituting probability values in equation 1 considering optimist values after human performance factor improvement we have:

$$
\begin{aligned}
HEP = \ & [P(\bar{C}|\bar{T},\bar{P},\bar{S}) \times P(\bar{T}) \times P(\bar{P}) \times P(\bar{S})] \\
& + [P(\bar{C}|T,\bar{P},\bar{S}) \times P(T) \times P(\bar{P}) \times P(\bar{S})] \\
& + [P(\bar{C}|\bar{T},\bar{P},S) \times P(\bar{T}) \times P(\bar{P}) \times P(S)] \\
& + [P(\bar{C}|T,\bar{P},S) \times P(T) \times P(\bar{P}) \times P(S)] \\
& + [P(\bar{C}|\bar{T},P,\bar{S}) \times P(\bar{T}) \times P(P) \times P(\bar{S})] \\
& + [P(\bar{C}|T,P,\bar{S}) \times P(T) \times P(P) \times P(\bar{S})] \\
& + [P(\bar{C}|\bar{T},P,S) \times P(\bar{T}) \times P(P) \times P(S)] \\
& + [P(\bar{C}|T,P,S) \times P(T) \times P(P) \times P(S)]
\end{aligned}
$$

$$HEP = [1 \times 0{,}3 \times 0{,}5 \times 0{,}1] + [0{,}9 \times 0{,}7 \times 0{,}5 \times 0{,}1] + [0{,}9 \times 0{,}3 \times 0{,}5 \times 0{,}1] + [0{,}7 \times 0{,}7 \times 0{,}5 \times 0{,}1] + [0{,}3 \times 0{,}3 \times 0{,}5 \times 0{,}1] + [0{,}2 \times 0{,}7 \times 0{,}5 \times 0{,}1] + [0{,}4 \times 0{,}3 \times 0{,}5 \times 0{,}1] + [0{,}02 \times 0{,}7 \times 0{,}5 \times 0{,}1] = 0{,}89\%$$

In general terms the main Bayesian Networks coins are:

- The relation between performance factors effect on HEP (Human Error Probability) can be calculated by conditional probability;
- If applied Bayesian network software, the HEP calculation became easier;
- Mostly, Bayesian Network is easy to understand graphical when applied to human reliability problems;

The drawbacks are:

- Difficulties in obtaining conditional probabilities in the data bank;
- As higher as a number of performance factors which take influence on HEP (Human Error Probability) harder to get reliable information from specialist opinion;
- Mostly, Bayesian Network is easy to understand graphical when applied to human reliability problems, but hard to calculate human error probability;

Whenever we access tasks which involve hazard we realize that omission and commission error can lead to unsafe conditions and trigger an accident. Such errors are highly influenced by human performance factors which affect the human performance in different tasks from different industries.

The preventive approach to prevent accidents is carrying out Human reliability analysis of such tasks that are identified some hazard in order to avoid accidents.

In order to decide which tasks mostly are carried out a Human Reliability analysis is necessary to establish a criterion that can be defined based on risk analysis (PHA, FMEA, HAZOP, ETA, FTA BOW Tie), accident and incident historical data or even for input from different.

Despite important, Human reliability analysis has some limitation because is hard to predict the human behavior. Despite a lot of different approaches and methods to deal with human error is hard to establish a method to predict the physiologic mechanism which takes influence on human error like we have the root cause or physical failure causes that explain equipment failure modes.

Regardless of what the limitation, Human Reliability Analysis must be carried out because have a high influence on system performance and safety.

DISCLOSURE

"Part of this chapter has been previously published in Gas and Oil Reliability Engineering Chapter 5 – Human Reliability Analysis 2013, Pages 349–419".

REFERENCES

Bell, J & Holroyd, J. (2009). Review of human reliability assessment methods. Retrieve from: www.hse.org.uk. The Health and Safety Laboratory for the Health and Safety Executive, Reseach report 679.

Calixto, E. (2012). Gas and Oil Reliability Engineer: Modelling and Analysis. Imprint: Gulf Professional Publishing, ISBN: 9780123919144, 07 Nov 2012.

Drouguett E.Lopez; Menezes R. da Costa Lima. (2007). Análise da confiabilidade humana *via* redes Bayesianas: uma aplicação à manutenção de linhas de transmissão. Produção, v. 17, n. 1, p. 162-185, Jan./Abr. 2007.

Embrey, D. E., Humphreys, P., Rosa, E. A., Kirwan, B., and Rea, K. (1984b), "SLIM-MAUD: An Approach to Assessing Human Error Probabilities Using Structured Expert Judgment, Volume 2: Detailed Analysis of the Technical Issues", NUREG/CR-3518, Brookhaven National Laboratory, Upton, NY.

Evaluation of Human Reliability Analysis Methods against Good Practices. (2006). NUREG-1842,September 2006.

Everdij M.H.C., Klompstra M.B., Blom H.A.P. and Fota O.N. (1996) MUFTIS work package report 3.2, final report on safety model, Part I: Evaluation of hazard analysis techniques for application to en-route ATM, NLR TR 96196L, (MUFTIS3.2-1)

FIRMINO, P. R. & DROGUETT, E. L. (2004). Redes Bayesianas para a parametrização da confiabilidade em sistemas complexos. Engenharia de Produção, Universidade Federal de Pernambuco, Centro de Tecnologia e Geociências.

FIRMINO, P. R.; MENÊZES, R. C. & DROGUETT, E. L. (2005). Método aprimorado para quantificação do conhecimento em análises de confiabilidade por redes Bayesianas. XXV Encontro Nac. de Eng. de Produção – Porto Alegre, RS, Brasil, 29 out a 01 de nov de 2005

Grozdanovic, M. (2005). Usage of Human Reliability Quantification Methods. International Journal of Occupational Safety and Ergonomics (JOSE) 2005, Vol. 11, No. 2, 153–159.

J. W. Hickman. (1982). PRA Procedures Guide: A Guide to the Performance of Probabilistic Risk Assessments for Nuclear Power Plants. The American Nuclear Society and NRC Grant No. G-04-81-O00.The Institute of Electrical and Electronics Engineers. NUREG-CR-2300.december 1982.

MANNAN, Sam.(2005). Lees' loss prevention in the process industries. 3rd ed. New York: Elsevier, 2005.

Menezes, R. da Costa Lima. (2005). Uma metodologia para avaliação da confiabilidade humana em atividades de Substituição de cadeias de isoladores em linhas de transmissão. Dissertação de Mestrado. UFPE. Recife, Junho / 2005.

SWAIN, A D. (1987). Accident Sequence Evaluation Program Human Reliability Analysis Procedure. NUREG/CR-4772.February 1987.

Swain A. D. & H. E. Guttmann. (1980). Handbook of Human Reliability Analysis with Emphasis on Nuclear Power Plant Applications, draft, NUREG/CR-1278, October 1980.

NUREG/CR 4772

NUREG/CR-6883,INL/EXT-05-00509.

Ronald L. Boring.; David I Gertman. (2005). Advancing usability evaluation through human reliability analysis. Human Computer Interaction International, 2005.

Shamus P. Smith and Michael D. Harrison. (2002). Blending Descriptive and Numeric Analysis in Human Reliability Design. Lecture Notes in Computer Science Volume 2545, 2002, pp 223-237.

SILVA, V. A. (2003). O planejamento de emergências em refinarias de petróleo brasileiras: um estudo dos planos de refinarias brasileiras e uma análise de acidentes em refinarias no mundo e a apresentação de uma proposta de relação de canários acidentais para planejamento. 2003. 158 f. Dissertação (Mestrado em Sistemas de Gestão).Universidade Federal Fluminense, Niterói, 2003

SWAIN, A D & GUTTMANN, H. E. (1983). Handbook of Human Reliability Analysis with Emphasis on Nuclear Power Plant Applications. US Nuclear Regulatory Commission. Washington.

Spurgin, Anthony J. (2010). Human Reliability Assessment: Theory and Practice.CRC Press. Taylor & Francis Group.2010

Vestrucci P. (1990).Modelli per la Valutazione dell'Affidabilità umana" Franco Angeli editore.

Willian,J.C.(1988). A data-based Method for Assessing and Reducing Human Error to improve Operational Performance",Proceeding of IEEE Fourth Conference on Human Factors in Power Plants, pp.436-450,Monterey,CA.

http://www.hse.gov.uk/comah/sragtech/casepemex84.htm. Accessed on 1170172015.

CHAPTER 8

Safety Standards

Abstract: In order to face different occupational risk, different industries such as Aerospace, Chemical, Oil and gas, Railways, Nuclear has established standards guidelines. In fact, some of such standard is a guideline for all industries like OHASA 18001 which address the Safety and Occupational Health management or IEC 61508 which address the safety integrity level. By the other way round, the specific standard area applied to risk management in different industries like SCR05 which addresses risk management in oil and gas industry or the nuclear safety case that also focus on risk management. In general terms, all those standards are a good guideline to support safety and occupational management in different industries. Even though, it's important to have in mind that such standards do not reflect the best practices to mitigate the risk in many cases. Indeed, the companies all over the word must to make all effort to apply the best practices to achieve high effectiveness in their safety and occupational management applying the standards as a baseline. This chapter aims to demonstrate the main standards applied to the most relevant industries in term of risk.

Keywords: Risk, OHSAS 18001, SCR05, Safety case applied to the nuclear industry, EN 51026, Key program asset integrity, IEC 61508.

8.1. INTRODUCTION

In general, Safety and occupational health management is supported by different standards. Such standards are different depends on the industry and focus application. Despite similar concepts, it is usual to see difference in similar standards applied in different countries. Indeed, such difference can be explained by local law applied to safety and occupational health as well as the local necessity and industry requirements.

The most of industries all over the world apply specific safety and occupational health standards to support their management but some particular industry is referenced like:

- Aerospace,
- Chemical,
- Oil and gas,
- Railways,
- Nuclear.

Eduardo Calixto

The Aerospace industry all over the world is the high guide by safety requirement and such requirement take influence in most of decisions and other requirement and target like reliability, maintenance an investment as well. Indeed, reliability, maintenance and safety are deeply related because many of the failures in aircraft may lead to a catastrophic accident. Indeed, such particular characteristic make aerospace the best example, safety, maintenance and reliability, integration which is taken into account since the beginning of project concepts and goes through the whole asset life cycle.

Despite a big challenge, after decades of constant improvement, the aerospace has been internalized the safety concepts as priority and such concept has been confirmed in all standards and most of the procedures.

Even though, accident all over the world caused by human or equipment failures has been reported that proves the necessity to improve and apply best practices in order to have safe operation and avoid the incident and accident.

The Oil and Gas industry is another good example of a safety standard application and it is one of the most standardized industries with a lot of safety and occupational standards applied all over the world. Indeed, the reliability, maintenance and safety are not as integrated as we can see in the aerospace industry. Despite all effort to avoid incidents and accident, the production is still the main drive in Oil and Gas Industry and depends on a country or company, safety and occupational health issues has more importance or not than production. For such industry, depends on accident, there are environmental impact and such incident start to be a concern for the last decades.

Despite all accident and environment impact on last decades, the Oil and gas industry has been implemented considerable improvements to avoid such events. In facts, it is necessary to take into account the process complexity as well as the constant new technology that is implemented to allow adaptation to the new competitive environment regarding all society requirements as well as engineering new challenges like deep water, limit toxic gas emission and sustainable profits.

One of the most important issues that must be taken into account as a challenge that Oil and Gas industry faces is the asset complexity. The different risk poses on different process and equipment that is comprised of refinery plants, platform, drill facilities and subsea system that demands a huge technological and capability. In addition, the huge number of assets that operates in different life cycle phase is also a big challenge in term of safety and occupational management. Indeed, despite all challenges and complexity, we can say that Oil and gas industry is a very good reference in terms of safety and occupational health management. Some organization provides support by producing standards for Oil and Gas industry like the API (American Petrol Institute) that has been providing a lot of standards which support safety and occupational health management being a reference all over the world. Whether we consider all Oil and Gas assets all over the world and all improvement implemented for the last decades, reducing the incident, accidents and environment impacts we get into the conclusion that it is in fact a reference in safety and health management.

The Chemical industry also can be considered a reference in terms of safety an occupational health because all effort and contributions for the last decades. Despite all fatal accidents on last decades, the bibliography developed in order to avoid such accident has given a great support to all industries in order to understand the operational and storage conditions of chemical products and process management. Some organizations like AICHE (American Institute of Chemical Engineering) and CCPS (Center for Chemical Process Center) has been promoting safety and occupational health for years supporting with books, procedures and organizing conferences.

The Railway industry has been improving their standards and procedures as well as the safety and occupational health management. For such industry, the big challenge is to develop training and railways equipment, safety enough in project phase and maintain such safe operational condition to avoid catastrophic accident. In trains and railway equipment the reliability and safety in many cases are related because some systems and equipment unsafe failures lead to an accident like breaks, bogie, TCMS (Total Control Management System), fire system detector and external doors. In order to support the safety and risk management of such equipment the EN 50126 standard series provide a very good guideline for risk management throughout train and railway equipment life cycle. Basically, all safety activities are defined along different train and railway equipment asset phase that is well described by V Diagram.

The nuclear industry is also a good reference in terms of safety and occupational health management. Indeed the risk management carried out by such industry is very well structured and also the human reliability procedures to avoid human error which lead an accident with catastrophic consequences or even improve human performance in emergency situations. Like others industry, the nuclear industry is very standardized and controlled by international authorizes which provide guidelines, training and procedures for most of activities and process.

Indeed, it is one of the most international regulated industries in terms of safety and risk management. In the USA, the United Stated Regulatory Commission has been developing a set of standards related to safety and human reliability known as NUREG (Nuclear Regulation).

In general terms, after several decades all such industries has been developing their own procedures and has been influenced one each other in order to achieve excellence in safety and occupational management as well as risk management. Nevertheless, a common problem that poses nowadays for all such industries are the natural disaster and terrorism attach. Therefore, it is necessary to come together to discuss different approaches in order to reduce the vulnerability that such industries face nowadays. Indeed, the international conference like ESREL, PSAM, Working on Safety and others has been providing a good chance for an international forum to discuss such issues in the last decades. In addition, some specific standards like OHSAS 18001 and ISO 31000 has been a good example of such effort to supply references and guidelines for the whole industry. In fact, such guidelines and standards allow different industries to achieve high performance by implementing best practices.

The first step to understand the particular challenges for each industry is to understand their own standards and the standards that can be applied for all of them like ISO standard series. Despite particular issues, all industries face common safety, occupational and risk management problems that can solve by the best practices and methods.

This chapter will present some examples of standards and guidelines most applied all over the world for the most critical industries in terms of safety, occupational and risk management.

The first standard to be discussed will be the **OHSAS 18001** that provides a guideline to implement an Occupational and Safety management System. Such standard is based on PDCA (Plan, Do, Check and Analyze) concept which defines specific items in each PDCA phase. In 90th decades the OHSAS 18001 standard was one of the most demanding standards for different organizations in order to get the certifications like other ISO series and was a very good baseline for many companies in different industries develops their own Safety |management System.

The second standard to be discussed is the **ISO 31000** that focuses on Risk Management. Despite a good risk management guideline, different from others ISO series, there was not a massive demanding for certification based on this standard. Even though, it does not reduce the importance of such standard as a guideline for many industries as well as small companies.

The third standard to be discussed is the offshore safety case **SCR05** (which applies to offshore facilities like platform in Oil and Gas Industry). Such standard defined by the HSE in UK, focuses on Risk management regarding the main asset phases which encompass hazard identification, risk analysis, risk assessment and risk mitigation. Such risk analysis comprises qualitative and quantitative approaches as well as emergency plans to mitigate accident consequences.

The fourth standard presented in this chapter is the **EN 50126 (**which presents the risk management applied to railway industries concerning the risk management ad defines risk assessment methods as well as the main safety issues and the link between safety and reliability. The railway industry has very established and structured standards with a defined task for whole assets life cycle phases based on V-Diagram definition. Indeed, such standard has different series that focuses on different subjects like Reliability, Availability, Maintainability and safety (EN 50126-1 -1999).

The fifth standard discussed in this chapter is the **nuclear safety case** proposed by HSE in UK, which the main objective is to demonstrate safety assurance along the nuclear plant life cycle which include asset operation and modification. Therefore, it's necessary to provide a written demonstration that risks have been reduced as low as reasonably practicable (ALARP).

The sixth standards are the "Key Program Asset (KP3)" which starts as a request from the UK Secretary of State for Work and Pensions afterwards after a debate about the Piper Alpha disaster.

The main program objective is focused on offshore installations on the United Kingdom regarding the maintenance of safety-critical systems which are relevant in the case of major accident.

The last standard discussed in this chapter is the **IEC 61508**. Indeed, the principal Safety Integrity Level standards used worldwide are: ISA S84.01, IEC 61508 and IEC 61511. The IEC 61508, addresses the application of Safety Instrumented Systems, with an approach that puts attention on the field devices as well as the logic solver as integral components of the SIS (Safety Instrumented System). That means that this document is an industry-specific standard. The important aspect of this standard is that it established a complete risk management throughout SIS life cycle in order to assure that safety requirements are achieved. This standard has been used since its release mainly in USA, although a lot of countries have adopted it for the implementation of their SIS.

8.2. OHSAS 18001

In order to support the worldwide Safety and occupational management, the OHSAS 18001 standard was created by effort from a number of the world's leading is safety and occupational management.

An international Project Group comprised of different specialists called The Occupational Health and Safety Advisory Services (OHSAS) was formed to create a single unified standard in order to support the safety and occupational management.

The Series consisted of two specifications: 18001 provided requirements for an OHS management system and 18002 gave implementation guidelines. By 2009 more than 54,000 certificates had been issued in 116 countries to OHSAS or equivalent OHSMS standards.

The main objective of occupational health and safety management system (OHSMS) is to provide a guideline that support organizations to accomplish:

- Occupational Health and safety Management high performance;
- Risk mitigation as well as incident and accident reduction;
- Legislation compliance.

The OHSAS 18001 implementation, follows the same principal and the structure of the existing ISO 9001 and ISO 14001 management systems. Historically, some organizations try to integrate these different standards in order to improve the management system efficiency. By the other hand, some organizations look at implementing two or all three standards at the same time, which can be apparently cost-effective and minimizes disruption. In fact, that is a tremendous challenge by the implementation point of view. The main barrier to implement such ISO management System all together are time and culture. Even when existing the favorable culture, its required time to train all employees and they also need time to assimilate such standard concepts and incorporate on their daily work routine. In some cases, it is necessary years to achieve that for only one specific standard.

Depends on organization, one standard is easier than others to be implemented and maintained. That happens because the importance and necessity as well as the employee's culture and leadership support for such management system is higher comparing with others.

In addition, what can be seen in many organizations throughout the world is that maintain the system effectiveness is the real challenge because it requires for employees real assimilate such standard concepts and implement that on their daily work routines.

Unfortunately, many organizations need parallel management approaches to support their Occupational and health management based on OHSAS 18001 because it's not internalized in employee's routine.

Indeed, many organizations implement such System Management in order to get the certification as costumer requirement and not in order to achieve occupational and safety excellence.

By the other hand, it is possible to see many organizations throughout the world that has achieved a successful OHSAS 18001 implementation and are able to maintain a high performance in occupational health and safety management.

Indeed, is more applicable to the organization which faces several hazards in their process which requires an occupational health and safety management. The key subjects that will be assessed by OHSAS certification are:

- System structure, responsibilities and documentation;
- Risk management;
- Employees training and awareness;
- Communication of safety management systems;
- Emergency response plan;
- Monitoring and continual improvement.

Such subjects are comprised into the PDCA (Plan, Do, Check and Analyze) philosophy that intends to achieve the continuous improvement a long time. Indeed, each PDCA phase comprises different activities that must to be taken place on OHSAS 18001 system. The Fig. **1** shows the PDCA cycle in OHSAS 18001.

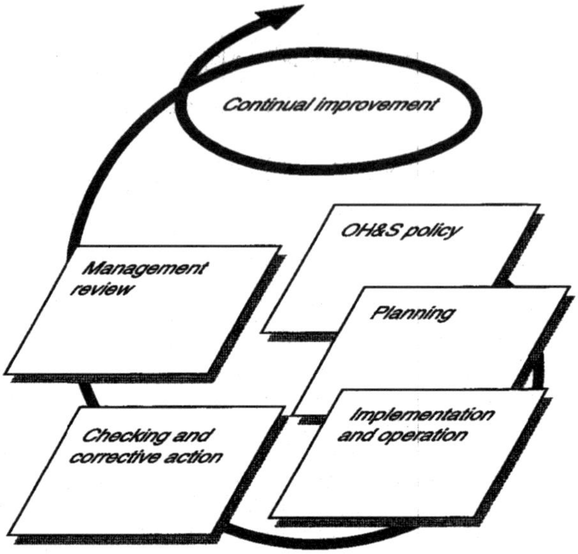

Figure 1: Elements of Successful OH&S management (Source: OHSAS 18001 standard, 1999).

- The first step to implement the OHSAS management system is to define a policy that's based on OHSAS 18001 must attend at least seven requirements that are:
- To be related to the nature and scale of organization's occupational, health and safety risk ;
- Commitment to continuous improvement;
- Commitment to comply with current applicable OH&S legislation;
- To be documented, implemented and maintained;
- Communicate to all employees with the intent that employees are made aware of their OH&S obligation;
- To be available to interested parties;
- To be reviewed periodically to ensure that it remains relevant and appropriated to the organization;

Depends on the organization, the OH&S policy is more complex in order to achieve all interested parties and comply the OHSAS 18001 requirements. An example of an OH&S policy of XX Group can be seen below.

"The XX Group is committed to minimizing in our processes the risk of injury or ill health to people, damage to property or the environment. The XX Group has a The XX Group fully accepts their obligations and legal responsibilities, which will be achieved by:"

- *Including a commitment to continual improvement in OH&S management and performance;*
- *A commitment to comply with all applicable legal requirements subscribes that relate to its OH&S hazards;*
- *Providing and maintaining methods that minimize the risks to health, Safety and welfare;*
- *Ensuring all employees are provided with adequate information, instruction, training and supervision to carry out their activities complying OH&S requirements;*
- *Maintain an open communication with stakeholders and community with all aspects related to safety;*

It is clear that all OHSAS 18001 requirements for the policy were stated on The XX group OH&S policy. The remarkable point in OH&S policies is to be stated only what is required for OHSAS 18001 standards. Any additional statement must be complied and will be checked during the audit process. That is a point that companies must be aware about the OH&S policy statement and improve and increase its scope on time.

Indeed, the OH&S policies are the first step to implement an OH&S management system like OHSAS 18001. For companies that intend to be certified, it is really necessary to be aware about the certification proposals as well as the cost involved in training, implement all standard requirement and audit process. In general terms, the OHSAS 18001 specification is applicable to any organization that wishes to achieve client compliance, demonstrate the OH&S standard application and achieve high performance in OH&S.:

The certification process can be summarized in three phases that are initial risk assessment, manual prescription and presentation for certification. The first step requires an effort to identify all hazards, assess the risk, evaluate the risk and mitigate them. All risk mitigation must comply with OH&S policy. In some cases, after risk assessment the policy must be revised. Based on risk assessment, all objectives and goals to mitigate such risk must be taken place. That means, define procedures, training employees and define responsibilities. The second step is to write an OH&S manual that describe the organization process and their hazards. Such manual must be sent to audit organization. All audit process will be planned based on such organization manual. Whether it's been necessary additional information, the audit organization can requires for that. The final step is a presentation and certification that is the audit process which has the main objective to verify the compliance based on standard. Once the compliance level is satisfactory the Organization get the certificate. The Fig. **2** summarizes the three certification steps.

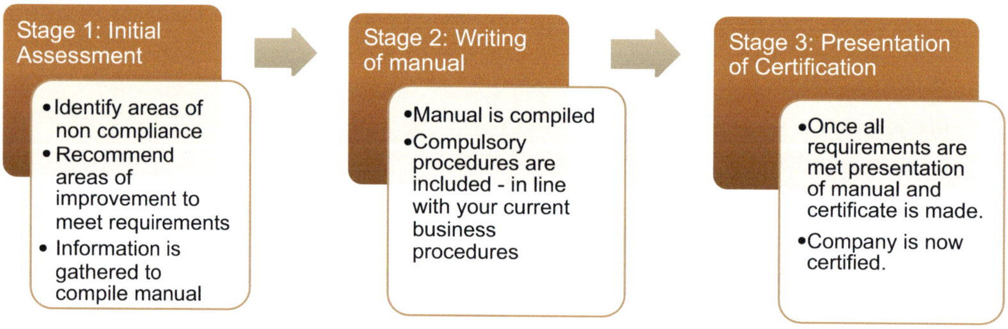

Figure 2: Certification Process (Source: http://www.isoqsltd.com).

In fact, the certification process is a long term process that requires a huge effort in terms of risk assessment, training, procedures and responsibilities definition that is well defined in different phases in OHSAS 18001 standards.

After the OH&S policy definition the next step is planning. The planning phase comprises risk assessment, objectives and goals definition as well as program to achieve such targets. Based on OHSAS the planning phase is defined by (OHSAS 18001):

- Planning for hazard identification, risk assessment and control (item 4.3.1);
- Legal and other requirements (item 4.3.2);

- Objectives (item 4.3.3);
- OH&S management programs (item 4.3.4).

The Planning for hazard identification, risk assessment and control must comprise all hazards into organization processes that might cause some damage to employees, subcontractors and visitors. In order to identify such hazards and assess risk it is advisable to implement a well-structured tool that enable identify hazards and assess risk in all activity tasks. In order to assess the risk it is necessary to define the qualitative risk matrix that enables to classify the hazards frequency and severity based on legislation and company policy. In chapter 2 are defined some examples of risk matrix that will not be discussed here. In addition, all routine and non-routine activities must be contemplated in such hazard identification and risk assessment. The Table **1** shows an example of hazard identification and risk assessment based on the preliminary risk analysis methodology. In addition, it is necessary to define the objective, goals, mitigation action and legislation related to each particular hazard.

In addition, it is necessary to document the responsibilities of all employees in such OH&S management system regarding all organization levels. Thereby, the OH&S management programs is a very important step to system implementation and it is usual to define an employee to represent a high management level to maintain the OH&S management system high performance.

Table 1: Occupational Health and Safety risk analysis

			Occupational heath and safety Risk Analysis								Local System:				
Process:						Acessed by:	Aproved by:								
Identification							Assessment, Evaluation and mitigation								
PROCESS	Activity	Task	Safety and Health hazard	Rotine	Not Rotine	Consequence	Frequency	Probability	Severity	Risk	Legislation	Objective	Goals	Mitigate Action	Resposabilities

The next step is the implementation and operational phase that comprises:

- Structure and responsibility (item 4.4.1);
- Training, awareness and competence (item 4.4.2);
- Consultation and communication (item 4.4.3);

- Documentation (item 4.4.4) ;
- Document and data control (item 4.4.5) ;
- Operational control (item 4.4.6) ;
- Emergency preparedness and control (item 4.4.7).

The ***structural and responsibility*** is one of the most important steps in order to implement an effective OH&S management system. In many cases, organization leaders define only one employee to be responsible for the whole system and that has been shown a big mistake in many organizations. That happens because such employee has no authorities to prior improvement actions to the OH&S management system and in most of cases, such employee has also his own work routine to carry out. In order to maintain a continuous improvement is advisable to define a board with the main organization authority's managers to be responsible for the OH&S management system. In this case, such border must have a constant agenda meeting and be supported by specialist in order to achieve all OH&S goals and targets. Unfortunately, in many cases, organizations define only one employee to be responsible for the whole OH&S management system and what happen in reality is that such objectives and target are reviewed some months before audit process just to maintain the organization OHSAS 18001 certification.

The second most important action when implementing OH&S management system is ***the trainee, awareness and competence*** that must be achieved and performed for employees in all levels in order to have a successful and sustainable high performance OH&S management system. Indeed, the trainee must allow awareness about the OH&S management system importance and whenever enable competence of employees carry out their activities regarding OH&S management system requirements on their daily work routine. In order to employees achieve competence it is required knowledge as well as practice with such requirement. Obviously that competence in such subject requires time to understand, exercise and implement in daily routine work.

The huge mistake concerning trainee is that in many organizations when implementing OH&S management system, they trainee employees about the concept and procedures but not take into account the awareness and competence. The audit process is presented one list that certifies that all employees were trained, but in some cases when some of them are asked about OH&S issues they give not a correct answer or have no idea how such OH&S issues are related to their work routine.

In order to avoid such a problem, it is important for organizations to be aware that awareness and competence about OH&S management system takes time and it is a continuous improvement process that must be supported by leaders by reinforcing improvement actions giving the requested resource as well as personal support.

That is one of the most serious questions about implementing more than one Management System on the same time like ISO 900, ISO 14001 and OHSAS 18001. The main problem here is that employees at all levels must achieve the awareness and competence in a short period of time.

In addition, the further problem involving trainee is in many cases, there is not a specific and effective trainee dedicating for leader in all levels. Whether leaders not achieve awareness and competence about their duties to lead OH&S management system, employees will certainly not achieve as well.

The ***"consultation and communication"*** is another important issue in an OH&S management system that takes a high influence on such system performance. Indeed, the communication is not well done from strategic to operational organizations level in many cases. That happens because in many cases employees have not the same communication channel than managers like access some email. The second point is that the language used by the manager is not clearly understood for many employees. By this way, many actions are not well implemented and that compromise the OH&S management system performance. In fact, communication is always a challenge for all organizations and in order to translate the goals and targets from the strategic level to operational level the middle managers, coordinators and supervisors must be involved in such process. Depends on how many people are involved in such process, it can take long time. In contrast, of all difficulties to communicate the target and goals, one successful approach is encouraging leaders like supervisors and manager to carry out a daily safety dialog with employees and discuss OH&S management system issues. Therefore, it is also a challenge task that requires employee's involvement. Indeed, the mistake in some cases is not carrying out such communication face to face. In contrast, some leaders send some information by radio in operational ground or only make a brief communication with their team and get their signature to present as evidence during the audit process. In order to get employees' involvement it is important to listen their opinion because they have a lot of solutions for organization problems they only must to be listened.

The following step is ***the documentation*** if the OH&S management system is in most of case well done in electronic form but not in paper form. That happens because the OH&S issues that must be part of operational procedures is not discussed with supervisor and operators and maintenance technicians and by this way, such procedures do not represent the real steps that are carried out in many cases.

The document and data control reflects the documentations philosophy and once again, in operational level, it is necessary to discuss how to storage, when such documents must be updated and how control such documents. Indeed is usual in many organizations to see that is hard to locate documents and, in some cases, what operators, says is not what is described in the document. That happens because leaders did not listen or involve enough supervisors, coordinators and operators in the documentation process. It seems clear that involve coordinators and supervisor do not mean that operators are involved because they have specific issues that in many cases are not listened for their coordinator and supervisors and because of that they must be involved as well.

The operational control is very important to drive action to mitigate risk and such activity must involve operators to identify correctly the hazards and specialist to carry out the risk assessment and mitigation action together with operators. In many organizations, only specialist carries out such activities based on their background and historical data. Indeed, no matter how experience such specialist area, the operators must take part in such activity because they know the real operational situation in terms of risk that change from time to time. In fact, employees must be involved in risk assessment process in order to understand the importance of risk mitigation and control action. In addition, the specialist mostly carries out such risk assessment on the beginning of the OH&S management system, but they are not involved in the risk mitigation process and control after OH&S management system implementation. In fact, all employees in different level must understand the risk involved in their processes and be prepared to take place, their responsibility to mitigate and control such risk in order to avoid an incident and accident.

The ***emergency preparedness and control*** is set up by risk assessment based on accident scenarios that requires an emergency preparedness, response in order to mitigate accident scenarios consequence. The big challenge here is to keep a higher level of performance in something that rarely will take place. Even though, it's necessary to develop procedures and train employees to enable them to take place emergency response whenever it's necessary. In order to achieve such high performance in emergency response it is essential to apply emergency response exercise that can involve the complete resources involved in one specific accident scenarios or part of them. Indeed, the emergency response exercise requires economic resource and mobilizes employees from their usual routine and such activities must be well planned. Specific details about emergency response in different levels can be seen in chapter 3.

The next step is the check and a corrective phase that comprises:

- Perform measurement and monitoring (item 4.5.1);
- Accident, incident, non-conformance and preventive and corrective actions (item 4.5.2);
- Record and records management (item 4.5.3);
- Audit (item 4.5.4);

In order to analyze the OH&S management system performance it is necessary to establish ***the perform measurement and monitoring*** process. Such process must be defined based on OH&S management system requirement preventively in order to enable preventive actions and improvement. The preventive index can be defined like mitigation action index (number of mitigation implemented/number of mitigation action planned), trainee index (number of employees trained/number of total trained, planned), document index (number of document under compliance/Total number of document).

By the other hands, it is also necessary to have a reactive index to measure and control internal audit non-compliance, external audit non-compliance, incident and accident.

The main issue in defining representative performance index that is able to drive improvement action in the OH&S management system. It is not necessary to define many indexes and that is a common mistake that requires additional work to measure and control without any additional contribution to OH&S management system improvement. The possible solution is defining verification and control index. The index must be defined in different organizational levels, which means strategic, tactical and operational. The verification strategic indexes are related to control index that are verification index in tactical organizational level. For managers in tactical level, those strategic control indexes are their verification index and their control index are verification index in operational level. In operational level supervisors and coordinators can also define control index for their teams. At the strategic level, as an instance, can be established the OH&S effectiveness index that comprises by control indexes like mitigation action index, trainee index, internal audit non-compliance, external audit non-compliance, incident and accident. Such indexes are verification index in tactical level that might include other verification indexes like document index. Therefore, in operational indexes there will be others control indexes.

"The accident, incident, non-conformance and preventive and corrective actions" management is reflecting of index and plan established at the beginning of an OH&S management system. Indeed accident is a result to different cause combination. In reality, many incidents are not taken into account seriously in order to eliminate their root causes and in time accidents happen. It's very important to understand that incident does not cause health damage to employees in many cases but it would be. Therefore, it is important to deal with incidents like accident implementing a root cause analysis like was discussed in chapter 6 in order to avoid that such incident turn out into accident. In many cases, simple actions are taking place when an incident happens and the real solution is postponed due to high cost or even the time that is necessary to be dedicated to solve the problem.

The record and records management of OH&S management system documentation is mostly not a problem, but even though it is necessary to create an easy and effective procedure that establish how to record all information about the OH&S management system to be available during audit process as well as daily for employees consultancy.

The audit process is one of the most important to certify the OH&S management system effectiveness because is an independent analysis about such management system. In order to be successful in the external audit process many companies establish an internal audit process before the external audit. Depends on the situation, the organization need to establish more than one internal audit during one year in order to verify the OH&S managing system performance and drive the correct resource to improvement actions. Definitely it is essential to understand that audit process is an essential tool to check the OH&S management system effectiveness and by this way, such audit must be representative in terms of processes verified and the number of people involved. In addition, it must be clear that the main audit process objective is to look for compliance in order to prove OH&S management system effectiveness and not look for non-compliance. In order to implement such philosophy the leaders and auditors who lead such process must be aware about that.

The final step is the ***"management review"*** that must be taken place after an internal or external audit or other periodic verification in order to enable improvement actions on the OH&S management system. The main issue in this phase is that if leaders did not take part in the whole process, they will not be able to understand why there are some gaps to be improved and why such gaps exist. Indeed, to implement all improvement actions it is necessary to establish a plan with responsible for each action as well as the necessary resource. In this case, the leaders will be convinced to give resource for such action only if they are committed to OH&S managing system performance and understand its importance.

In general terms, the OHSAS 18001 is an excellent standard that allows companies from different industry like Oil and gas, chemical, aerospace, train, transportation, nuclear and others to establish the minimum required necessary to implement the OH&S management system. It is not necessary for certificate, but it is always necessary to manage the OH&S in the best way to avoid health damage and accident. Thereby, all companies are responsible to define which the best practices are in order to fulfill such requirements, based on their resources, culture and process characteristic.

Indeed, no matter how much economic resource, complex process or technology are involved, the organization has to guarantee employees health and safety applying the best practice to achieve such goal and maintain high performance in the OH&S management system a long time.

8.3. ISO 31000

In general terms, the ISO 31000 standards scope is the risk management that can be applied to different types of organization which face the biggest challenge that is managing their process risk.

The standard concept is based on the continuous improvement philosophy that comprises different activities along the PDCA cycle in order to mitigate the risk and maintain such risk under acceptable level. The general continuous improvement cycle is described in Fig. **3**.

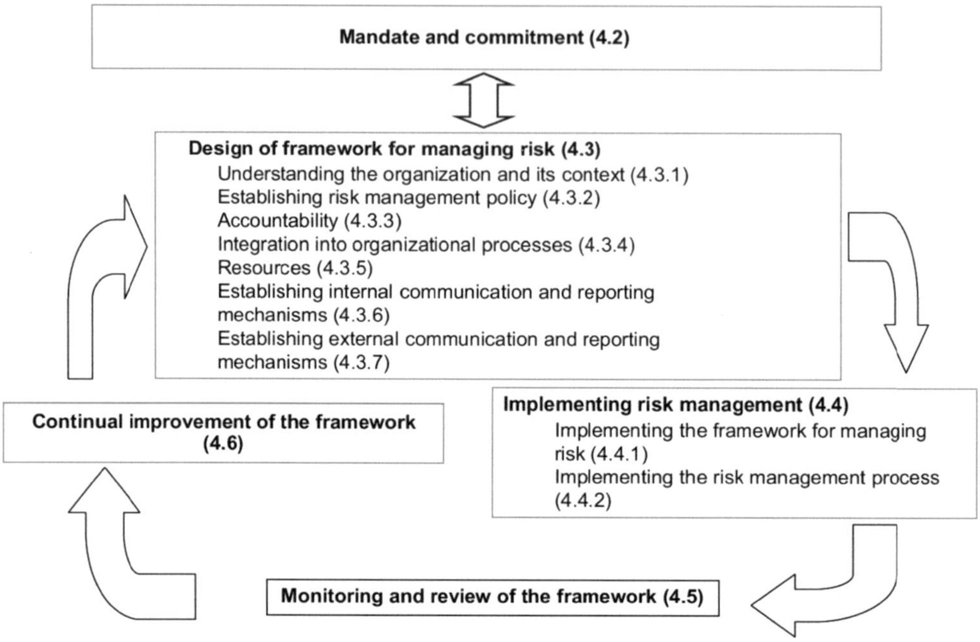

Figure 3: Relationship between the components of framework for managing risk (Source: ISO 31.000, 2009).

The first step on PDCA cycle is to establish bases on standard is the ***"mandate and commitment"***. Such first step requires the risk management police establishment in Strategic level and such policy will drive all goals and index throughout the organization in order to achieve high performance in risk management. Indeed, most of organization which defines risk management policy is very well aware about their process risk as well as internal or external factors that would affect the risk management performance. In fact, on most of cases, the risk management is a complement of OH&S management with the clear objective to be more effective in managing the process risk to avoid major accidents.

The second step is ***"The design of a framework for managing risk"*** which will support the first one more effectively and will guarantee that the risk policy is adequate to company taking into account the internal and external organization's context. The framework and communication process must be very well establish to achieve high performance in risk management and in many cases the OH&S management framework can be used as well for such proposal. The external communication is always a big challenge in terms of risk management because involve different stakeholders with different requirement and expectation about risk policy. The most challenging is the society that is affected in case of major accident. Therefore, more than communicate the risk it is necessary to establish a trustful relationship in order to be able to carry out evacuation trainee and take place such procedures whenever is necessary to guarantee society safety in case of an accident.

The third step is ***"implementing risk management"*** that comprise four phases which are Risk assessment, risk treatment, risk communication and risk monitoring and review as shows Fig. **4**. Indeed, all such phases are very important risking management and whether one of them is missing or mistreated the risk management performance is compromised.

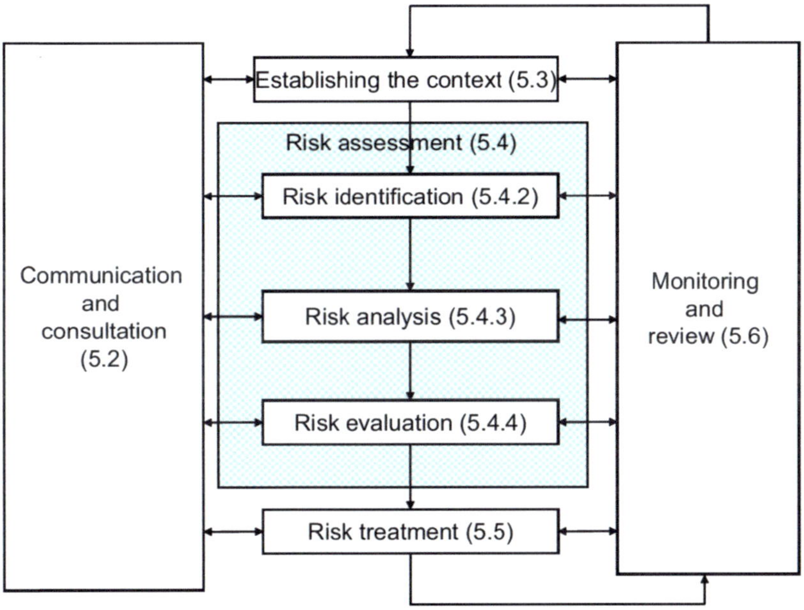

Figure 4: Risk management Process (Source: ISO 31.000, 2009).

The most common errors in risk management are:

- Focus too much effort on risk assessment specially in risk analysis and do not dedicate effort to others risk management phases;
- Do not dedicate effort on risk treatment in order to certify that most of risk mitigation recommendations are taken place on appropriated time;

- Do not dedicate effort to communicate the risk for employees and community;
- Do not dedicate effort to monitor and review the risk during the asset life cycle;

Indeed, the more technical challenge is the risk analysis because requires different qualitative and quantitative method's application that must be implemented along the asset life cycle in order to obtain the best results in terms of mitigate the risk. All qualitative and quantitative methods are well described in chapter 3. However, by the risk management point of view the main concern is to implement such risk analysis in the correct time along asset life cycle in order to be able to implement the risk analysis recommendation on adequate time. The Table **2** shows the correct asset life cycle phase to apply the risk analysis.

Table 2: Risk analysis methods along asset life cycle

	Asset Life cycle phases				
	Pre-Feed	Feed	Executive	Operational	Abandon
Risk Analysis Methods	What if				
	PHA	PHA	PHA	PHA	PHA
		HAZOP/HAZID			
		FMEA		FMEA	
		FTA		FTA	
		ETA		ETA	
		Bow Tie	Bow Tie	Bow Tie	Bow Tie
		QRA		QRA	QRA
		SIL		SIL	

On Table **2**, the methods highlighted in blue is qualitative and the red ones are quantitative. During the pre-feed phase the project has a very preliminary concept, but even though, it is important to take into account risk issues in order to take better decision and select safe technologies. In this phase, previous risk analysis that is similar to the project in discussion may support in a way to recognize the risk involved in the new project. Whenever the process is known and there is not a new technology involved some risk analysis likes what if and PHA (**P**reliminary **H**azard **A**nalysis) are the most common methods applied in such a phase.

During the feed phase, many project information came out and it is more appropriated to take place different qualitative and quantitative like PHA (**P**reliminary **H**azard **A**nalysis), HAZOP (Hazard Operability), HAZID (Hazard Identification), FMEA (Failure Mode Analysis), FTA (Fault Tree Analysis), ETA (Event Tree Analysis), Bow Tie, QRA (Quantitative Risk Analysis) and SIL (Safety Integrity Level). All those methods have the main objective to identify the hazards quantify the risk and provide information to compare the risk level required as a target in order to define risk mitigation actions.

On Execute phase, the risk in such process is about construction and for this specific process the PHA and Bow tie are the most common methods applied in such a phase.

On an operational phase, the operational routine faces, their daily process risk and in order to support decision to control and mitigate such risk the PHA (**P**reliminary **H**azard **A**nalysis), FMEA (Failure Mode Analysis), FTA (Fault Tree Analysis), ETA (Event Tree Analysis), Bow Tie, QRA (Quantitative Risk Analysis) and SIL (Safety Integrity Level) are the most common methods applied in such a phase.

The last phase is abandoning or decommissioning. In such phase the risk involved in decommissioning activities must be identified to avoid the incident and accident. Depends on the case, some quantitative methods might be necessary to support decision to how disassembly or store products and equipment. Therefore, PHA (**P**reliminary **H**azard **A**nalysis), Bow tie and QRA (quantitative risk analysis) are the most common methods applied in such a phase.

In order to mitigate the risk, all recommendation defined for risk analysis methods must assessed and a criterion to establish which one must be taken place have to be well defined. Indeed, it is necessary to define the recommended plan is taking into account when, who, where and how such recommendation will take place. In addition, it is really necessary to follow up such recommendation plan and define specific indexes for that. It is possible to find a different criterion to prior risk mitigation action recommendations like risk value, severity and it includes the cost and feasibility to implement such recommendations. In fact, two important drivers will take high influence in such decision those are law obligation and available budget. The first one is mandatory and in many cases, there is no discussion about that otherwise, the organization does not get their operating license or is forbidden to operate. By the other way round, most of such recommendation are not mandatory and depends on the budget they will take place or not. Because of that, managers and leader must be involved in the risk management process to guarantee that all necessary resources, include financing will be available to mitigate the risk.

The ***"monitor and review process"*** is part of risk management and in many cases is neglected for many organizations. In fact, monitor and review the risk is a hard task because in many cases, different employees work in different asset phase and do not follow up the risk assessment along asset life cycle or are not aware about which approaches took place in the previous asset life cycle. That situation is usual for safety engineer who work in operational phase and do not take part in the risk management in project phase or safety engineer that works on project phase and will not follow up the asset risk management in the operational phase. Therefore, the organization must define a structure that is enabled to deal with such issues, but even though, it is complex because in many cases there is more than one asset to manage the risk and such assets are in different life cycle phases.

Nevertheless, most organizations recognize the importance of monitoring and review the risk despite all challenges and difficult that they face. Indeed, one of the most important issues related to review the risk is related to process modification in the operational phase. In many cases, such modifications are not assessed in terms of risk impact and any incident and accident occur because of lack of awareness about the new process conditions. Therefore, it is important to establish a risk management process that enables detect such process change and update risk assessment.

The final Risk management phase is the ***"continuous improvement of the framework"*** that enables organizations to adapt their risk management to new external requirements, society demands or even to achieve better performance. Indeed, all management process must be dynamic and flexible to be adapted to external requirements in order to survive with high performance for a long period of time. Once again, to promote such continuous improvement it is necessary the managers and leaders involvement in all organizational levels.

8.4. SAFETY CASE APPLIED TO OIL AND GAS INDUSTRY (SCR05 - HSE)

The HSE safety case regulation **"SCR05"** applied to offshore assets has the main objective to establish good risk management practices along offshore asset life cycle in order to avoid a major accident. In other words, that means the major accident risk is, or will be, avoided based on risk management program throughout offshore asset life cycle. In general terms, organizations most provide evidence that:

- Identification of all hazards with the potential to cause a major accident;
- Assessment and evaluation of risks of major accident;
- Implementation of measures to mitigate the major accident risk.

In order to certify that such requirement, it is necessary to implement the risk management assessment that requires qualitative, semi-quantitative or quantitative risk analysis and further risk evaluation. The type of risk assessment used should be appropriate to the magnitude of risk as shows Fig. **5**. (Q=qualitative, SQ=Semi-quantitative, QRA=quantitative risk analysis).

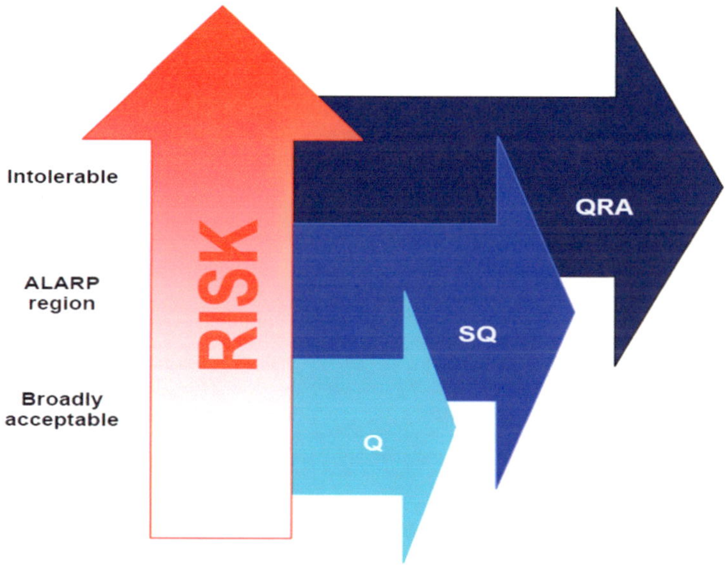

Figure 5: Proportionate Risk Assessment (Source: SCR05, 2005).

In terms of risk assessment, the risk under broadly acceptable level may be analyzed by qualitative risk analysis methods like PHA, FMEA and HAZID. In fact, the PHA and HAZID are the most applicable because identify the hazard clearly and based on risk matrix, qualify the risk level. Based on such assessment the hazard classified in intolerable level will require a quantitative risk analysis. Whether risk poses under ALARP region a semi-quantitative analysis is required and in this case methods like FTA, ETA, SIL analysis must be an option for such assessment. In the worst case, whenever risk poses under intolerable risk level, it will be necessary to carry out the Quantitative Risk Analysis, that means, consequence and effect analysis carried out by software as well as individual risk and PLL (Potential Loss of life) evaluation.

Indeed, the main point in risk assessment is to be careful to sub estimate some hazard's consequence effects and do not treat such risk adequately. In many cases, organizations are aware about their risk, but avoid classifying risk as intolerable in order to avoid hard argue with authorities about risk mitigation actions.

Nevertheless, the authorities have similar safety cases to compare, and in case of experience professional are reviewing such safe case they are able to detect such sub-estimate risk. Furthermore, the main issue to be discussed in risk assessment is the risk target imposed by authorities.

The risk evaluation in SCR05 is based on "ALARP" concept as discussed on Chapter 3 and need to be provided as evidence to be assessed. In order to evaluate the risk and take decisions for mitigating it other aspect like benchmarking, regulators and stakeholders must be taken into account.

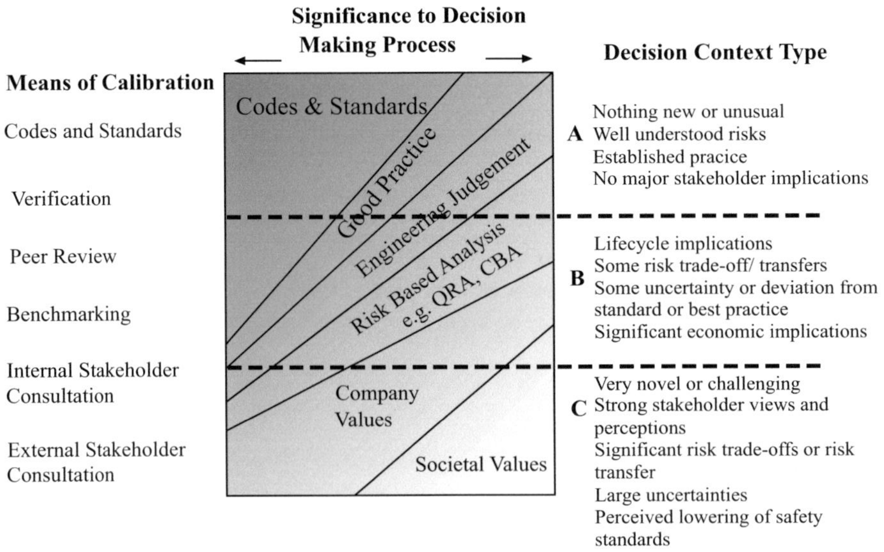

Figure 6: A Decision Support Framework for Major Accident Hazard Safety (Source: www.hse.org).

In addition, the decision context that defines the level of knowledge about the technology assessed and certain about the risk as well as implications in case of major accident must also to be taken into account. The Fig. **6** summarizes the decision process based on risk level.

Based on **"SCR05"**, the cost benefit analysis(CBA) must be taken into account when risk mitigations take place. Despite it´s not an easy task, in some cases, it´s important to have a precise valuation about the consequence in terms of economic value in order to justify investments to mitigate such risk. Indeed, the risk criterion is the first one to prior which recommendation to mitigate the risk must be implemented first and the CBA will support better such decision.

Despite high focus on risk management, the SCR05 has other requirements that must be attended in order to allow the organization go ahead with their entrepreneur proposal. Indeed, such requirement is different depends on the asset life cycle and specific cases that are characterized by different "schedules". Those schedules are specifically for design, operation, decommissioning phase as well as for non-production installations and combined production situations. Thereby, depends on the type of schedule, there will be different requirements. Basically the requirements are:

- Company identification;
- Process characteristic;
- Major accident analysis;
- Particulars.

The **"company identification"** requires all legal information as name, local of installation from such company which is responsible for the asset.

The **"process characteristic"** requires all description about process that will take place like production, installations, equipment and such information must be provided by drawing, layout and technical document that clarify the asset characteristics.

The **"major accident analysis"** is about how a company identifies the hazards, analyze the risk and evaluate it in order to mitigate the risk of major accidents. Indeed, such requirement is extensive because involve all risk management steps discussed above.

The **"particulars"** is about specific information such as vulnerabilities that might lead to a major accident like meteorological conditions, operational limits, combined operation.

The Table **3** summarizes the SCR05 schedules describing the information necessary for each type of requirement.

Table 3: SCR05 scheduler requirements

		SCHEDULES				
		Schedule 1 (Design)	Schedule 2 (Operation)	Schedule 3 (Non-production installation)	Schedule 4 (Notification of combined operation)	Schedule 5 (Dismantling)
INFORMATION REQUIRED	Identification	The name and address of the operator of the installation	The name and address of the operator of the installation	The name and address of the operator of the installation	The name and address of the operator of the installation	The name and address of the operator of the installation
	Project Characteristic	Chosen design concept, including suitable diagrams, and a summary of the other design options which were considered.	Review if necessary	Review if necessary		A description of how the proposed arrangements, methods and procedures for dismantling the installation and connected pipelines take adequate account of the design and method of construction of the installation and its plant.
	Process Characteristic	The principal systems on the installation,the installation layout, the process technology to be used, the principal features of any pipeline, any petroleum-bearing reservoir intended to be exploited using the installation and the basis of design for any wells to be connected to the installation.	The principal systems on the installation,the installation layout, the process technology to be used, the principal features of any pipeline, any petroleum-bearing reservoir intended to be exploited using the installation and the basis of design for any wells to be connected to the installation.	The principal systems on the installation,the installation layout, the process technology to be used.	The principal systems on the installation,the installation layout, the process technology to be used.	The principal systems on the installation,the installation layout, the process technology to be used.
	Major Accident	The Risks assessment and evaluation providing informations that the potential to cause a major accident are reduced to the lowest level that is reasonably practicable.	Review or preparation of the safety case	The Risks assessment and evaluation providing informations that the potential to cause a major accident are reduced to the lowest level that is reasonably practicable.	A description of how the management systems for the installations involved in the combined operation will be co-ordinated so as to reduce the risks from a major accident to comply with the relevant statutory provisions, a description of any activities during the combined operation which may involve hazards with the potential to cause a major accident on or in connection with an installation and a description of any risk control measures introduced as a result of that review.	A summary of how any safety representatives for that installation were consulted with regard to the revision of the safety case and a description of arrangements made for protecting persons on the installation from toxic gas at all times other than during any period while they may need to remain on the installation following an incident which is beyond immediate control
	Particulars	The meteorological and oceanographic conditions to which the installation may foreseeably be subject,the properties of the sea-bed and subsoil at its location, Particulars of the types of operation, etc.	The previous information from design phase and in addition, the maximum number of persons, Particulars of the plant and arrangements for the control of well operations, A description of any pipeline with the potential to cause a major accident,A description of arrangements made for protecting persons on the installation from toxic gas, the emergency plan and particulars of any combined operations which may involve the installation at all times other than during any period while they may need to remain on the installation following an incident which is beyond immediate control.	The previous information from design phase and in addition, the limits of the environmental conditions beyond which the installation cannot safely be stationed or operated, identifying the routes and locations of pipelines, wells and other subsea equipment; and a summary of the arrangements in place for co-ordinating the management systems of all duty holders involved in any such combined operation, the plant likely to be used during any such combined operation, the likely impact any such combined operation may have on the installations involved.	Particulars of any plant to be used in connection with the combined operation but which is not described in the current safety case for any of the installations involved in the combined operation.	

8.5. EN 51026 (RISK MANAGEMENT APPLIED TO RAILWAY INDUSTRY)

The EN-40126 -2 is a specific standard for risk management application addressed to railway industry, which define RAMS process activities throughout the asset life cycle as well as the concepts and risk analysis methods.

The remarkable characteristic that the reliability is also a safety performance index. In such trains, for example, some equipment such as breaks, buggies, external doors, TCMS can have unsafe failure which cause accidents. Therefore, reliability, availability, maintainability and safety are integrated in the same standard, which describes the RAMS process by the "V-Diagram" concerning all asset life cycle phases as show Fig. **7**.

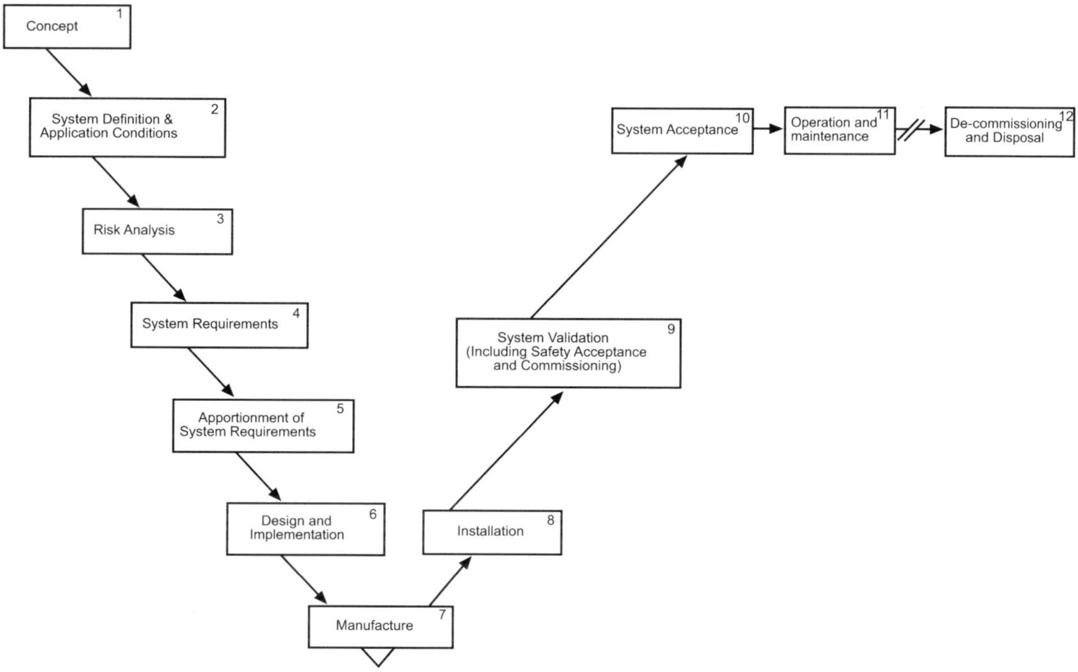

Figure 7: V-Diagram (Source: EN 50126-1).

In general terms, each phase has a specific task to be carried out in order to guarantee the best performance on safety as shown in Table **4**.

In **"concept phase"** it is required to look for similar projects and similar safety policies as well as requirement.

During the **"System definition and application condition"** is required to carry out preliminary risk analysis in order to identify the main hazards as well as established an overall safety plan and define risk tolerability and criterion.

The next step is **"Risk analysis"** to carry out the in system level as well as risk management.

After risk analysis phase, the following step is **"System requirement"** where has defined system safety requirements as well as a safety acceptance requirement and functional requirements. In addition, in this phase the safety management must be defined and established in order to follow up the requirement achievement and assure that all actions necessary to certify high safety performance will be implemented.

The next phase is the **"Apportionment of system requirement"** that's defines the subsystem and component safety requirements as well as their acceptance safety criterion.

On **"Design phase"**, the supplier data related to risk analysis must be assessed in order to certify the safety requirement assurance. The safety case must be implemented in order to certify that all requirements are achieved based on risk analysis methods defined on procedures and requirement.

Table 4: Safety Lifecycle phases (Source: EN 50126-1)

LIFE CYCLE	PHASE RELATED GENERAL TASKS	PHASE RELATED RAM TASKS	PHASE RELATED RAM TASKS
1. Concept	• Establish Scope and purpose of Railway project • Define Railway project concept • Undertake financial analysis & feasibility studies • Establish Management	• Review previously achieved RAM performance • Consider RAM implication of the project.	• Review previously achievement, safety performance • Consider Safety implication of the project. • Review Safety policy & safety target
2. System definition and application conditions	• Establish a system mission profile • Prepare system description • Identify operation & maintenance strategy • Identify operation	• Evaluate past experience data from RAM • Perform preliminary RAM analysis • Set RAM policy • Identify long term op & Mtce Conditions • Identify the influence on	• Evaluate past experience data for Safety • Perform Preliminary Hazard analysis • Establish Safety Plan (overall)

	conditions • Identify maintenance conditions • Identify the influence of existing infrastructure constraints	RAM of existing infrastructure constraints	• Define tolerability of risk criteria • Identify the influence on RAM of existing infrastructure constraints
3. Risk analysis	• Undertake project related Risk Analysis		• Perform System Hazard & Safety Risk Analysis • Set-up Hazard Log • Perform Risk Assessment
4. System requirements	• Undertake requirement analysis • Specify system (overall requirements) • Specify Environment • Define system demonstration & acceptance criteria (overall requirement) • Establish a validation plan • Establish Management, Quality & Organizational requirements	• Specify System RAM requirement (overall) • Define RAM acceptance criteria (overall) • Define system functional Structure. • Establish the RAM program • Establish RAM management	• Specify System Safety requirement (overall) • Define Safety acceptance criteria (overall) • Define system functional Structure • Establish the RAM program • Establish RAM management

5. Apportionment of system requirements	• Apportion system requirement
	  • *Specify subsystem & component acceptance criteria*
	  • *Define sub-system & component*
		• Apportion system RAM requirement
		  • *Specify subsystem & component RAM acceptance criteria*
		  • *Define sub-system & component RAM acceptance criteria*
			• Apportion system safety, target & requirements
			  • *Specify subsystem & component safety requirement*
			  • *Define sub-system &*

	acceptance criteria		*component safety, acceptance criteria* • Update system safety plan.
6. Design and implementation	• Undertake requirement analysis • Specify system (overall requirements) • Specify Environment • Define system demonstration & acceptance criteria (overall requirement) • Establish a validation plan. • Establish Management, Quality & Organizational requirements.	• Implement RAM program by reviewing, Analysis, testing and Data assessment, covering: • *Reliability & Availability* • *Maintenance & Maintainability* • *Optimal maintenance policy* • *Logistic Support* • Undertake programme control, covering: • *RAM program management* • *Control of sub-contractors, supplier*	• Implement Safety Plan by reviewing, Analysis, testing and Data assessment, addressing: • *Hazard log* • *Hazard analysis & risk assessment* • *Justify safety related design decision.* • Undertake programme control, covering: • *Safety management* • *Control of sub-contractors, supplier* • Prepare generic Safe case control, covering: • Prepare (if applicable) generic application safe case

On **"manufacturing phase"**, the system, subsystem and equipment must keep their safe performance predicted on the design phase. By this way, the supplier must adapt their manufacture process to avoid non-compliance in such phase.

On **"installation phase"**, the system, subsystem and equipment must keep their safe performance defined at design and maintained on manufacturing phase. Therefore, the supplier must follow instruction based on Safety plan to avoid mistakes that compromise systems, subsystem and equipment reliability. The human error is usual in this phase, thereby; whenever some critical installation process is carried out the "Human reliability analysis" might be necessary to avoid such human error.

On **"System validation"**, which starts the system commissioning and all safety requirement states in safety case as well as risk analysis recommendation must be accomplished.

The next phase is **"Acceptance"**, which all safety requirement states in safety case as well as risk analysis recommendations must be assessed and evaluated. Whether further actions are required it must be done in order to have system approved before to start the operational phase. Unfortunately, that´s not happening in many cases and unsafe failures happen more frequently and earlier than expected during the operational phase which increase the risk of accident.

On **"Operation and maintenance phase"**, safety and occupational management must be established in order to follow up system's safety performance. In addition, it also important to define well the maintenance program focus on safety functions, layers of protection and all equipment that in case of unsafe failure trigger an incident.

On operational phase the **"performance and monitoring"** is the key activity in terms of follow up system safety performance. Therefore, indexes must establish and measure in a defined frequency of time.

The other important activity is operational phase is the **"modification and retrofit"** that is required whenever some process modification is performed and such modification may affect the safety with a new unsafe condition or increasing the risk of accident. In such situation, some risk analysis must be carried out in order to verify the impact of the new modification.

The final asset phase is **"decommissioning and disposal"** where the risk analysis must be carried out in order to define critical activities which would lead to an accident. The Table **4** summarizes all activities that must be performed in different asset phases.

Table 4: Safety Lifecycle phases (continuation) (Source: EN 50126-1)

LIFE CYCLE	PHASE RELATED GENERAL TASKS	PHASE RELATED RAM TASKS	PHASE RELATED RAM TASKS
7. Manufacturing	• Perform Production Planning • Manufacture • Manufacture and Test Sub-assembly of components • Prepare documentation • Establish training	• Perform Environmental stress screening • Perform RAM improvement testing • Commence failure reporting and corrective action system (FRACAS)	• Implement safety plan by: review, analysis, testing & data assessment • Use Hazard Log
8. Installation	• Assemble System • System Install	• Start maintainer training • Establish spare parts and Tools provision	• Establish Installation Program • Implement the installation program
9. System Validation (including safety, acceptance and commissioning)	• Commission • Perform probationary period of operation • Undertake training	• Perform RAM demonstration	• Establish commissioning program • Implement commissioning program • Prepare application specific safety case
10. System requirements	• Undertake acceptance procedures, based on acceptance criteria • Compile evidence for acceptance • Entry into service • Continue probationary period of operation (if appropriate)	• Assess Ram demonstration	• Assess application specific safety case
11. Operational maintenance	• Long Term System Operation • Perform on going maintenance	• Ongoing procurement of spare parts & tools	• Undertake ongoing safety centered maintenance • Perform ongoing

	• Undertake ongoing training	• Perform ongoing reliability centred maintenance, logistic support	safety performance, monitoring and hazard Log Maintenance. •
12. Performance	• Collect operational performance statistics • Acquire, analyze and evaluate data.	• Collect, analyze, evaluate and use performance RAM statistics.	• Collect, analyze, evaluate and use performance & safety statistics.
13. Modification and retrofit	• Implement change request procedures • Implement modifications and retrofit procedures	• Consider RAM implications for modification retrofit	• Consider safety implications for modification & retrofit.
14. Decommissioning and disposal	• Plan Decommissioning and disposal • Undertake decommissioning • Undertake disposal	• No activity for RAM	• Establish Safety Plan • Perform Hazard Analysis & Risk Assessment • Implement Safety Plan

By Safety point of view, the critical aspect of safety management along the asset life cycle is the risk assessment process that requires the risk target definition in order to lead system to safety high performance.

The EN – 50126 2 (2007) establish the qualitative and quantitative risk assessment methods and concepts in order to evaluate the risk and verify if acceptable levels are achieved or if it´s necessary additional mitigation actions. In fact, the risk can be evaluated qualitatively or quantitatively. In the first case, the risk matrix and qualitative risk methods are applied and the second case, quantitative methods as well as a quantitative risk index such as individual risk are applied. In fact, the first step of risk management is to define a risk target. Whenever qualitative methods are applied, the risk matrix must be defined. Based on EN 50126, a risk matrix six per five as must be applied in Railway industry.

The qualitative risk analysis method such as PHA (Preliminary Hazard Analysis) and FMEA (Failure Mode Analysis) enable to identify the hazard and evaluate the risk based on risk matrix. The Fig. **8** shows an example of risk matrix with risk index.

Frequency		Risk			
Frequent	6	6	12	18	24
Probable	5	5	10	15	20
Ocasional	4	4	8	12	16
Rare	3	3	6	9	12
Unprobable	2	2	4	6	8
Unexpected	1	1	2	3	4
		1	2	3	4
		Insignificant	Marginal	Critic	Catastrophic
		Consequence			

Figure 8: Risk Matrix.

On Fig. **8** above is possible to identify different risk levels by different colors. The red regions means intolerable level and all risk in such region must be mitigated. The orange and yellow region mean tolerable and moderate, respectively, and theoretically it is necessary to mitigate such risk as much as practicable and on green region that is not necessary any mitigating action.

The **"ALARP"** concept is applied to evaluate risk based on tolerable risk region and in case of intolerable risk, such risk must be mitigated as well as tolerable and moderate risk must be mitigated as much as practicable in terms of return on investment and feasibility.

The **"MEM"** (Minimum Endogenous Mortality) concept incorporates the lowest natural death rate and uses this to assure that the total additional technological risk will not be greater than such natural death risk. The tolerable individual risk is defined by the number of deaths per year as shows Fig. **9**.

The other additional Risk concept is **"GAME"** and principle states that a new system should be globally at least as safety as the current system, including an element of continuous improvement (Calixto, 2013).

No matter Risk principles establish per regulator authority, to define **"individual risk"** is necessary to identify the risk scenarios that have death consequences like "derailment" for example and estimate the number of deaths as well as frequency of occurrence. Such index is known as individual risk and in order to calculate such individual risk for the whole system like a train for example, it is necessary to sum risk for each accident scenario as shows equation below.

$$RI = \sum_{i=1}^{n} C_i X f_i$$

where:

f = frequency of accident scenario (year)

C = consequence of accident scenario (deaths in plant area)

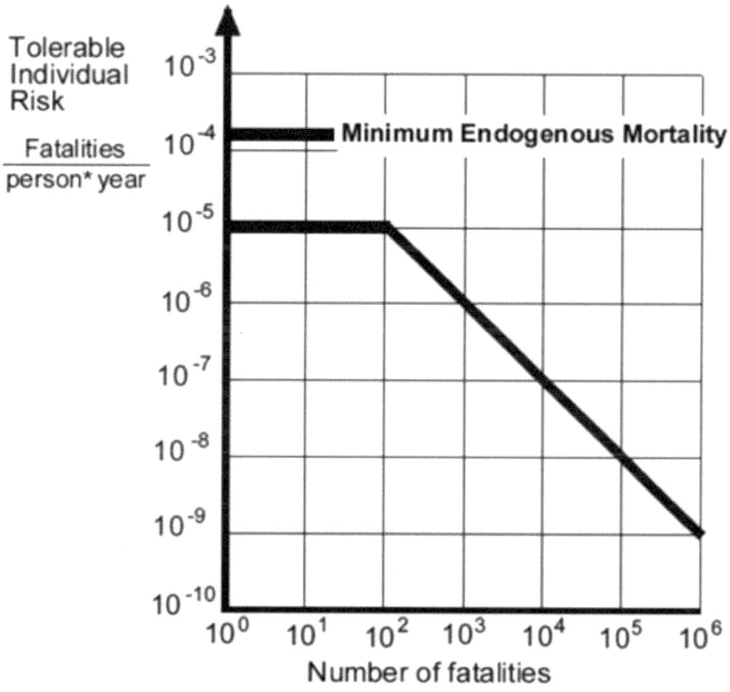

Figure 9: MEM Individual Risk Criterion (Source: EN 50126-2, 2007).

The Risk management process requires different types of analysis in different asset phase, but in general term risk management can be summarized as shown the Fig. **10**. Indeed, the EN-50126 define additional Qualitative Risk analysis approach like Hazard log and C-Hazard as well as some quantitative methods like FTA, ETA, Bow tie and SIL analysis must be applied along enterprise's phases in order to define more precise risk and help to mitigate it.

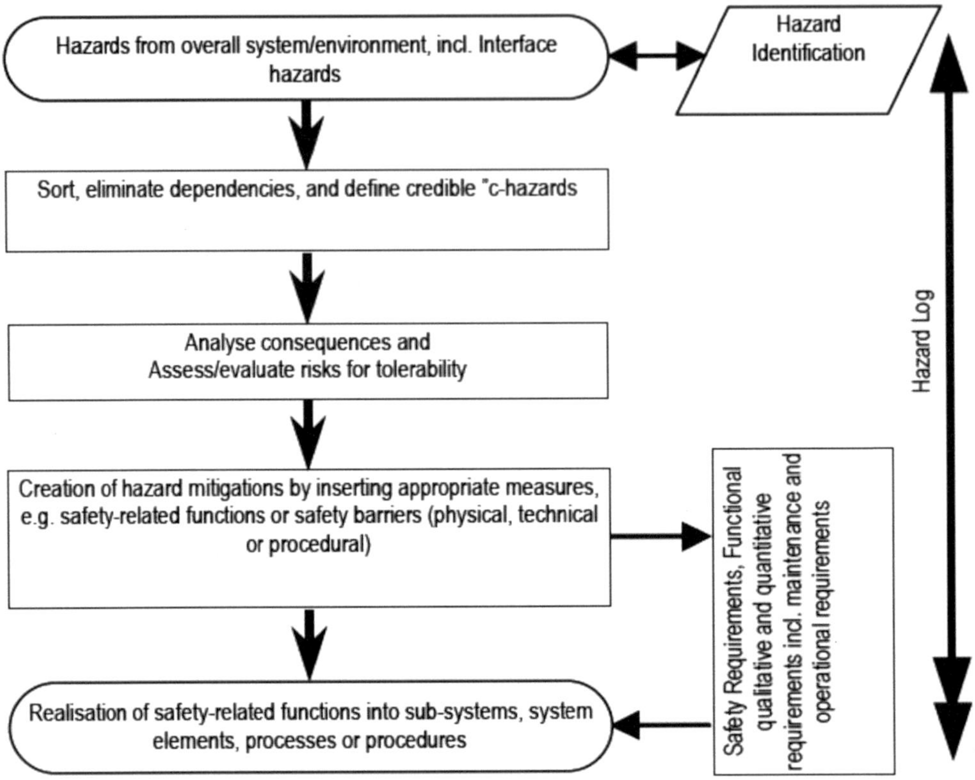

Figure 10: Risk Assessment (Source: EN 50126-2, 2007).

Once the safety requirement has a high relation in many cases with the RAM requirement both processes must be integrated.

The important thing is to be on mind that such standard is a guideline and do not limit the methods to risk assessment. In fact, the best practice must be applied to mitigate the risk as well as allow equipment with unsafe failure achieve a high level of reliability.

In addition, preventive maintenance and inspections have an important whole in risk assessment and the best approaches performed to achieve high operational availability must be applied to achieve high integrity and safe condition.

8.6. SAFETY CASE APPLIED TO NUCLEAR INDUSTRY (HSE)

The Health and Safety at Work Act 1974 establish the employer responsibility about safety in the workplace which is reinforced by Act 1965 (NIA) 4. In the

UK, in order to have a Nuclear plant, it´s necessary to get a license from Health and Safety Executive which requires a safety case. The safety case has the main objective to demonstrate in written form that risk is under tolerable level and a risk management process will be established to maintain such acceptable level throughout the nuclear asset life cycle. In addition to efficient risk management is expected that safety case demonstrate also the safety principles, safety practices, standards compliance and safety procedures as well as a safe control during both normal operation and fault conditions.

The licensee is legally responsible for the safety case. However, it is those employees of the licensee who have direct responsibility for delivering safety who should have 'ownership' of it.

The licensee submits the safety case for approval based on UK and international standards, thereby the Nuclear Installation Inspectate (NII) perform the safe case assessment in order to approve it. The assessment of safe case is a continuous process because the NII carries out periodic inspections at all nuclear sites to ensure that the license conditions and regulatory requirements are all met. Such inspection and audit process focus on critical process and hazards which might lead a major accident in case of failure.

The Fig. **11** below presents an overview of the nuclear safety regulatory framework and identifies the key stakeholders and their roles.

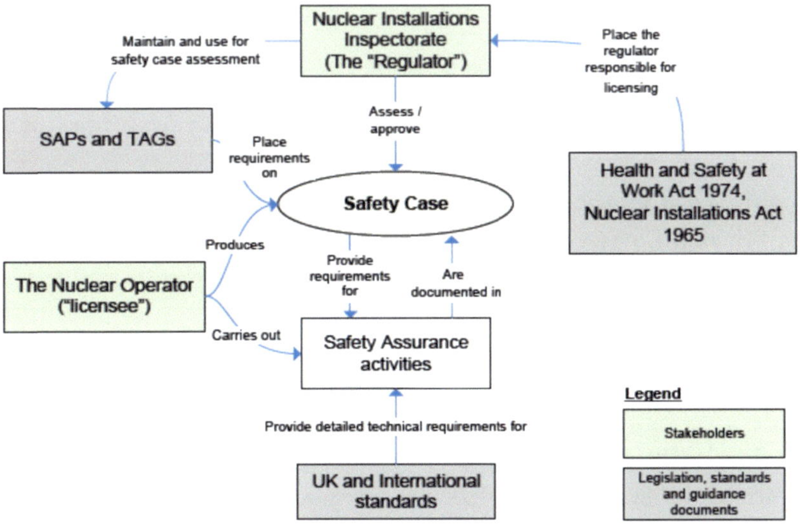

Figure 11: The UK regulatory framework for nuclear safety (Source: HSE, 2014).

In order to have the safety case approved by NII it´s necessary to take into account three different strategies such as goals based approach, standards or vulnerability.

Similar to other safety cases, it´s necessary to demonstrate compliance with UK and international standards such as IEC 61508, IAEA NS-R-1 or IEC 6088035.

The goal-based approach requires safety evidence demonstration related to risk management which enables to operate the nuclear plant under acceptable risk level even in crisis situations.

The vulnerability-based requires to demonstrate the copy capacity to mitigate different internal and external threats. The Fig. **12** shows the safety justification triangle.

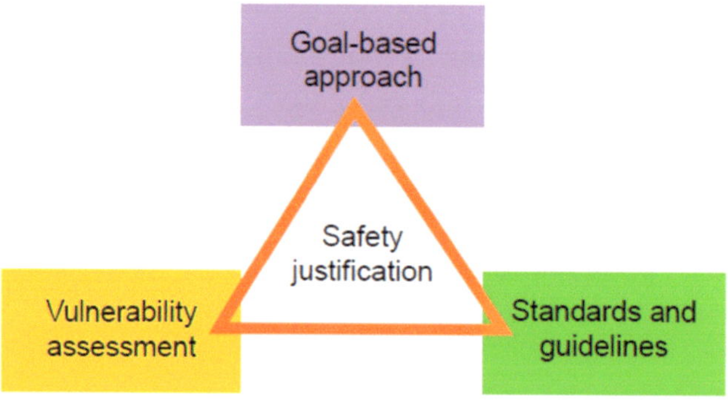

Figure 12: Safety justification triangle (Source: HSE, 2014).

The nuclear safety case is defined by the HSE as:

"... The totality of documented information and arguments which substantiates the safety of the plant, activity, operation or modification in question. It provides a written demonstration that relevant standards have been met and that risks have been reduced as low as reasonably practicable (ALARP)."

The ALARP concept is mitigating the risk as low as reasonably possible. That means the tolerable risk may also be mitigated based on cost benefit assessment. In fact, the tolerability does not mean accepting, it refers to a willingness to live with a risk to yield certain benefits so long as there is confidence that it is properly controlled.

In order to understand the ALARP concept, it´s necessary to be clear about the risk concept.

The **"Individual risk"** is a chance of death that an individual or group of people has when they are located in one vulnerable region and exposure to some hazard into the operational ground (industrial area). The individual risk is usually expressed in terms of ISO-Risk curve or ALARP region. In Nuclear Industry case, the accident consequence may be either in the form of 'early effects' or 'late effects'. Early effects will occur if the radiation dose is very high and will result in direct death. With regards to late effects, the greatest concern is cancer.

The **"Societal risk"** is a chance of death that communities outside Plant area has due to be exposed to industrial hazard sources. The societal risk is usually represented by an F-N curve that shows expected number of fatalities on each frequency level. The Fig. **13** shows the ALARP tolerable risk representation.

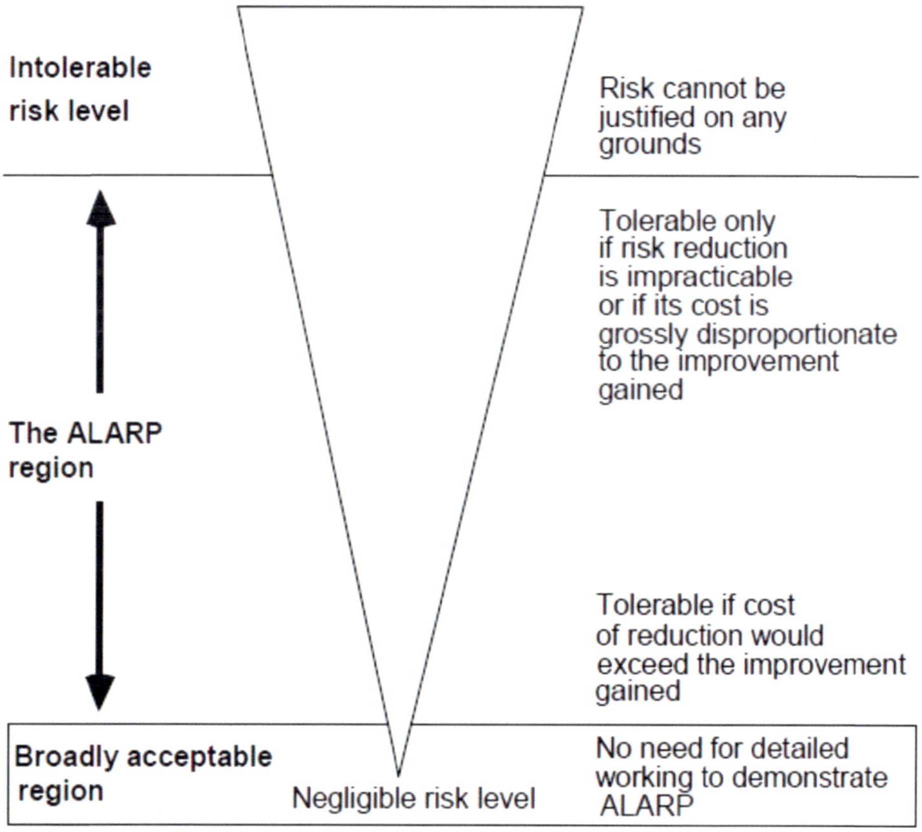

Figure 13: ALARP (Source: HSE, 2014).

The safety case provides written evidence of risk management compliance like others safety cases from different industries. The remarkable difference in Nuclear industry is the recognition of human factors influence on safety.

The human factor starts from design, concerning human machine interaction and its influence on safe state like controls, alarms, indications and instrumentations.

Indeed, several standards related to human reliability analysis (HRA) have been developed and implemented in Nuclear Power Plants projects include operational and maintenance processes. Some of HRA methods were described and exemplified in Chapter 7.

In addition, such standards and procedures provide a background to take into account such human factors when risk analysis methods are performed.

Likewise, human reliability analysis has supported nuclear power plants during projects, operation and maintenance task in order to avoid human error. The control room design is an example of human factor application in many Nuclear plant projects during the design phase. Such HRA methods have been applied by other industries such as Oil and Gas, and Transportation.

8.7. KEY PROGRAM ASSET INTEGRITY (KP3)

The **Key Program Asset Management** come out as a request from the UK Secretary of State for Work and Pensions afterwards a Parliamentary debate on 2 July 2008 that marked the 20th Anniversary of the Piper Alpha Disaster.

Despite the Key program asset management is not a standard, it´s important for company which have processes that can lead a major accident.

In this specific case, the main program objective was focused on offshore installations on the United Kingdom Continental Shelf, and revealed significant issues regarding the maintenance of safety critical systems used in major accident concerning the period between 2004 and 2007.

The **Asset integrity** is defined by HSE as the ability of an asset to perform its required function effectively and efficiently while protecting health, safety and the environment.

The KP3 subjects can be summarized in process safety management, leadership, learning, communication, competence, safety culture and workforce involvement.

In order to verify compliance is established the audit process which verify compliance of such subject described above. However, it´s necessary to define the assets and critical process to prioritize the audit based on risk classification. Such approach is also used for other safety standards audit process and the main objective is to start from the most critical processes which may lead a major accident in order to certify that all measures are taking place to mitigate the risk.

Likewise, other audit process, it´s necessary to collect evidence of compliance about the subject defined by KP3 program. The Fig. **14** shows an example of an audit check list applied for the maintenance of safety critical elements.

<table>
<tr><td>Offshore D</td><td colspan="3">Maintenance of safety critical elements (SCE)</td></tr>
<tr><td colspan="4">1. Does the maintenance work order for a SCE contain a statement of or reference to the relevant SCE performance standard?
2. Does the work order describe any tests to be conducted prior to re-commissioning, to demonstrate that the relevant performance standards has been met?
3. How is the result of this test recorded (e.g. pass / fail / remedied)
4. What do you do if the test doesn't meet the acceptance criteria?</td></tr>
<tr><td colspan="4">RELEVANT LEGISLATION
PFEER 5 Assessment - establish appropriate performance standards
SCR 2 (5) SCEs remain in good repair and condition
PFEER 19 Suitability and condition of plant
PUWER 5 Maintenance</td></tr>
<tr><td>NON COMPLIANCE / MAJOR FAILING</td><td>ISOLATED FAILURE / INCOMPLETE SYSTEM</td><td>IN COMPLIANCE / OK</td><td>NOT TESTED / NO EVIDENCE</td></tr>
<tr><td></td><td></td><td></td><td></td></tr>
</table>

Figure 14: KP3 Audit guide (Source: HSE, 2014).

The Integrity Asset Management program must be integrated with other systems, programs and methods in order to lead efficient high performance in asset integrity. The usual standards which can be integrated with asset integrity are Six Sigma, ISO 14.001, ISO 9001, OHSAS 18001. In addition, management process and methods such as Process Safety Management (PSM) and RAM (Reliability, Availability, Maintainability) can also support the asset integrity management as shows Fig. **15**.

Indeed, that is a big challenge, but a good opportunity to succeed the programs all together and drive organizations to achieve high performance by different aspects.

The success of such implementation like all programs depends too much of leadership involvement and awareness as well as the organizational culture in favor of integrity management.

In order to achieve high performance it is always necessary to learn with past and implement new ideas otherwise the gaps is still on organizations.

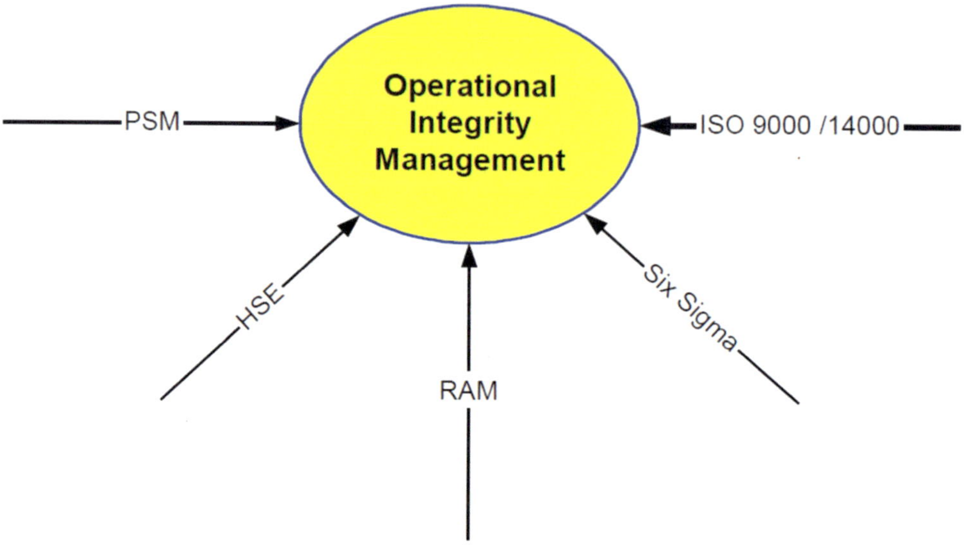

Figure 15: Operational Integrity Management (Source: Sutton Technical Books, 2005).

The Operational integrity management is a track to achieve operational excellence in the future that means high asset performance, high quality, low incident and accident and low environmental impacts.

The operational excellence track started on occupational safety in 19th and them standards, procedures, methods were implemented, but we still have a lot of gaps that must be fulfilled in order to achieve the operational excellence as shows Fig. **16**.

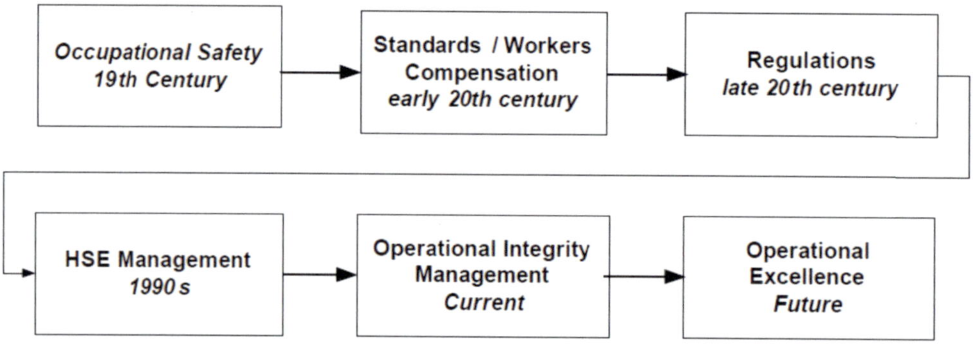

Figure 16: Operational Excellence (Source: Sutton Technical Books, 2005).

Nowadays, the asset integrity management takes into account not only risk management, but also human factors and reliability. Indeed, all elements discussed previously are comprised in such three pillars. The chapter 9 will give an example of Asset integrity implemented during the design phase.

The importance of human factors is demonstrated by major accident history where many of them have human error as one of the root causes. In fact, the human factors must take into account since the design phase. A good design provides system which technological human factors have low influence on human error. As discussed in chapter 7, there are human performance factors such as psychological and social organization that has not too much influence. By the other hands, technological human performance factors are under organization control and it´s possible to apply the best practice to mitigate such factors influence on human error.

The reliability is the chance that a system, equipment or component works property during a period of time. In order to achieve such desired reliability it´s necessary to implement different reliability engineering methods since the design phase throughout the asset life cycle.

Unfortunately, low attention is being given to reliability of equipment with safety function or even equipment which unsafe failures that can lead to a major accident. Despite many improvements has been implemented in the last decades, it´s necessary to integrate safety and reliability together during design and also operational phase.

The maintenance has also an important hole in asset integrity because enables to detect unsafe failures and hazard condition that can be eliminated before an accident happens. The big challenge considering maintenance is to establish a maintenance program for equipment with safety function or equipment which has unsafe failures that lead to an accident.

8.8. IEC 61508 (SAFETY INTEGRITY LEVEL STANDARD)

The principal standards that are used worldwide: ISA S84.01, IEC 61508 and IEC 61511. The standard ANSI/ISA S84.01 "Application of Safety Instrumented Systems for the Process Industry" was the first published by the three (ANSI/ISA, 1997) in 1996, although the committee in charge of writing ICE 61508 was already working in that direction during the same year. This standard appeared for addressing the application of Safety Instrumented Systems, with an approach that puts attention on the field devices as well as the logic solver as integral components of the SIS, specifically in the realm of the process industry. That means that this document is an industry-specific standard. The revolutionary aspect of this standard is that it established a complete life cycle for the development of the SIS, ensuring the safety requirements are compiled when implementing the system. This standard has been used since its release mainly in the USA, although a lot of countries have adopted it for the implementation of their SIS.

The very recently, in the year 2003, the specific-sector standard IEC 61511 has been released. It has been developed in the frame of IEC 61508, to cover specifically the implementation of Safety Instrumented Systems using E/E/PE technology in the process sector.

The IEC 61508 is concerned with safety systems that use electrical or electronic or programmable electronic (E/E/EP) technologies. It is not specific to any industrial sector: therefore it can be applied any sectors using E/E/EP safety systems, such as nuclear, transport, process, manufacturing, *etc*. One of its objectives is to provide a basis for the development of standards for specific sectors and the safety during assets life cycle based on the safety systems assurance. The Fig. **17** shows the main focus of those main different standards.

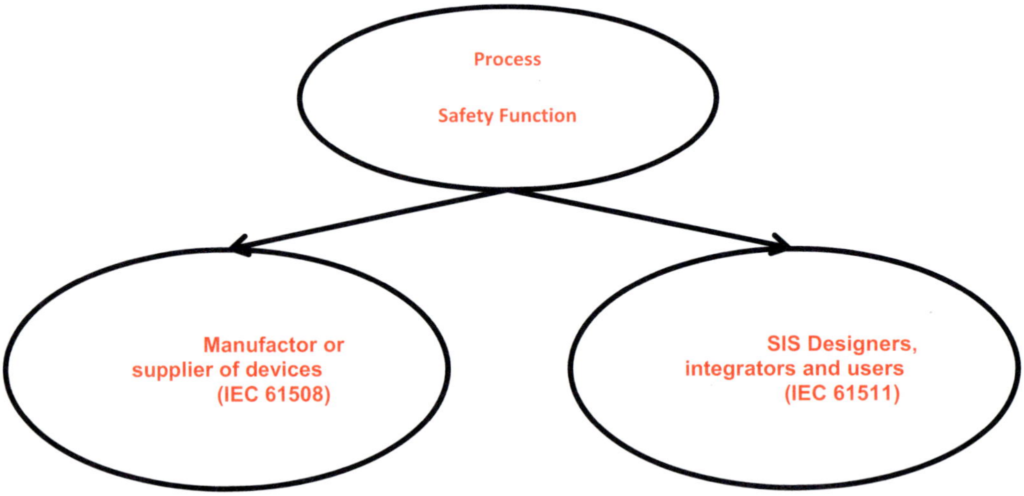

Figure 17: SIL Standards (Source: Echeverria, 2003).

The IEC 61508 is comprised in seven parts such as general requirements (Part 1), Hardware and Software requirements and definitions (Part 2,3 and 4), methods for the determination of SIL (Part 5), guidelines on the application (part 6) and overview of measures and techniques (part 7).

In order to achieve an acceptable risk level, the Safety Integrity Level (SIL) assessment is applied in order to mitigate the risk when it´s necessary. As a result, the Safety Instrumented Function (SIF) can be implemented as a layer of protection.

The SIF has a SIL number classification which varies from 1 to 4 and such levels can be translated as failure on demand. Indeed, as higher is SIL number lower is the probability of failure on demand, in other words, high reliability is required to achieve risk acceptable level.

Basically, each SIL number is related to one SIF and each SIF comprise one or more initiate element (sensor), logical element and final element (actuator and valve). In fact, SIF can comprise more than one of such elements as shown in Fig. **18**. At higher levels, there is an SIS (Safety Instrumented System) which comprises more than one SIF (Safety Instrumented Function) as shows Fig. **18**.

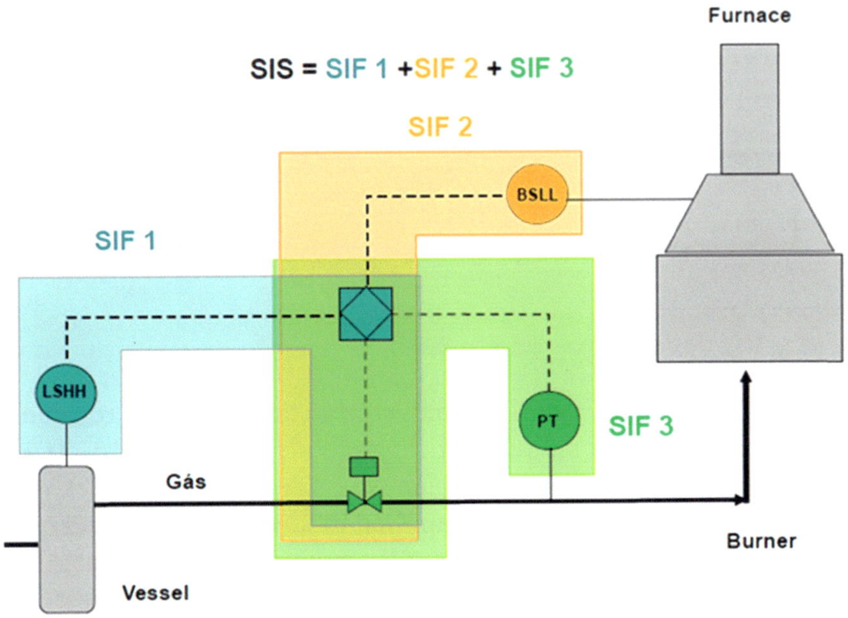

Figure 18: Safety Instrumented System (Source: Calixto E, 2012).

In order to apply such requirement, IEC 61508 it´s necessary, adequate competency that is assured for training, knowledge, experience and qualifications. In addition to SIL assessment methods it´s also necessary for other methods to provide reliability prediction such as RBD or FTA model and Monte Carlos Simulation.

The Safety Life Cycle, defines three different phases like to analyze, realization and operation in order to assure safety throughout SIS life cycle as shown Fig. **19**.

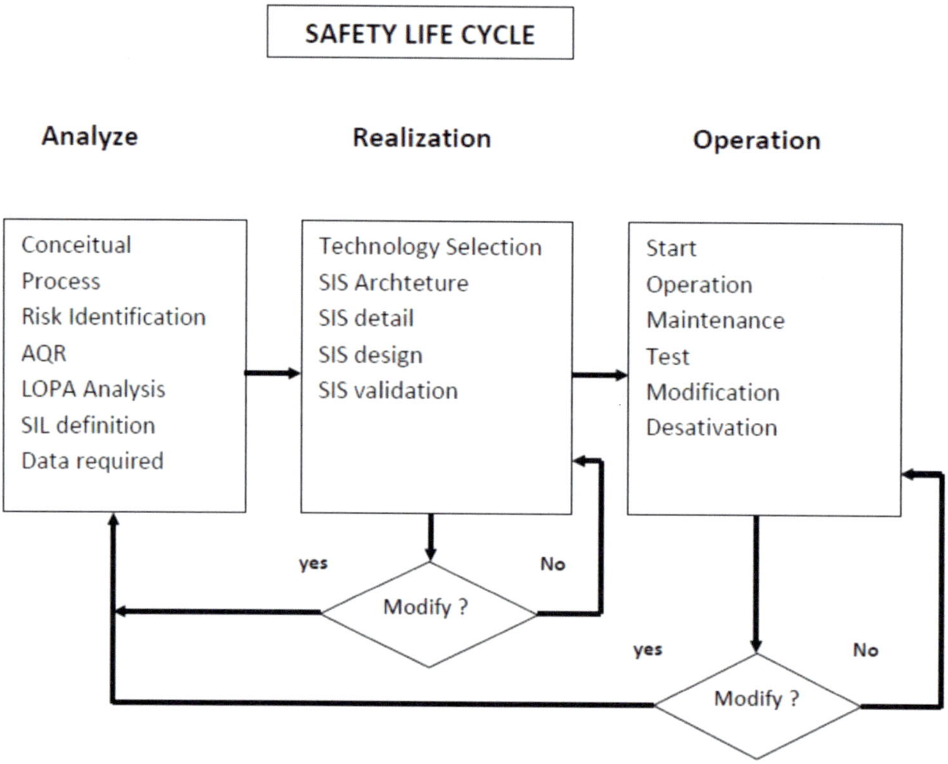

Figure 19: Safety Life Cycle (Source: Schartz, 2002).

During the Analyze phase, the risk analysis methods have a relevant hole to support SIL selection. The qualitative risk methods such as PHA, HAZOP and FMEA and quantitative risk methods such as FTA, ETA, LOPA, Consequence analysis are the most common methods applied to identify, assess and evaluate the risk. Based on such method results, the SIL methodologies are also applied to define the SIL level required to mitigate the risk in acceptable levels. The SIL analysis methodologies are those discussed in chapter 3 such as Hazard matrix, Risk Graph, Frequency Target and Individual or Societal Risk.

The SIF (Safety Instrumented Function) is related to SIL values which range from 1 to 4 that give a reference in order to predict the risk mitigation whenever a SIF take place as a layer of protection as shows Table **5**.

Table 5: SIL Classification (Source: Schartz, 2002)

Safety Integrity Level (SIL)	Average Probability of Failure on Demand (PFD)	Availability* in %	Risk Reduction Factor
1	10^{-2} to $<10^{-1}$	>90 to 99	>10 to 100
2	10^{-3} to $<10^{-2}$	>99 to 99.9	>100 to 1000
3	10^{-4} to $<10^{-3}$	>99.9 to 99.99	>1000 to 10,000
4	10^{-5} to $<10^{-4}$	>99.99 to 99.999	>10,000 to 100,000

Thereafter SIL definition, it is necessary to define SIS technology and configuration. It´s important to assure such reliability requirement and in this case, during the SIF design phase, accelerated life test **(ALT)** and high accelerated test **(HALT)** can be performed to predict SIF reliabilities and robustness.

The DFMEA, PFMEA and SFMEA has an important hole to drive the SIF performance throughout the design phase and also to avoid error during the manufacturing and installation phase.

During the operational phase its also important monitor SIF in order to verifies if such system achieves the reliability (SIL) requirement defined by **"warranty terms"** as well as establish an **"Inspection & Test program"** to certify that SIF are available to fulfil its mission when required. One of the most challenging tasks in Inspection and Test program is to define the best time to perform such inspection and test. Therefore, the reliability engineering methods such as lifetime data analysis and RAM analysis can support such decision by defining the time related to the reliability level required and also the effect of such intervention on SIF operational availability and reliability.

REFERENCES

BS EN 50126. (2007). PD CLC/TR 50126-2:2007 Railway applications. The specification and demonstration of Reliability, Availability, Maintainability and Safety (RAMS). Part 2: Guide to the application of EN 50126-1 for safety.

BS EN 50129. (2003).Railway applications. Communication, signaling and processing systems Safety related electronic systems for signaling The European Standard EN 50129:2003.

David J Smith, Kenneth G L Simpson. (2001). Functional Safety: A Straightforward Guide to applying IEC 61508 and Related Standards Second edition. Elsevier 2001. ISBN 0 7506 6269 7.

Echeverria, Alejandro Carlos Torres. (2003). Practical application of a sil analysis to a process plant the university of sheffield faculty of engineering department of chemical and process engineering. September 2003.

http://www.hse.gov.uk/offshore/programmereports.htm

http://www.hse.gov.uk/safetycasereports.

ISO 31000. (2009). Risk management.

Robin Bloomfield, Nick Chozos, George Cleland Adelard LLP. (2014). Safety case use in the nuclear industry. Retrieve from: London http://www.hse.gov.uk/safetycasereports.

CHAPTER 9

Safety and Occupational Health Management

Abstract: The safety and Occupational Management has been a huge challenge faced for different organizations throughout the world. Indeed the first challenge is to implement the best methods in the different safety subject, such as occupational risk, risk management, accident analysis, human factor analysis, emergency response. The second challenge is to manage such information and activities in order to produce a high effectiveness in Safety and Occupational Health (SOH) Management. This requires an effective management model as well as considering different aspects which impact on Safety and Occupational Health performance, such as safety culture, leadership, organizational learning and safety economic valuation. Those aspects are the pillars of SOH management, which must also be integrated with Integrated Management System and Asset Management. However, all this effort does not achieve their objective if a proper management model support (SOH). Finally, in order to achieve the highest performance in (SOH) the organizations must understand that the main focus of (SOH) are people who works in process and are effected for that.

Keywords: Leadership, culture, organizational learning, risk management, asset integrity management, integrated system management, safety and occupational health management.

9.1. INTRODUCTION

The Safety and Occupational Health Management effectiveness is a high influence of critical factors like safety culture, leadership, organizational learning and safety economy valuation. In addition, specific issues require an additional management focus like Management Model, Asset Integrity Management, Risk Management and Integrated Management System (IMS).

The management model also takes influence on such safety management efficiency because it dictates how the resources are allocated, coordinated, planed on time to achieve the targets and objectives considered by Safety. Indeed, for all organizations all over the world, to achieve high performance in safety requires a high effort, commitment and resources and the most important issue is to focus the safety Management on people which is the heart of SOH as show Fig. **1**.

The following items will describe all such aspects which take influence on the safety management effectiveness with real examples, discussion to try to understand safety management performance.

Eduardo Calixto

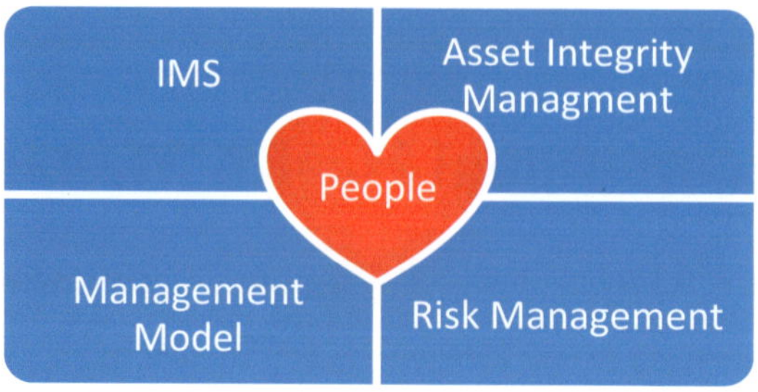

Figure 1: Safety Management concept.

9.2. SAFETY AND HEALTH MANAGEMENT CRITICAL FACTORS

9.2.1. Safety Culture

All organization has a safety culture, which can be described as strong or weak, positive or negative, which depends on influence on safety performance in a sense to prevent an incident and accident and preserve employee's health.

Indeed, quite few organizations have safety cultures which allow safety high performance over a long period of time. As Reason puts it, "like a state of grace, a safety culture is something that is striving for but rarely attained" (Reason, 1997; Parker, 2006).

Therefore, understanding how organizational cultural works can provide insights into ways which organizational cultures need to be modified to give a higher priority to safety. Indeed modify culture is always a big challenge once culture is related to peoples' values.

Although clearer than safety culture, organizational culture has itself been defined in a variety of ways. Such variety provides a useful summary of the way the concept of culture has been used by various writers like: observed behavioral regularities, group norms, espoused values, formal philosophy, rules of the game, climate, embedded skills, habits of thinking, shared meanings and root metaphors (Schein, 1992). It will be noted that some of these usages focus on values and attitudes as the key element of culture, while others stress behavior. Cooper sees this as the crucial distinction. "The main difference between such definitions (he says) appears to reside in their focus on the way people think, or on the way people behave" (2000:112).

Perhaps the best known definition of organizational culture, "the way we do things around here" (Deal and Kenney, 1982), is clearly behavior focussed.

Schein himself has at times referred to organizational culture simply as "the way we do things around here" (Schein, 1992), although his formal definition is more complex. Moreover Hofstede, after discussing whether it is better to focus on values or practices in defining organizational culture, concludes that "shared perceptions of daily practices should be considered to be the core of an organization's culture" (Hofstede, 1997). It is simply a question of emphasis. "The way we do things around here" carries with it the connotation that this is the right, or appropriate or accepted way to do things.

In order to understand how the safety culture work in organizations its better to start looking at successful organizations in terms of Safety, Occupational and Health Management performance. In reality, successful means high performance in avoiding an incident and accident as well as preserve employee heath in the work place.

Indeed, looking at such organization, it's possible to see that to maintain a safety culture is always a challenge because depends on people who works on the organization. In fact, we need to take into account the different holes that different employees have in the organization considering safety culture. Basically, there are leaders and followers and both are important to establish and maintain the safety culture.

The leadership in safety is highlighted as one of the most important aspects to lead organization to achieve safety high performance. By the other way out, leadership by itself do not enable safety high performance if the other employees do not support their leader by acting preventively to avoid the incident and accident. Consequently, not only the leadership, but the whole team must realize the safety importance and act preventively daily to achieve high performance in safety.

It is important to state that a leader is someone who are able to influence a person's behavior into such organization. Therefore, in an Organization we can see formal and informal leaders. The formal leaders are the supervisors, managers, directors and CEO. The informal leader are, those are able to influence on a person's behavior, values and culture.

Despite it is clear the importance of leadership in rigorous safety culture, what it is seen in many cases is that different leaders into organization have different focus and target and not all of them promote the safety culture.

Indeed, in many organizations the highest safety performance is realized as a safety managers target, but that must be all managers target. Therefore, it is possible to understand in many cases why is it hard to achieve high safety performance by safety culture. In fact, in many organizations, the maintenance or production leaders do not take into account safety as their own responsibility and do not recognize as one of their objectives.

In order to solve this problem the leaders must be involved and promote a safety culture by dedicating time to discuss and promote a safety culture, invest the necessary resources and recognize safety professional effort as a strategic organizational target.

Remarkably, to discuss when concerning safety culture and leadership influence is that leaders are not alone.

Indeed, the leaders will not able to implement a safety culture and maintain over a long period time if the employees do not accept or cooperate by preventive attitudes.

By this point, come out a real discussion that is how to get employees cooperation. Indeed, once the employer has safety as a value, which means, preventive actions which drives to avoid incident and accident, such attitude must be reinforced by leaders. Such attitude might be reinforced by training, promotion and more than it support all preventive action daily and avoid punishment for mistakes.

In fact, if we think about the human nature it is clear that nobody has an intention to get hurt in the workplace. Thereby, arise the main question: what happens that makes people to perform unsafe actions or do not perform preventive actions?

In order to answer this question it is necessary a deeply understand about individual behavior as well as their motivations and social context in the workplace. In fact, it is quite impossible to explain why some employees that even with long experience, hard training and in some cases after many preventive actions committed unsafe behavior or do not perform preventive action that in some cases lead to an incident or accident.

Despite, that is such difficult answer, some tips are clear to understand that by practicable point of view.

Based on my experience on operational ground, what I see in many cases is that conflict target lead to unsafe actions like finish a maintenance task as fast as possible. In another case, do not perform a preventive action is related to the misperception of risk and how such preventive action can avoid an incident or accident.

The worse case is when employees are not encouraged to report any unsafe condition or human error to be afraid of punishment.

The last one is the reducing risk perception caused by the lack of incident or blind vision of high experiment professional does not suffer an accident.

Unfortunately, all such issues influence on safety performance and describe safety culture over time. Because of that, safety must be rigorous daily and a leader must support employees to carry out preventive action constantly.

Despite such effort, leader and employees change from time to time and such employees population change also take influence on safety culture.

The safety is a constant and hard challenge that must be seen as a real organizations strategic objective and promoted by all leaders' level, which will reinforce their employees' preventive actions constantly, no matter how is the loss of production, service deliver delay or investment required.

Safety culture is something that must be built up and maintain for all employees in different organizational levels, positions and holes into the organization.

9.2.2. Leadership

Whatever situation into organizations, all processes are influenced for leadership. A simple understanding of leadership is the capacity that an individual influence on others individual's attitutes. Therefore, leadership becomes a very important driver in Safety and Occupational Health Management because independent of hierarchy defined by the organization, many employees are able to influence on other behavior positively or negatively.

Considering the formal hierarchy, definitely such leaders, which have a manager function into the organization will influence on individual's behavior and it is necessary to understand the leadership styles in order to select the best leaders for managing Safety and Occupational Healthy Management. The Table **1** shows the different schools of leadership.

Table 1: Leadership Schools (Source: Bolden, R., Gosling, J., Marturano, A. And Dennison, P, 2003)

LEADERSHIP SCHOOLS	CHARACTERISTIC
Trait Theories - Great Woman or Man Theories	Based on the belief that leaders are exceptional people, born with innate equalities, destined to lead. The lists of traits or qualities associated with leadership exist in abundance and continue to be produced.
Behaviorist Theories	This approach concentrates on what leaders, in fact do rather than on their qualities.
Situational Leadership	This approach sees leadership as specific to the situation in which it is being exercised. For example, whilst some situations may require an autocratic style, others may need a more participative approach.
Contingency Theory	This is a refinement of the situational viewpoint and focuses on identifying the situational variables which best predict the most appropriate or effective leadership style to fit the particular circumstances.
Transactional Theory	This approach emphasizes the importance of the relationship between leader and followers, focusing on the mutual benefits derived from a form of contract through which the leader delivers such things as rewards or recognition in return for the commitment or loyalty of the followers.
Transformational Theory	The central concept here is change and the role of leadership in envisioning and implementing the transformation of organizational performance.

Table 2: Leadership Schools (Source: Bolden, 2003)

TRAIT	SKILL
•Adaptable to situations	•Clever (intelligent)
•Alert to social environment	•Conceptually skilled
•Ambitious and achievement-orientated	•Creative
•Assertive	•Diplomatic and tactful
•Cooperative	•Fluent in speaking
•Decisive	•Knowledgeable about group task
•Dependable	•Organized (administrative ability)
•Dominant (desire to influence others)	•Persuasive
•Energetic (high activity level)	•Socially skilled
•Persistent	•Technical knowledge
•Self-confident	
•Tolerant of stress	
•Willing to assume responsibility	

The **"Trait Theories"** are supported for many people in organizations, but in fact in real life is very hard to find, out someone into organizations with so many special traits and skills as shows Table **2**. In fact, what it has seen is different levels of such traits and skills in different employees.

All traits and skill listed in the table are important, but, concerning Safety and Occupational Management, the most important skill is "technical knowledge" and the most important traits are **"willing to assume responsability"** and **"adaptable to situations".** That's because the challenges faced by safety professionals in different situations, that can even change in some hours, have required always technical knowledge to go in the right direction as well as to be adaptable to situations and take responsibility for decisions. Mostly, Safety leaders must take a decision based on technical knowledge and need to know profoundly the technical tasks of the team members to support them and provide them the required resource. The big mistake associated with leadership in Safety Management is to choose a leader only by their trait and forget the "technical knowledge competence". Definitely, a Safety Manager must be a Safety Specialist no matter engineer or technician, and of course with some essential trait listed above like **"willing to assume responsibility"** and **"adaptable to situations"** that enable them to lead their team).

The **"Behaviorist Theories"** defends the idea that leadership can be learned and are based on leaders' behavior which focus on tasks and people. Concerning Safety Management, to perform the tasks is much more important than understand the individual which performs such task because the most important thing is to avoid an accident or a health damage. The main point here is understanding that the employees must be prepared to perform their task successfully and the leader must support them whenever is necessary. Indeed task and people focus on leadership are not contradictory, such issues are complementary.

The **"Situational Leadership"**defends the idea that leaders are able to be adapted accordly to each faced situation that they need to influence their team member. Concerning the Safety Manangement, most of leaders must focus on task and efficiency no matter how autocratic, democratic and laissez-faire is their approach. The difficulty to be adaptive is that the leader's personality defines the way that they pursued their participative leadership and such characteristic must fit on their team member. The big problem here is that team members are individuals and each one of them have their own characteristic which requires a certain level of participative decision support.

The **"Contingency Theory"** defends the idea that a leader with specific characteristics is a good leader in any specific situation or context but is not a good leader in different situation and context. Concerning the Safety Management, leader must focus on task efficiency as well as supporting the team members to perform such tasks successfully. As mentioned before, the technical knowledge is an essential skill in Safety Management. Therefore, a Safety Manager must have specific trait and skills which enable them to lead a team of specialist that is totally different context comparing with maintenance, operational and marketing department. In fact, each context requires a specific leader, a good human resource management leader will not be a good safety management leader and the other way round. Unfortunately, many organizations do not pay attention to the context and believe that their leader are able to lead different teams in different subjects under different context which cause a lot of problems.

The **"Transactional Theory"** support the idea that leaders is under total control of the work process by defining the work requirement and take influence on their team members by reward and punishment as stimulus. Concerning Safety Management, no matter the reward or punishment, the safety professional must be committed to fulfill their tasks successfully because it can avoid an accident or health damage. Such context must be taken into account for safety managers because keep their team members motivated to perform their job is totally different than keep than stimulated to fulfill their tasks by reward and punishment. In fact, rewards and punishment must be taken into account to stimulate employees, but it is necessary to have in mind that it is not always possible to reward all employees all the time. In addition, punishment does not solve all cases of human error and in many cases only makes employees to be afraid to perform specific tasks because they can make some mistake and be punished for that.

The **"Transformational Theory"** regards that team member will follow a leader who inspire them and such leader with passion and enthusiasm will be able to contaminate their followers by energy and enthusiasm. Concerning the Safety Management, the team members must be aware of their important tasks which avoid accident and health damage. Therefore, such team members must be motivated all time by themselves and not depends on specific leader to get enthusiasm to achieve high performance in their tasks.

Finally, we get into conclusion that independent of what organizations understand leadership and how they believe that such a leader must apply as trait and skill to lead different teams, concerning safety and occupational health management the most important skill is "technical knowledge" and the most important traits are **"willing to assume responsability"** and **"adaptable to situations"**. In addition, the leader must be adaptive to different situation and context and pursue their participative leadership depends on each team members necessity. By the other hand, team member must be motivated to perform their tasks no matter how enthusiastic is the leader. Indeed, what team member really needs is support to take decisions and have to the resources available whenever is necessary.

9.2.3. Organizational Learning

The Organizational learning is also an important factor in safety management because allow employees take place the bad practices by good ones in their routines as well as reinforce the preventive behavior that avoid the incident and accident. However, individual learning does not necessarily lead to organizational learning (Ikehara, 1999).

In order to understand the organizational learning it is necessary to first understand the individual learning process.

There are different theories which try to explain the individual learning process as shows the Table **3**.

Table 3: individual learning schools

TYPE	MODEL	FOCUS	MAIN LEARNING MODES
Behaviorism	Stimulus Response	Reinforcement	Experimental learning
Cognitive	Environmental and expectancy	Mental act	Rational learning
Social Cognitive	Stimulus organism response	Symbolizing Forethought Self-regulatory Self-reflective	Observational learning Enactive learning Self-efficacy
Gestalt	Pattern of wholes	Balance of cognitive, physical, emotional and spiritual factors.	Experimental learning

The behaviorism theory can be divided into:

- The Classical Conditioning Theory;
- The Operant Conditioning Theory;
- The Reciprocal Inhibition Theory;
- The Incubation Theory;

The Classical Conditioning Theory attributes, learning process to stimulus and response (Pavlov, 1927). Based on such theory the employees will have a safe behavior based on reward and punishment. That means, they will use the individual protect equipment in order to avoid punishment or will keep all safety documentations updated in order to get points for their promotions.

The Operant Conditioning Theory state that behavior is shaped and maintained by its consequences that is shaped by environment. Such theory is very similar to the previous one, but regards that different environment shapes, different consequence and consequently shapes different behavior. For example, whether we consider two similar Operational Management with different environments, we can expect different employee's behavior. Considering the previous example, when employees go to perform some service at **"Operational Management A"**, they always use their individual protections because the manager required that and punish all employees that do not do that. By the other way round, when such employee work on "**Operational Management B**", they do not use such individual protection frequently because they know the manager do not punish them.

The Reciprocal Inhibition Theory regards that two elements are involved in the inhibition of a response during extinction: reactive inhibition, which describes an inhibitory state dissipating with time, and negative conditioning, which leads to a permanent decrease in response probability (Wolpe, 1958). The theory behind this type of reciprocal inhibition theory is that, with sufficient repetition, the old, undesirable response can be unlearned, and a new behavioral pattern can be permanently established. A good example applied to safety is an employee which fall down from a scaffold because he was not using the personal protective equipments now is being trained to work on a scaffold with such equipment to learn that such workplace is safe whenever such personal protective equipments are being used.

The Incubation Theory observes that behavior followed by negative consequences is not eliminated, which cannot be explained by the Operant Conditioning Theory (Eysenck, 1976). In many cases, extinction does not fail to occur. In addition, there is an incremental enhancement effect, so the unreinforced conditioned stimulus may produce increases in anxiety. In the previous example, the operator that falls down from a scaffold, sees the other similar accident and he is now more afraid to work on scaffolds than before.

The Cognitive Theory recognizes learning through association between the environmental cues and the expectancy (Luthans, 1998). Applying to safety issues, a safe behavior, learning occurs when certain cognitive cues associated with the choice actions may eventually lead to a goal or a reward, like high safety management performance achievement. In this case, once managers have safety performance goals an they will be assessed for such performance they will effort actions to achieve such goals.

The Social Cognitive Theory integrates both social and cognitive processes to understand motivation, emotions and action. The social cognitive process occurs not only for assessing individual rewards or consequences, but also by observing other people's behavior and consequences. Considering the previous example, once the managers realize that other managers are not rewarded they may not feel motivated to perform such effort to achieve safety management goals because such effort will not reward as expected.

The basis of Gestalt theory is "that" human nature is organized into patterns or wholes, that it is experienced by the individual in these terms, and that it can only be understood as a function of the patterns of wholes of which it is made (Perls, 1973). The control room operator sets up a wrong valve and opened it because the numerical sequence of such control induced him to a wrong sequence interpretation. The consequence was a toxic product loading in a Tank under maintenance. There was not employees' injuries because the operators was not in the tank at this moment. Indeed, the panel controls design must take into account the gestalt theory which explain the tendency that human have to complete and generalize their visions of pictures.

The organizational learning is a social learning and take place by observing other employees' behavior when the results of such behavior is something wanted for who observes. In addition, the context and organizational environment are also observed in such learning process (Ginter, 1982).

The organizational learning takes place also by the own employee experience. Indeed the employee behavior is reinforced when such action is promoted and supported by leader or achieve some employee benefits like promotion, professional acknowledgment. By the other hands, the employee behavior is avoided when that behavior has bad consequences like punishment, bad reputation, bad workplace social acceptance.

Therefore, based on such experiences, the employees broad over their mind, start to define concepts and generalized them and finally test such concepts in different situations as shows Fig. **2**.

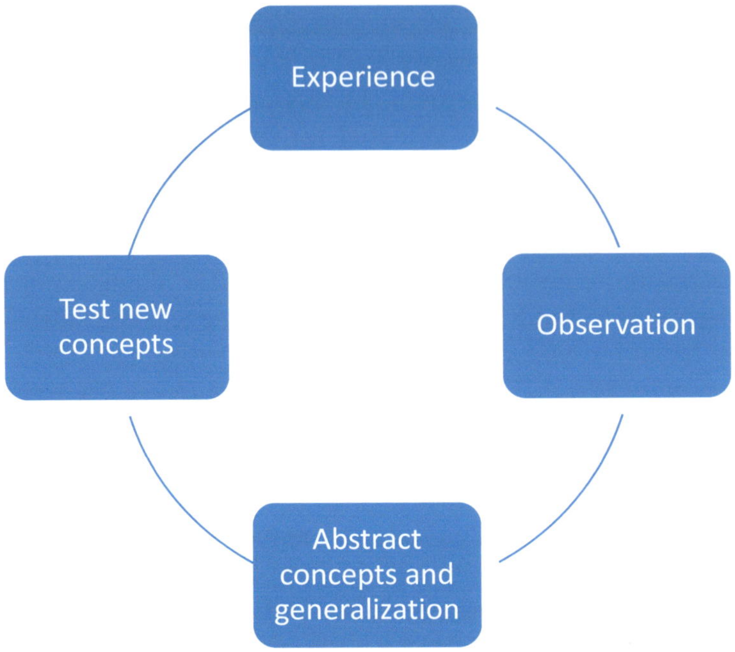

Figure 2: Organizational learning process (Source: Rubin & Irvin,1974).

Despite the complexity of individual learning, the organizational learning is much more complicated process to understand and apply in organizations. Indeed, it considers not only the capacity of each employee to learn, but also a systematic method to retain and spread out such knowledge all over the organization.

Similar to the previous case, the organizational learning has different approaches that consider process learning, culture, knowledge management, continuous improvement and innovative and creative.

The Table **4** summarizes the organizational learning philosophies. The first one is those theories are **"Individual learning"** which consider that the organizational learning depends on each individual learning into the organization as stressed in different example previously.

The second theory is **"Process or system"** which regards how organizations manage their experience (Glynn, 1992). In terms of Safety and Health Management, many organizations implement the lesson learning based on audit process as well as other assessments in order to avoid the previous mistakes as well spread out as the best practices that enables to achieve high performance.

Table 4: Organizational learning types (Source:Wang, 2002)

TYPE	THE CONCEPT OF ORGANISATIONAL LEARNING	PRACTICES
Individual learning	"Organisational learning occurs by individuals".	Staff training & development
Process or system	"Organizations understand and manage their experiences".	Enhancement of information processing and problem solving capability
Culture or metaphor	"A learning organization should be viewed as a metaphor rather than a distinct type of structure".	Creation and maintenance of learning culture
Knowledge management	"Involves knowledge acquisition, dissemination, refinement, creation and implementation".	Facilitation of interaction and strengthening of knowledge base
Continuous improvement	"A learning organization should consciously and intentionally devote to the facilitation of individual learning in favor of the whole organization".	The adoption of TQM practices
Innovation and creativity	"In the hyperdynamic business context, organizational learning is the process by which the organization constantly questions existing product, process and system".	Facilitation of knowledge creation; focus on creativity quality and value innovation

The next theory is **"Culture or Metaphor"** which regards employees learning as a conscious communal process for continuously generating, retaining and improving individual and collective learning to improve performance of the organizational system. Considering the lessons learning process different management in the same organization can learn the best practices in order to achieve high performance in their Safety And Health Management System.

The **"Knowledge Management"** regards knowledge acquisition, dissemination, refinement, creation and implementation. In this way, organizations develop a process of training people in order to spread out such knowledge and apply it. Considering the Safety methods applied to avoid incident and accident as an instance, the behavior audit, there will be an established and define process to give such knowledge for employees in order to apply it in order to avoid and prevent incidents. The point is that some activities are based on skills and in this case, it is not possible to write down in procedures for example

The **"Continuous Improvement"** concerns that organizational learning should consciously and intentionally devote to the facilitation of individual learning in order to continuously transform the entire organization and its context (Pedler, 1991). Such context is too linked to give resource and encourage people to bring knowledge and continuous learning through the organization. Once again, many safety methods applied in organizations to improve safety performance are based on Knowledge and skills that depends on person's involvement, perception and skills to carry out such task like combat fire, bring down an equipment to a safe state like a furnace. That shows how big is the organizational learning process.

The **"Inovative and Creativity"** understand organizational learning as strategic process which the organization constantly questions existing product, process and system, identify strategic position. Thereby, the precise learning process is established in order to achieve sustained competitive advantage. Therefore, Safety and Occupational Health management is linked to organizational strategy by defining safety goals and indexes in order to achieve sustainable performance based on Knowledge and skills. In fact, in some specific situations that depends not only of people involvement and perceptions but also skills to carry out such task related to safety like combat fire, bring down an equipment to a safe state like a furnace. That shows how big challenge is the organizational learning process.

9.3. SAFETY ECONOMIC VALUATION

The Safety economic valuation fundamentals are also an important tool to support the Safety and Occupational Management because helps to understand the required investment to mitigate the risk and avoid accidents and health damages.

Despite an important issue to be discussed in safety management, it is not usually applied to many organizations as a system management tool to support the decision process related to risk mitigation.

Indeed, the first remarkable application of economic analysis in the accident was in an Exxon Valdez Accident in EUA. In order to evaluate the environmental cost of such accident, the environmental economic valuation was except for EUA congress as the best methodology to be applied to define the economic value in local activities losses and environment resources losses. Such methodology can also be applied to provide an economic safety valuation of preventive actions and layers of protection.

The basic concept of such methodology is that the social welfare has an economic value. Therefore, whenever there is a loss in social welfare, it must be compensated. In fact, there's a good risk level associated with health which this welfare is not disturbed, because the social cost compensate the social welfare. Below some risk level, considered by the specific safety level, there's welfare lost and it's must be compensated.

Therefore, it's important to define the safety valuation taking into account different aspect like direct and indirect social cost. The safety valuation regards ***"use value"*** and ***"not use value"***. The first one comprises direct use value, indirect use value and future option of use value. Not used value comprises only existence value.

The ***"direct use value"*** associates safety with risk or any loss of welfare caused by the accident or health damage like medical assistance cost for instance.

The ***"indirect value"*** is associated with some aspect of welfare that you can lose indirectly like loss of leisure time caused by an accident.

The **"future option value use"** is associated with some welfare that is not being used in the present, but it may be used in the next future like your free leisure time when you'll be retired. The "not used value" comprises existence value and it's associated with some moral issue.

In order to calculate each specific value is necessary some methodologies as Function Production method and Function Demand method.

The first one, Function Production Method, regards that a specific welfare has a direct relationship with safety and if that safety is not available the welfare value change, so it's possible to estimate the safety economic value.

The second method, the Function Demand Method, regards the relationship between safety and the welfare that can be understood as quality of life and health. Thus, the welfare economic value depends on the quality of life and health. Considering health, the economic value depends on willing to receive of each employee whom is affected by a specific hazard in the workplace. Indeed, employees are willing to receive a higher salary or a monetary compensation to work in a place with such hazards. In many cases, the WTA (willing to accept) is more applied than WTP (willing to pay), because the last one is limited by the individual's economic resources like salary. In order to predict the WTA value it is necessary to interview people to know the economic value they willing to accept to face a specific risk in the workplace. In a real WTA calculation is necessary to interview many people to define a WTA function. Indeed, that is the hardest task in safety valuation because there are many issues involved that influence on the reliable WTA valuation like the probe selected to such research, the interview influence and the clarification of questions.

Indeed, the Value of Statistical life (VOSL) estimates the existence or contingent value, which concerns the WTA (willing to accept) or WTP (willing to pay) as well as a discount index on time as demonstrated on equation below.

$$VOSL = \frac{\sum_0^i WTA_i \Delta r_i}{\sum_0^i \Delta r_i}$$

Where:

WTA = Willing to accept

r = risk

The VOSL calculation is a complex method and it is not the main objective here to describe the whole process but highlight its importance. The Table **5** shows examples of VOSL calculation performed by different researchers (Pearce, 2000).

Table 5: VOSL Studies (Source: Pearce, 2000)

Study	Original studies	VOSL
Van den Bergh *et al.* 1997	10 US and 1 UK wage-risk studies	\$3.86 million ('most reliable estimate)
Desvousges *et al*, 1998	28 wage-risk and 1 CVM study, USA	\$3.6 million, with confidence interval \$0.4 to 6.8 million
Day, 1999	16 wage-risk studies = 10 USA, 2 Canada, 4 UK	Best estimate of \$5.63 million

In addition to estimating the VOSL, is also possible in estimating the VOLY (Value of life in years) which regard the annuity which is discounted over the remaining life span of the individual at risk would equal the estimate of VOSL. Thus, the VOLY is calculated by equation:

$$VOLY = \frac{VOSL}{A}$$

where:

$$A = \frac{[1 - (1 + r)^{-n}]}{r}$$

An example to demonstrate the VOSL estimation is related to traffic accidents. In this case, considering that £1.5 million relates to traffic accidents where the mean age of those involved in fatal accidents is such that the average remaining life expectancy would have been 40 years and "n" is years of expected life remaining and "r" is the utility discount rate. The Table **6** shows different values of VOLY for n = 40 years (Pearce, 2000).

Table 6: VOLY Studies (Source: Pearce, 2000)

	n = 40 years					
	r = 0.3%		r = 1%		r = 1.5%	
VOSL	A	VOLY	A	VOLY	A	VOLY
1000	37.600	26.60	32.800	30.49	29.9	33.44
1500	37.600	39.89	32.800	45.73	29.9	50.17
2000	37.600	53.19	32.800	60.98	29.9	66.89
3000	37.600	79.79	32.800	91.46	29.9	100.33

The other model applied to safety economics valuation is the cost/benefit analysis (CBA) which in safety and occupational health management supports economic evaluations of the measures of risk treatment, was expected total costs are weighed against the total expected benefits in order to choose the best option.

On of important aspect when performing the CBA analysis is to identify the stakeholders that can have costs or receive benefits. A global CBA includes all the stakeholders. The direct and indirect benefits and costs to all relevant stakeholders on the options being considered are identified.

In a CBA in safety and occupational health model concerns also both tangible and intangible costs and benefits should be considered, as well as direct and indirect costs and benefits.

Inputs for CBA can include information on costs and benefits to relevant stakeholders and on uncertainties in those costs and benefits. Tangible and intangible costs and benefits should be considered. Costs include the resources expended and negative outcomes, benefits include positive outcomes, negative outcomes avoided and resources saved (Ramos, 2012).

The Direct benefits are those that flow directly from the action taken, while indirect benefits are those that are coincidental but might still contribute significantly to the decision. Examples of Indirect benefits include reputation improvement, staff satisfaction and "peace of mind", factors which are often weighted heavily in decision-making.

The Direct costs are those that are directly associated with the action. The Indirect costs are those additional, ancillaries and sunk costs, such as loss of the asset cause of an accident. When applying a cost benefit analysis to a decision on whether to treat a risk, costs and benefits associated with treating the risk, and by taking the risk, should be included (Ramos, 2012).

The Table **7** shows the social costs and benefits of two measure packages that were analyzed in Elvik's Norwegian study (Elvik, 2007). Before the packages were compiled, the costs and benefits of the separate measures were determined. The package 'Optimal use of road safety measures' consists of measures of which the individual benefits were estimated to be higher than the individual costs. The package 'Strengthening present policy' is an intensified continuation of measures that are already being taken in Norway. These two project alternatives were compared with the null alternative in which these measures are not applied. The period under consideration is 2009-2020.

Table 7: Cost benefit analysis (Source: Elvik, 2007)

	Options	
	Optimal use of road safety measures	**Strengthening present policy**
Benefits		
Road safety	10.042	8.471
Travel time	-816	1.591
Transport costs	184	-240
Environment	121	-17
Public health	66	80
Increase of mobility	8	70
Total benefits	9.604	9.953
Cost	6.472	11.042
Profitability		
Benefits -cost balance	3.132	-1.088
Benefits-cost ratio	1,48	0,9

The CBA analysis aggregates the monetary value of all costs and all benefits to all stakeholders that are included in the scope, adjusted for the different time periods in which costs and benefits occurs. Thus, the decisions to be made can be based on the analysis of the Net Present Value (NPV), considering the internal rate of return (IRR) or the ratio between the present value of benefits and present value of costs.

The Net Present Value (NPV) which is produced by a CBA analysis becomes an input into decisions about risk. A positive NPV associated with an action would normally mean the action should occur.

The Table **8** shows an example of NPV application applied to Safer Island Program in Maldives. The Maldives is a small island nation comprised of coral atolls, located in the central Indian Ocean. While it is comprised of over 1,000 islands, approximately 200 of these are inhabited. The real island vulnerability is that 80 percent of its islands are less than 1 meter above sea level. The Maldives is considered one of the country's most vulnerable to the predicted consequences of global climate change. The expected increase in sea level rise and in the intensity of extreme weather events and the seriousness of their adverse consequences has necessitated the Maldives to consider climate change and disaster management in all aspects of its future development. While a tsunami of the magnitude experienced in December 2004 is extremely rare, the event heightened awareness of the vulnerability of the Maldives islands as it provided a 'snap shot' of a potential future dominated by sea level rise.

Table 8: Net Present Value Analysis (Source: UNDP Maldives, 2009)

Severe Tsunami in Maldive Island	
Limited Safe Island Protection	**Value cost**
Coastal Protection	$504,000.00
EPZ around Island	$420,000.00
Resilient Habour	$2,238,730.38
Houses & buidings retorfitted	$450,000.00
Drainage in rainfall flood prone areas	$14,169.96
Total Investment	**$3,626,900.34**
Estimated losses	**Value cost**
Severe Tsunami	$3,780,200.40
Discount rate	7.50%
Project Lifetime	50 years
Net Present Value	$1,411,753.14

In a cost/benefit analysis, when all tangible and intangible costs and benefits have been identified, a monetary value is assigned to all costs and benefits. Indeed, a cost-benefit analysis should permit to answer to the following questions:

What OHS investments should be made?

How much should be spent on preventive measures?

When should we make a given investment?

What business value can we expect from a prevention investment?

Even performing such methods explained above it is always difficult to precisely how much accuracy is such economic value, but in fact, the main objective is to obtain an economic value dimension to support decisions. Despite all limitations, take into account safety, the economic values of the decision are always better than do not consider such values.

9.4. SAFETY AND OCCUPATIONAL HEALTH MANAGEMENT

The Safety Management Clinic approach is based on Management Clinic theory proposed by Neto, Cerqueira (Cerqueira, 2003). The Management Clinic treats organizations like patients when compared to a medical clinic, where, depends on the disease, the patient need to go to a specific clinic to have treatment. In case of Organizations, many of them have specific safety performance problems that are related to low performance in safety external environment, safety internal environment, safety strategic thought, safety management, safety routine or safety improvement actions. Thereby the organizational leader must to have a future vision and resources to lead the organization to higher performance by going though the different clinic buildings and implement the action and resource support leaders and teams. The Fig. **3** shows the Safety Management Clinic elements and their interactions. Each one of those six elements is a clinic building where the organization has health problems and must be treated.

In order to go though the management Clinic, the organizational leader must have future vision to lead the whole organization to the objective target in future as well as resources to invest in teams and leader to enable them to achieve their target. In case of Safety and Occupational Health management, the future vision might be benchmarked and safety in their industry for example.

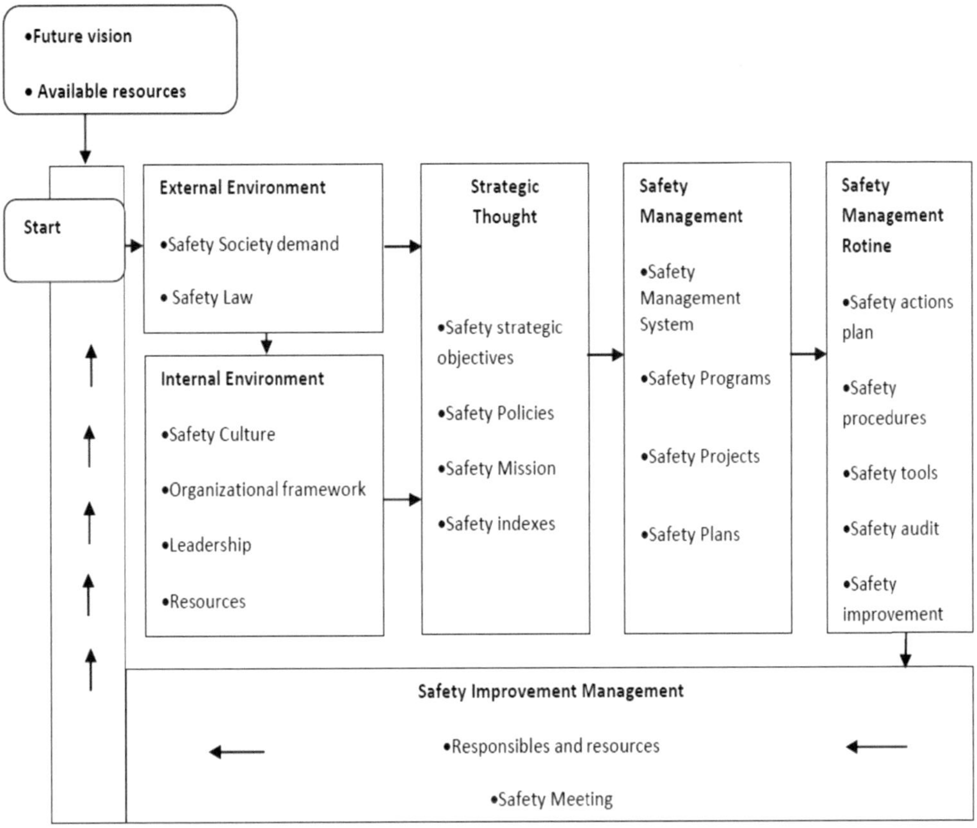

Figure 3: Safety Management Clinic (Source: Adapted from Cerqueira Neto, 2003).

The most important about it is that the main leader regards the internal and external organizational environment to be realistic and establish feasible objectives.

The external environment comprises all safety laws and society requirement that is defined by the low number of incident and accident, acceptable risk and no environment impact cause by an accident. The internal environment concerns the safety culture, organizational framework, leadership and resources as explained in previous item. The external environment change over time and that is out of organizations control. Thereby, it is necessary that the organization be flexible and adaptive to external environment and requires that leader have a good external environment perception to adjust his future vision and target and invest their economic resources in order to achieve high performance in safety. Therefore, it is always to have in mind that it is necessary to apply the best approaches and methods to achieve high performance in safety and occupational health because the law became more restricted and the society's tolerance for accident and environmental impacts as well.

Indeed, the main challenge is to maintain a favorable internal environment over time because employees change as well as leader, which requires a constant train about safety and occupational health.

The internal environment is also important and must be taken into account when the future vision as well as objectives and indexes are established. The organizational environment concerns the safety culture, organizational framework, the leadership and resources as discussed previously. Indeed, the main challenge as well as an external organizational case is to face the internal changes over time. In fact, the leaders change, the employee changes as well as the resources available to invest in an organization. In order to face such challenge the main leader must to have in mind such changes and make the organizations flexible enough to be adapted to such changes.

The next step in Management clinic is the strategic though, where the main leader will define the safety strategy over a long period of time and break it in objectives and indexes for middle and short time. The strategy is part of future vision and is defined by strategic objectives for example:

- Minimize the incident and accident occurrence;
- Mitigate all risk to lower than tolerable level;
- Mitigate all risk to lower than tolerable level;

Indeed, such strategic objective must be defined in terms of tatic and operational objectives as well as indexes to follow up the progress of action that enables such objectives achievement. Therefore, the Safety management that is the next builds of management clinic will provide such information detailed in projects, programs and action plans. In order to establish feasible targets it is important to apply process risk analysis to understand clearly the level of risk of each activity and task, which risk must be mitigated and which procedures must be taken place to maintain the risk in acceptable level as well as identify new potential risk.

Indeed, the process to identify hazards, risk assessment, risk evaluation, risk mitigation, risk communication and documentation is part of Risk management. Depends on the situation, the Risk Management is a program or even a

Management system into the Safety and Occupational and Health Management. Depends on process complexity and risk involved the particular Risk management program or system is required. Such programs are very important in project phase when the risk can be mitigated by design options and equipment and product improvement. When the project phase finishes all recommendations and risk will be part of safety and occupational health management.

The further step is to establish a safety management routine that comprises all safety tools like

- Safety Dialog;
- Safety Occurrence Reports;
- Preliminary Hazard Analysis;
- Behavioral Audit;
- Safety Process Risk Analysis.

In addition, it is necessary to establish action plans for improving safety performance and achieve the target in short, middle and long terms.

The safety procedures must also be established in order to allowance of continuous high performance in safety based on guidelines. The importance of procedures is to record the best practices and knowledge and improve over time in order to establish a routine and guideline for employees achieve the safety objectives.

This last Management Clinic build is the Safety Manangment Improvement which the main objective is to perform improvement based on the evidences of a safety audit, safety occurrence report, incident reports, accident reports, risk analysis recommendations or project requirements.

The main organizational leader must to be in mind that in some cases, some improvement actions will be required to adapt the organization to the new external requirement and that must be clear for all leaders in different organizational levels.

9.4.1. Safety and Occupational Health Management Case Study

In order to demonstrate the application of Safety And Occupational Health Management (SOHS) the implementation of SOHS applied to mining industry will be presented. Such Safety and Occupational and Health Management was a big challenge due to the number of people involved as well as the process complexity. The organizational framework is defined in Fig. **4** which shows the organizational framework of the General Operational Management that is comprised for another four operational management, and specialized support teams like security, Medicine at work and human resource.

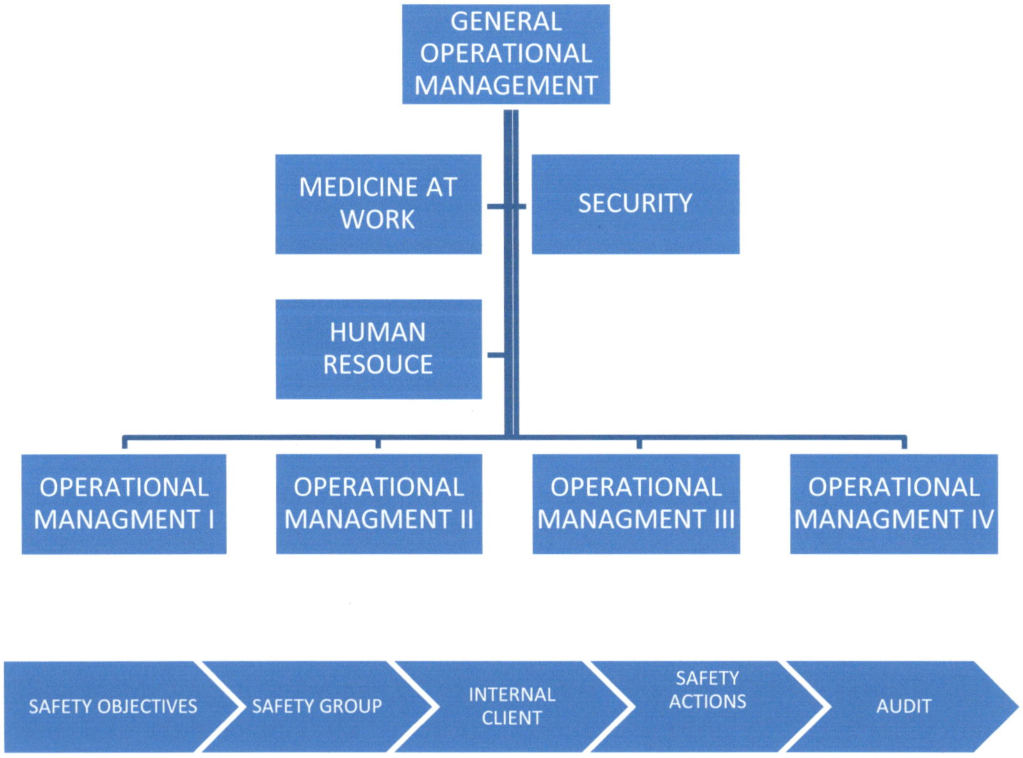

Figure 4: SOHS IMPLEMENTATION.

The SOHS implementation was driven by a Safety Group that had representatives from each Operational Management which are specialist in safety, occupational and health management. This group is sponsored by the main leader, the general manager, which support the Safety group with resource and define the objectives, target and indexes as well as approve the improvement actions in order to achieve the safety objectives.

The leader has a future vision:

"To become a benchmark in in safety performance in mining industry"

In order to achieve such objective the leader got into conclusion that was necessary to implement the Safety, Occupational and Health Management System and the main objectives were:

- Minimize the incident and accident occurrence;
- Mitigate all risk to lower than tolerable level;
- Anticipate the incident and accident;

Such objectives came out of future vision as well as the external requirement to improve performance on safety. The external parties, as well as government and society were not satisfied with this organization safety performance mainly because the increasing number of incident and fatal accident.

The other important issues to be taken into account was the internal organizational environment that includes safety culture, organizational framework, leadership and resource.

Based on such objective the main indexes target was stable like:

- Safety System implementation, which is calculated by safety tools correct usage measure by the safety audit in term of percentage;
- Percentage of unsafe condition reported are solved;
- Percentage of mitigation action implemented on time;
- Percentage of behavioral audit implemented on time;
- Number of incidents;
- Number of accidents;
- Percentage of law requirement document delivered on time.

Most of Safety indexes are preventive in order to avoid an incident and accident which require a preventive safety culture more than everything.

In this case, the safety culture was a big challenge because there was not an established preventive safety culture so far. The other issue is that the production culture was predominant, that means, the production is more important than safety and environment. Such predominant culture would be detected when some maintenance used to be carried out and safety was stepping aside to perform such maintenance fast to make the operational plant available.

Therefore, the main internal challenge was to develop a safety culture internally by change the reactive behavior for a preventive behavior which start for the leader examples in their daily decisions.

By the other hands, the leader had resource to invest in training, technology and additional resources to implement the SOHS. In addition, the leader defines the Safety Group tat is comprised of leaders, which were responsible to implement the SOHS in their operational management. In this case, those leaders work together in such SOHS implementation and define the same agenda and action to be implemented in each operational management. That enable to know the best practice and common problem and enable a faster implementation.

Indeed, to implement the SOHS the Safety group faced a huge challenge because was necessary to perform massive training for around 1200 people in a couple of eight month as well as apply all safety tools, define all safety procedures, audit, propose and implement improvement actions.

In addition, there were more five companies that perform specialized internal operational service like cleaning and maintenance that was included in SOHS implementation.

The aftermath, the next step was to go to the third Clinic Management that is the Strategic Though. Once the Future vision and strategic objectives was defined, it is also important to define the, mission and safety policy. Indeed the organization already had the quality and environment policy.

The safety mission is defined as:

"To produce pellets with operational excellence concerning quality environment preservation and employes safety and occupational health".

The next step is to define the SOHS policy that is defined as:

⇒ To apply the best practices when producing pellets concerning to preventive actions to avoid quality non-compliance, environment impact, incident and accident;

⇒ Safety and Occupational has been always prioritized;

⇒ Attend the Safety and Occupational laws and terms;

⇒ Establish communication with employees, union, government, client and society to discuss issues related to safety and occupational health;

⇒ Establish the improvement process to achieve always high performance in safety and Occupational and health performance;

⇒ Trainee and aware employees to achieve high performance in safety and occupational health;

The next step is to establish the SOHS indexes as well as definite projects, programs and action plan. In order to establish such actions it is necessary to be clear about the risk involved as well as the effort necessary to implement all safety management systems based on SOHS tools.

The first SOHS tool is the process risk analysis. Such process analysis is a qualitative risk analysis, which concerns the risk in all activities and task for process and maintenance. Similar to the preliminary risk unless there a risk matrix that defines the occurrence and severity for each hazard as shows Fig. **5**.

Such process risk analysis was carried out in the whole process in order to mitigate and control the risk. As the results of such process risk analysis many actions had to be taken place

The process risk analysis sheet is a guideline for operators and they are more aware about such risk they face daily in their activities. Indeed, for different processes the process risk analysis sheet was delivered to operators and is on their operating room to consultancy. All operators were trained in order to be aware about such risk in order to avoid an incident and accident. In order to support the Safety Occupational and Health Management, other safety tools were implemented when SOHM was implemented.

		Occupational Risk Analysis								Local Pellet Plant		Area Operational
Process: Pelletizing						Acessed by: Eduardo Calixto	Aproved by:					Review
		Identification										
PROCESS	Activity	Task	Safety and Health	Rotine	Not Rotine	Consequence	Frequency	Probability	Severity	Risk	Control measure	Preventive Action
1-Pelletizing	Pelletizing disc operation	Monitor, control and operate disc	Arm or hand trapped			Death or loss of body part	1		3	3	Check the disk operation condition before any intervention take place	Stop the disc and certify that is locked in safe position whenever physical contact will be necessary
			Noise exposure			Loss or disturbance or audition capacity	2		2	4	Monitore noise as well as the use of ear protection.	Use ear protection in pelletizing operational area
			Dust exposure			Damage to respiratory capacity	2		2	4	Monitoring the use of mask as well as the pelletizing disc particules emission	Use mask protection in pelletizing operational area
			Phisic Ergonomic factor			Muskolosesquel etal disorder	2		2	4	Check ergonomic conditions on operator room regarding operator phisical and	Ajust ergonomic conditions to operator phisical and biodinamic conditions

Figure 5: Process Risk Analysis.

The other usual tool is the **"Safaty Dialog"** which the main objective is to promote discussion about safety issues daily and increase the employees' awareness. In order to increase the participation each employee is encouraged to present a specific topic for discussion. Thereby, different topics are proposed like ergonomic, safety laws, safety procedures and even incidents and accident. Mostly, in operational management, such safety dialog is carried out daily.

The other SOHM tool is the **"Safety Occurrence Report"** which the main objective is to report whatever unsafe condition detected by any employee. Such report has made written the unsafe condition in a small paper file and such file is stocked in a specific box. Some employees are responsible to implement some action during 24, 48 or 72 hours depending on risk of unsafe condition related. Such actions must be reported to the employee which open the occurrence and such preventive actions are managed and assessed by managers.

On the first four months a lot of occurrence were related and a huge implementation plan was prepared to eliminate such unsafe condition. Indeed, some action required investment like change pipelines or small civil rebuilding. Even though, all of them took place and were able to avoid accidents. Despite the preventive effort, many of such “Safety Ocurrence Report (was a result of the incident and some cases of accidents)”.

The **“Preliminary Hazard Analysis”** is another safety preventive tool which is applied whenever some not routine task or specific task with high risk is performed. The Preliminary Hazard Analysis has the main objective to aware the executor about the hazard involved in their tasks and avoid unsafe conditions and action. Therefore, it is necessary to implement all recommendations defined by PHA before performing such tasks.

The **“Behavioral Audit”** is another preventive tool which the main objective is to identify safety gaps when employees performing these tasks like lack of procedure, not use equipment protection or even perform any unsafe actions and create unsafe conditions. The remarkable characters of the Behavioral Audit is that such audit was supported by a Safety volunteer group called **“CONSEG”.** Such Group has the main objective to promote safety discussion among employees and come out with solution to implement safety improvement. Indeed, this group was very effective because had employees in different management, but mainly operational and maintenance management. One of the most contributions was to support all Programmed Maintenance in order to avoid accidents.

The **“Safety Inpection”** was a more informal tool that must be performed periodically by all leaders' level in one specific operational plant. The Director performs such audit once per year, the general manager each three months, the Operational and Maintenance manager each month and the supervisor each week. Moreover, each issue related in such audit has to be treated and improved in order to avoid incident and accidents.

The additional safety tool is the **“Work Refuse Law”** which is basically the option that all employees have to refuse to execute a specific task if they consider that safety condition are not enough. In the beginning, such tool brought a lot of discussion, but after some month was complete understood for all leaders and employees. In fact, whenever such Work refuse Law is applied it is necessary to write a report about the task signed up by the team leader and the employee an investigation must be performed by a group with employee representatives.

The **“Special Work Permission”** is a safety tool which the main objective is to prevent an incident and accident whenever a dangerous activity is carried out. Thereby, it is necessary to perform a preliminary hazard analysis and in addition to have a reported signed up by the team leader. Depends on specific case more assessment is required like "Activities in High place which requires a special form to fill out".

The **"Safety Board"** has the main objective to assess the audit results as well as SOHS performance in order to implement improvement actions. The Such Safety Board must present evidence of their periodic meeting and the improvement action plans. The safety board is comprised of the main organizational leaders.

The final safety tool is the **"audit process"** which has the main objective **is to evaluate the** effectiveness of safety tools implementation which is measured based on punctuations which regards different criterion like availability, quality, accuracy and consistency. The availability measure how much the safety procedure is available to be consulted. The quality measure how well done the safety tool is performed. The accuracy measure how good the report is filled out and the consistence measure how much the safety tool is implemented in real cases.

The Table **9** shows a Preliminary Hazard Analysis audit criterion as an example.

Table 9: Preliminary Hazard Analysis evaluation

Document	Criterion	Item	Punctuation	Remark
Preliminary Hazard Analysis (PHA)	Availability	1.1.1	0	The procedure is not available in paper or electronic version in any place.
			1	The procedure is available in paper or electronic version in only one place.
			2	The procedure is available in paper or electronic version in more than 70% of places.
			3	The procedure is available in paper or electronic version in all places.
	Quality	1.2.1	0	The PHA do not define the tasks and the mitigation action properly.
			1	The PHA define 50% or less of tasks and the mitigation action properly.
			2	The PHA define 70% or more of the tasks and the mitigation action properly.
			3	The PHA define 100% of tasks and the mitigation action properly.
	Accuracy	1.3.1	0	Less than 50%t of blanks are not filled out correctly.
			1	50% of blanks are not filled out correctly.
			2	70% of blanks ore more are filled out correctly.
			3	All blanks are filled out correctly.
	Consistence	1.4.1	0	No evidence of PHA applied in required activities.
			1	50% or less of evidence of PHA applied in required activities.
			2	70% or more of evidence of PHA applied in required activities.
			3	100% of the evidence of PHA applied in required activities.

The Table **10** shows how the audit process makes the assessment for each safety tool considering the criteria stated previously as well as evidence presented during the audit process. Indeed, each criteria has different weight and the score calculation is done by multiplying the score by weights and them sum the four partial scores having a total score for each safety tool. The next step is to calculate the "Safety, Occupational and Health Management effectiveness" by dividing the total achieved scores by the maximum achieved scores.

Table 10: Audit process score table

Safety, Occupational and Health Management System Evaluation						
Item	Safety Tool	Criterion	Score	Weight	Partial Score	Total Score
1	Preliminary Hazard Analysis	Availability		1	0	0
		Quality		3	0	
		Accuracy		2	0	
		Consistence		3	0	
2	Safety Inspection	Availability		1	0	0
		Quality		3	0	
		Accuracy		2	0	
		Consistence		3	0	
3	Special Work Permission	Availability		1	0	0
		Quality		3	0	
		Accuracy		2	0	
		Consistence		3	0	
4	Work Refuse Law	Availability		1	0	0
		Quality		3	0	
		Accuracy		2	0	
		Consistence		3	0	
5	Safety Report	Availability		1	0	0
		Quality		3	0	
		Accuracy		2	0	
		Consistence		3	0	
6	Safety Dialog	Availability		1	0	0
		Quality		3	0	
		Accuracy		2	0	
		Consistence		3	0	
7	Safety Board Committee	Availability		1	0	0
		Quality		3	0	
		Accuracy		2	0	
		Consistence		3	0	
Maximum Scores						189
Achieved Scores						
Safety, occupational and Health Management Effectiveness						

The first internal audit was performed to test the operational teams in order to check the effectiveness of safety tools implementations and understanding about the Safety and Occupational Management System. An example of audit result was the operational team C which achieved 76% of performance. Indeed, the minimum performance target was 80%. In this case, the main issues were discussed in order to promote the necessary improvement in this team as well as with other teams. Based on first audit process result the main problem was the "Preliminary Hazard Analyis" and "Safety Work Permission". Taking the Team C as an example, it is possible to see the punctuation losses on Table **11**.

Table 11: Socres losses in audit process.

Loss of Punctuation in safety audit process								
Criterion	**Preliminary Hazard Analysis**	**Safety Inspection**	**Special Work Permission**	**Safety Report**	**Refuse Work Law**	**Safety Dialog**	**Total Score Losses**	**%**
Availability	x	x	x	x	x	x	**x**	
Quality	6	x	9	6	x	x	**21**	**23%**
Accuracy	8	4	x	x	x	4	**16**	18%
Consistence	12	6	18	6	6	6	**54**	**59%**
Total Score losses	**26**	**10**	**27**	**12**	**6**	**10**	**91**	100%
%	**29%**	11%	**30%**	13%	7%	11%		

The table above shows concerning 24% of score losses during the audit process we can say that 23% are related to quality criterion and 59% is related to consistence. That means the tools are not applied correctly and there are not enough evidence of real implementation. The tools which have lower performance where the Preliminary analysis with 29% of score losses and the Special work permit with 30% of scores losses. Indeed, on the first case, the Preliminary Hazard Analysis was not well implemented cause by lack of preventive culture that is reflected in the attitude to carry all tasks as fast as possible. At the beginning, the preliminary Hazard Analysis considered a waste of time. In the second case, the Special Work Permission was also a big challenge to be implemented due the additional analysis required to have such reported signed up for team's leader and because of lack or preventive culture, such safety tools have a higher loss of scores at the beginning of Safety and Health Management System implementation.

After this first internal audit improvement action was established by the Safety Board like additional training, specialist support in tools implementation as well as additional internal audit to measure the improvements. By this time, on the second audit process the Team C achieved 88% of Safety system performance as showing the Table **12**.

Table 12: Audit process score table.

Safety, Occupational and Health Management System Evaluation						
Item	Safety Tool	Criterion	Score	Weight	Partial Score	Total Score
1	Preliminary Hazard Analysis	Availability	3	1	3	19
		Quality	2	3	6	
		Accuracy	2	2	4	
		Consistence	2	3	6	
2	Safety Inspection	Availability	3	1	3	21
		Quality	2	3	6	
		Accuracy	3	2	6	
		Consistence	2	3	6	
3	Special Work Permission	Availability	2	1	2	21
		Quality	2	3	6	
		Accuracy	2	2	4	
		Consistence	3	3	9	
4	Work Refuse Law	Availability	3	1	3	25
		Quality	3	3	9	
		Accuracy	2	2	4	
		Consistence	3	3	9	
5	Safety Report	Availability	3	1	3	27
		Quality	3	3	9	
		Accuracy	3	2	6	
		Consistence	3	3	9	
6	Safety Dialog	Availability	3	1	3	27
		Quality	3	3	9	
		Accuracy	3	2	6	
		Consistence	3	3	9	
7	Safety Board Committee	Availability	3	1	3	27
		Quality	3	3	9	
		Accuracy	3	2	6	
		Consistence	3	3	9	
Maximum Scores						189
Achieved Scores						167
Safety, occupational and Health Management Effectiveness						88%

Despite the improvement, like the previous audit result, the Preliminary Hazard Analysis as well as the Special Work Permission was the main problem for this team as for the others. The Table **13** shows the different teams performance in such internal audit process.

Table 13: Socres losses in audit process

General Operational Management	Team A	Team B	Team C	Team D	Team E	Average Performance
Operational Management I	81%	x	81%	79%	75%	79%
Operations Management II	84%	81%	91%	81%	86%	84.6%
Operational Management III	80%	95%	94%	90%	87%	**89.2%**
Operational Management IV	93%	93%	88%	91%	90%	**91%**
Total Average Performance						**86%**

After seven months the whole Operational Management was audited on tools like Process Hazard Analysis, High Workplace Permission, Safety Programs Required by Law, Incident and Accident Analysis and Comunication, Safety Contractor Management and Safety Indexes. Indeed, most of such tools were developed together with all teams along the previous month. The one which was more difficult to implement was the Safety Contractor Management but after some months all teams together with contractors were integrated. Finally the external audit was performed and the Safety, Occupational and Health Management were considered implemented by external auditors achieving 89% of effectiveness as shows Table **14**. Despite high performance achieved, where notice that some improvement is particular safety tools was necessary like Process Hazard Analysis, Preliminary Hazard Analysis, Safety Inspection, Safety Work Permission, High Workplace Permission and Work Refuse Law.

9.5. RISK MANAGEMENT

The Risk Process Management can be understood as a group of activities as plan risk analysis, identify hazards, assess and quantify risk, implement risk analysis recommendations, risk communication and follow up risk. Such action has the main objective to keep process risk under acceptable level throughout the enterprise life cycle. In order to succeed is necessary leadership and human, economic and technological resources.

Therefore, it must be a preventive management approach in order to avoid accidents which cause severe damage to employees' health and the environment.

Table 14: Final Audit process score table

Safety, Occupational and Health Management System Avaliation						
Item	Safety Tool	Criterion	Score	Weight	Partial Score	Total Score
1	Process Hazard Analysis	Availability	3	1	3	16
		Quality	1	3	3	
		Accurancy	2	2	4	
		Consistence	2	3	6	
2	Preliminary Hazard Analysis	Availability	3	1	3	19
		Quality	2	3	6	
		Accurancy	2	2	4	
		Consistence	2	3	6	
3	Safety Inspection	Availability	3	1	3	21
		Quality	2	3	6	
		Accurancy	3	2	6	
		Consistence	2	3	6	
4	Special Work Permission	Availability	2	1	2	21
		Quality	2	3	6	
		Accurancy	2	2	4	
		Consistence	3	3	9	
5	High Workplace permission	Availability	3	1	3	25
		Quality	3	3	9	
		Accurancy	2	2	4	
		Consistence	3	3	9	
6	Work Refuse Law	Availability	3	1	3	25
		Quality	3	3	9	
		Accurancy	2	2	4	
		Consistence	3	3	9	
7	Safety Programs required by law	Availability	3	1	3	27
		Quality	3	3	9	
		Accurancy	3	2	6	
		Consistence	3	3	9	
8	Incident and accident anaysis and communication	Availability	3	1	3	27
		Quality	3	3	9	
		Accurancy	3	2	6	
		Consistence	3	3	9	
9	Safety Report	Availability	3	1	3	27
		Quality	3	3	9	
		Accurancy	3	2	6	
		Consistence	3	3	9	
10	Safety contractors management	Availability	3	1	3	27
		Quality	3	3	9	
		Accurancy	3	2	6	
		Consistence	3	3	9	
11	Safety Indexes	Availability	3	1	3	27
		Quality	3	3	9	
		Accurancy	3	2	6	
		Consistence	3	3	9	
12	Safety Boad Comitee	Availability	3	1	3	27
		Quality	3	3	9	
		Accurancy	3	2	6	
		Consistence	3	3	9	

Maximum Scores	**324**
Achieved Scores	**289**
Safety, occuplational and Health Management Effectiveness	**89%**

The preventive risk management concepts can be summaries in a way to anticipate environmental impacts and health employee's damages by detecting possible incidents and process deviation before those events happens. In fact, many of those events would be avoid in project phase by implementing risk analysis recommendations.

The huge challenge is implementing such steps along assets life cycle and manage risk of several assets in different phases.

Such complexity turn into risk management a harder and challenge task for many specialist all over the world. Such complexity is observed in huge companies from some industries like Oil and Gas, Chemical, Aeronautic, Nuclear and Metallurgic. Despite the different types of risk on those industries, many of the accidents have as consequence deaths or environment impact and because of that risk must be managed constantly. In addition, such complexity, in huge companies, there are different managements which deal with risk management a long assets life cycle phases with different professionals and leadership that also influence on risk management effectiveness.

In order to carry out Risk Process Management is essential to apply risk analysis methods and considering when such methods is better applied along enterprises life cycle. Risk analyses are methods which the main objective is to define hazards and qualify or quantify risk. Risk is understood as a combination between probabilities (frequency) of an event with its consequence and can be measured qualitatively by risk matrix or quantitatively by some index like individual risk and societal risk.

The risk matrix reproduces a combination of probability (or frequency) and consequence which are defined qualitatively. Thus, risk matrix can have different combinations of level of consequences and probabilities.

The individual risk measures an expected number of death frequency and results of the frequency of accident and expected number of death cause by such accident consequence. The individual risk is applied as an index to measure risk into operational ground, in other words, into an industrial area. Nevertheless, an ISO - RISK curve that is a graphic representation of individual risk can also be applied to measure risk which the community who lives close to industrial area are exposed.

The societal risk shows graphically the cumulative expected number of deaths per frequency and results from a cumulative combination of frequency of accident and expected number of deaths. The societal risk is applied as an index to measure the risk that communities close to industrial area are exposed. The Societal risk is usually represented by an F-N curve, which means frequency and number of deaths. No matter if qualitative or quantitative, risk have to achieve acceptable values that are the main objective of Risk management, which means to keep risk in acceptable values along enterprises life cycle.

In order to verify if the risk is acceptable is necessary to apply Risk analysis methods. Such methods are qualitative and quantitative by concept. In the first case, a group of specialist identifies hazards and qualify risk based on their opinion. On the second case, despite hazard be identifies qualitatively based on specialist opinion, risk is calculated by mathematic methods. Independent to be qualitative or quantitative some of them, called deductive and other inductive. In Deductive Risk analysis case, the first step is to identify hazards or incident and then theirs causes the consequences. In Inductive Risk analysis, the first step is to define process deviation, equipment failures or events that turn into consequences like an accident.

We can get into conclusion that Risk Analysis methods, applications are very important, but the correct application time a long asset life cycle is also important as show Fig. **6**. Despite the importance, apply risk analysis do not mean Risk management and other issues are also important as will be shown in the next item.

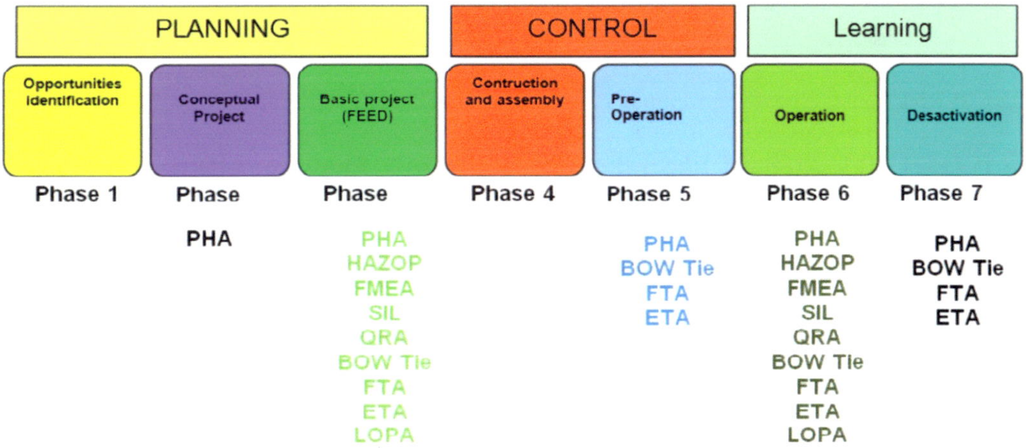

Figure 6: Risk Analysis a long enterprise life cycle (Source: Calixto, 2006).

The Risk Management based on ISO 31.000 defines steps to assess risk systematically which the main objective is to identify and minimize risk. Therefore, is important to establish controls to check if risk is under acceptable level constant. Despite the effectiveness of standard ISO 31.000, Risk

Management steps are subject to fail as shows Fig. **7** below. Such fail are layers in ISO 31.000 and a big challenge faced before, during and after Risk Management implementation.

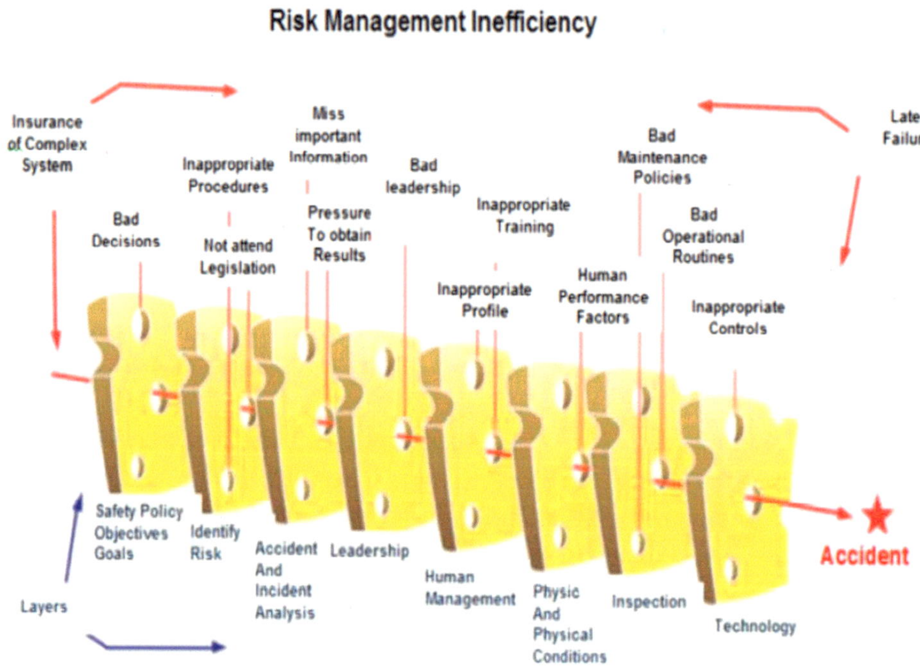

Figure 7: Risk management Inefficiency (Source: Moraes, 2010).

The ISO 31.000 standard like other ISO standards focus on efficacy to comply procedures and do not focus on other management factors like the result as Brazilian National Quality Awards. The Brazilian National Quality Awards have a straight ethic code and have trained professional to carry on the independent audit process for companies which are eligible for Quality award.

The Brazilian National Quality Awards are based on Deming Award and Malcolm Baldrige National Quality Award principles. The audit process analysis is based on a set of questions which punctuation regards performance factor like:

- Leadership – 110 Points
- Strategy And Plan – 60 Points
- Customer – 60 Points

- Society – 60 Points
- Information and Knowledge – 60 Points
- People – 90 Points
- Process – 110 Points
- Result – 450 Points

The main objective of Risk Process management focuses on best risk management practices in order to obtain excellence and achieve better results that means mitigate risk efficiently. One particular characteristic in proposal Risk Process Management is that despite punctuation, the final result will classify the audit management by the level of excellence in risk management. That means for each range of punctuation there will be an excellent classification. Such practice has the main objective to avoid competition among groups to achieve higher punctuation. In fact, one of the main objectives is promoting best practice knowledge among different teams. Another remarkable feature in such methodology is that each issue which takes influence in risk management is set up as a question to be audited. In addition, the methodology looks for best practices in order to obtain business results. The Risk Management factors analyzed in the audit process are:

- Planning
- Risk identification
- Recommendation
- Risk communication
- Risks follow up

On planning phase the main objective is to define a framework to support the Risk Process Management. In Addition to identify which assets must be assessed by specific risk analysis methods is also important to define resources like human, technology and finances in order to succeed process risk management. A final product of this phase is a real plan with all resource definitions. In order to check plan effectiveness some questions are preceding in audit process as well as evidence is required. Some examples of questions are:

1. All Plants are assessed and was defined which risk analysis must be carried on for each one and when?

2. To define which analysis must be carried out for each plant and in which frequency was used Company procedure?

3. Is there a formal definition about which employees will take part into all risk analysis?

4. For each risk analysis was defined the leader?

5. There are employees from Plants who are trained in Risk analysis?

6. Is there chronogram about plant risk analysis recommendations?

Each question is assessed based on follow score:

A – Total compliance (100% - know company procedures, has good practice established and have more than one evidence).

P – Partial compliance (75% - know company procedures, has good practice established and have only one evidence).

LC – Low compliance (25% - know company procedure, has good practice established but have no evidence).

N – No compliance (0% - Do not know company procedure and have no evidence).

NA – No Applicable.

An example of planning factor is shown in Table **15**, where a set of questions was proceeded by audit group and based on evidence and compliance criterion was defined final scores for planning phase.

The Risk identification has the main objective to guarantee that all risk analysis defined in the planning phase are not being performed too late as well as mitigation action defined as recommendations in such analysis. Is also important that whenever intolerable scenarios be identified as well as mitigation actions be established. Thus, as an example of questions about risk identification factor we have:

1. The Refinery has qualitative risk analysis for their plants (not those ones which take part into QRA)?

2. The Refinery has quantitative risk analysis of their plants?

3. The Refinery has integrated quantitative risk analysis for their plants?

4. The employees (who work on plants) take part in risk analysis?

5. Whenever is necessary to use risk matrix it follows company procedure?

6. In qualitative and quantitative risk were assessed person, environment, asset and image consequences?

7. In case of quantitative risk analysis was assessed intolerable risk?

8. In case of qualitative risk analysis was identified intolerable risk and was proposed recommendation?

9. In case of quantitative risk analysis (QRA), was identified individual or societal risk in the intolerable region (ALARP, F-N, ISO-Risk)?

10. For quantitative risk analysis, was identified risk in intolerable level and proposed mitigated actions?

11. In qualitative risk analysis the recommendations are identified by scenarios I, II, III e IV?

12. The accidental scenarios identified in quantitative risk analysis are taking into account on Emergency plan?

The third factor is Recommendation which the main objective is to guarantee that the recommendation be appropriated and be implemented on time as well as enough economic resources and responsible for their implementation be defined. The main questions on recommendation factor are:

1. All recommendations from risk analysis were assessed after the finish at maximum sixty day?

2. All recommendations approved to be implemented are in electronic system to be followed up?

3. Is there action Plan (PDCA) for all recommendations approved in electronic system?

4. Was established criterion for prior recommended implementation?

5. The recommendation planned to be implemented in this current year is an update?

6. The recommendation that was not implemented has clear justification and new data to be implemented?

Table 15: Audit planning evaluation

Questions	Local	Evidence required	Evidence	Mark							
				A	P	LC	N	NA	Weight	Total	Final
1 - All Plants are assessed and was defined which risk analysis must be carried on for each one and when?	EM	PDCA				x			3	7.5	
2 - To define such which analysis must be carried on for each plant and in which frequency was used Company procedure?	EM	Check up (PE-2AT-00023) and Table 1 (N-2782)				x			1	2.5	
5 - Is there formal definition about which employees will take part into all risk analysis?	EM	Check up plan		x					1	10	
6 - For each risk analysis was defined the lider?	EM	Check up plan		x					1	10	
8 - There are employees from Plants who are trainned on Risk analysis?	EM	Check up in Human Resource Management		x					1	10	
9 - Is there plan to all plant risk analysis?	EM	PDCA		x					1	10	
10 - Is there cronogram about plant risk analysis rcommendations?	EM	PDCA						x	1	10	
								TOTAL	9	60	6.7

The Risk communication criterion recommendation has the main objective to guarantee that risk analysis, recommendation and main results are aware among employees exposure to such scenarios or who are having a chance to affect it. The main questions about Risk communication area:

1. The risk analysis results are informed for community close to the refinery?

2. The risk analysis results are informed for refinery workforce?

3. The risk analysis results are informed for governmental agencies?

4. The risk analysis results are informed for companies which take part in the Local Emergency Plan?

5. The refinery performs a critical analysis about risk communication?

The last and no less important factor is Risk follow up that has the main objective to guarantee that recommendation and improvement action be implemented in order to mitigate risk whenever is necessary as well as risk analysis be updated whenever one modification in process be carried out. Some examples of questions in such criterion are:

1. The qualitative risk analysis is valid based on PE-2AT-00023?

2. The quantitative risk analysis is valid based on PE-2AT-00023?

3. The risk analysis (qualitative and quantitative) was assessed based on PE-2AT-00023?

4. The risk analysis (qualitative and quantitative) that was planned to be performed on current year is delayed?

5. The risk analysis (qualitative and quantitative) delayed have a new plan to be performed?

6. The risk analysis (qualitative and quantitative) is recorded in an accessible place for the workforce?

One remarkable characteristic of the audit process is to define assets in different life cycle phases that are managed for different management. In addition, at least 50% of all assets (Plants) in all enterprises phases must be audited in order to guarantee a reliable result in an audit. The Safety Management has always audited because their influence on Risk Process Management. As an audit result the refinery is classified in one of five levels of Process Risk management as shows Table **16**.

Table 16: Process Risk management level (Source: Calixto, 2012)

Category	Score	Process Risk Management Level
F	0 - 4,0	The Process Risk management practice are inadequate or was not demostrated in audit process
E	4,1 - 5,0	The Process Risk management practice are adequate for some company procedures
D	5,1 - 7,0	The Process Risk management practice are adequate for many company procedures
C	7,1 - 8,0	The Process Risk management practice are adequate for most of company procedures
B	8,1 - 9,0	The Process Risk management practice are adequate for all company procedures
A	9,1 - 10,0	The Process Risk management practice are adequate for all company procedures and best risk management practices are implemented

The case study has the main objective to test Risk Process Management Methodology in order to improve risk management in eleven Brazilian refineries of total twelve. The management audit was 70% operational and 30% Safety Management and Project Management. The process described here is the second audit process.

Thus, there was implemented some improvement in previous methodology that was only a set of general questions. The audit process was carried out in 2010 from September to November in such refineries. The audit group had two Safety specialists which took between three and four days to carry out audit process. The same auditors were audited for other groups, but in fact, when audit, the safety specialist only supports auditor's teams. The responsibility for each management was audited on Risk Process Management in their management.

In general terms was observed that risk analysis was carried out for contracted companies and most of such analysis are not spread out for the workforce, unless PHA and HAZOP. The other important point is that managers and team leader were more aware about Risk Process Management. One of the most critical point in oil refineries recommendations was not well manage with action plan defining responsible for each action, time to be implemented and economic resource. In addition was observed that there was not a standard criterion to prior recommendations implementation. The risk communication was another point to improve in most of refineries because there are some difficulties in communicates risk to the workforce and also to the community.

The final Company result considering all refineries was “B” that means most of Risk Process Management practice are applied for company. Some refineries achieved “A” and other “C” as shows Table **17**.

Table 17: Process Risk management, audit result (Source: Calixto, 2012).

	Refeneries												
Factors	A	B	C	D	E	F	G	H	I	J	J	Average Point	
Planning	8.1	9.0	9.4	7.9	6.3	9.6	9.5	10	9	10	9	8.9	B
Risk Identification	9.6	9.4	9.2	6.4	7.1	8.8	9.5	10	10	10	9	9.0	B
Recommendation	4.3	8.0	8.5	2.0	8.9	8.6	6.9	8	9	10	9	7.5	C
Comunication	5.0	3.6	6.9	10.0	6.8	5.4	4.6	8	8	6	9	6.7	D
Risk Follow Up	9.7	10.0	9.6	7.5	5.5	9.1	8.6	9	9	10	9	8.8	B
Process Risk management Result												8.2	B

Based on the audit result, some hypothesis came out as:

- As higher efficiency in planning, higher is efficiency on identification factor;
- As higher efficiency in planning, higher is efficiency of risk follow up factor;

On first case planning and identification factors of Risk Process Management are positively correlated (ρ=0. 9).

As expected, as higher as planning efficiency, higher is risk identification and analysis efficiency.

On second case planning and recommendation factors of Risk Process Management are also positive correlated (ρ=0.90). As higher as planning efficiency, higher is risk follow up factor.

By the other hand, identification factor has not a high correlation (ρ=0.67) with recommendation factor as well as risk follow up.

The correlation coefficient result means that if planning in Process Risk Management is well performed, risk analysis and risk follow up has a high chance to be well succeed because the leader and team are aware about importance of Risk

Process Management and there are resources planned to perform a risk analysis as well as risk follow up. By the other hand, perform risk analysis do not mean that the recommendations will be implemented and risk will be communicated for workforce and community. The Fig. **8** shows the correlation between Risk Process Management factors with regards point for each factor in the audit process.

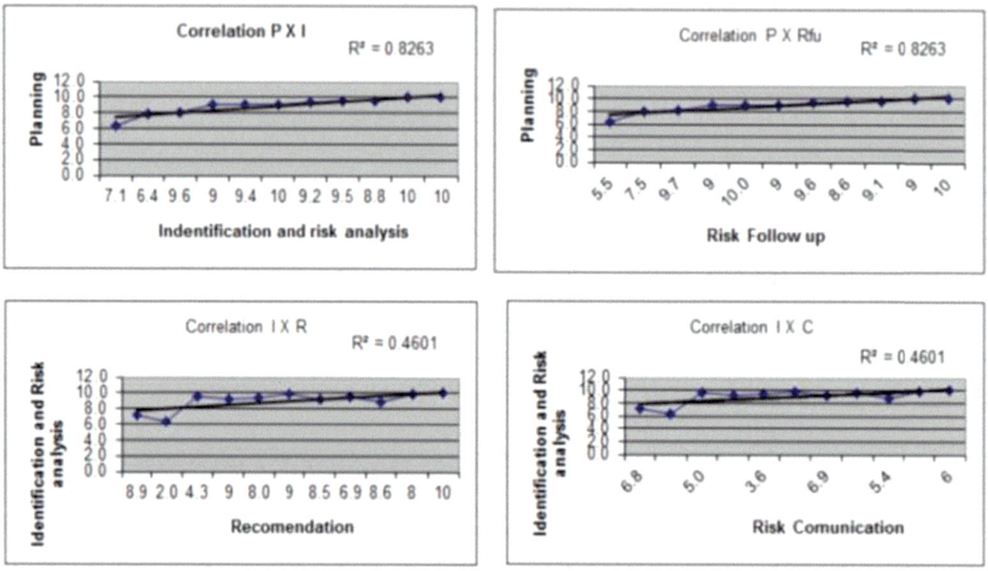

Figure 8: Correlation among Risk Process Management factors (Source: Calixto, 2012).

The Risk Process Management is a big challenge to the Company and must be carried on systematically. That means an audit process as well as best practice in risk analysis and management must be implemented constantly and must be part of operational routine. In doing so, is necessary human, technology and economic resources must be dedicated in order to achieve part of excellency in risk management.

One remarkable point is the awareness of most of managers and team leaders about Risk Process Management and now it seems to be clear that risk management is not only Safety management responsibility.

In general terms the Risk Process Management methodology requires time and human resource to be implemented and maintain high performance over a long period of time. The managers and leader are essential in such process and must stimulate cooperation among teams despite competition to achieve high compliance in the audit process. Consequently, it will be possible to have continuous learning all over the organization.

One remarkable point includes efficiency criteria considered by quality of actions like risk analysis in order to drive teams to succeed best practice and not make up evidences to audit the process.

Despite the good result some teams require more improve that other and due to exist many other management systems all over the company is necessary that such improvement actions as well as audit proceed annually.

9.6. ASSET INTEGRITY MANAGEMENT

In order to understand the Asset integrity philosophy is necessary to understand first the asset management concept. Asset Management is defined as the best practices applied along assets life cycle in order to achieve the best performance result. Indeed, such practices are carried out with the support of leaders in different organizational levels, which means, strategic, tactical and operational.

By safety point of view, unsafe failures must be avoided since design phase and requires specific reliability indexes be assured and maintained a long asset life cycle.

Therefore, different reliability engineering methods in different asset life cycle phase must be taken place as shows Fig. **9**.

The ideal asset performance is achieved when most of early life failures are eliminated on design phase that enable excellent performance during the operational phase, which is represented by the green bathtub curve in Fig. **9**, that means lower failure rate, in other words, high reliability.

In order to achieve such performance it is necessary to implement different methods a long asset life cycle. Indeed, all effort starts on design phase applying different qualitative (DFMEA, RCM, RBI, HALT, FRACAS, human reliability) and quantitative (RAM, ALT, Reliability Growth analysis and warranty analysis) methods. Such methods have the main objective to identify the early life failure during design and eliminate them whenever it is possible. On an operational phase, different qualitative (PFMEA, RCM, RBI) and quantitative (Lifetime data analysis and RAM analysis) methods must be taken place to maintain asset performance until the end of asset life when must be defined when decommissioning the asset (equipment) that is supported by ORT (Optimum Replacement Time), RAM analysis and Reliability Growth analysis.

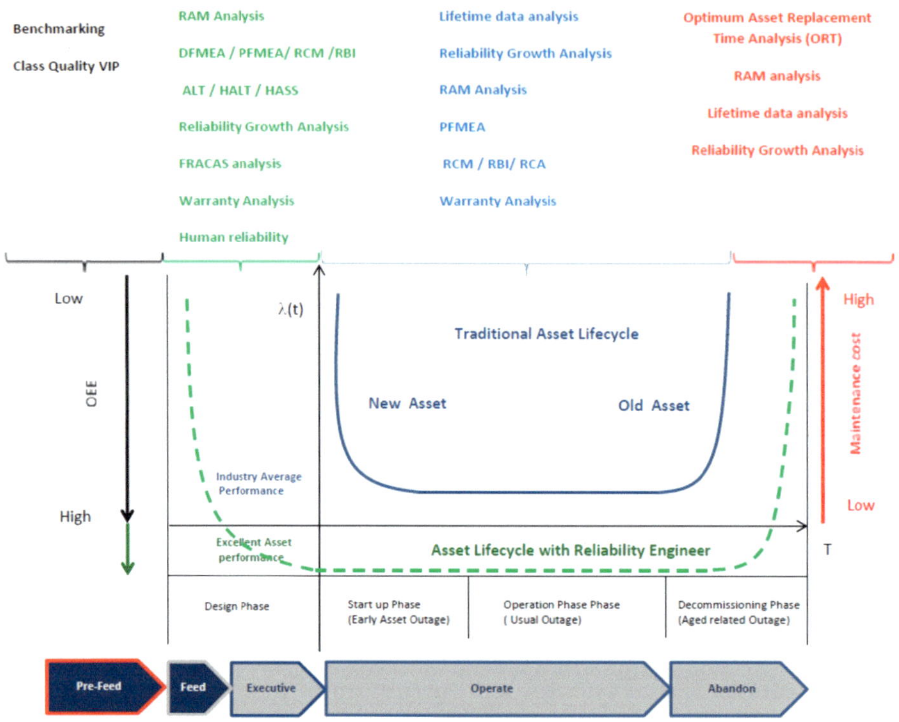

Figure 9: Asset Management life cycle (Source: Calixto, 2013).

Indeed, all such methods must be applied to avoid unsafe failures and consequently avoid accident. In such context, the Asset integrity must take place.

The **"Asset integrity"** is part of **"Asset Management"** but focus on asset with shows relevant impact on safety or environmental in case of failure. The Asset Integrity has the main objective to assure high asset performance considering safety a long asset life cycle. Based on KP3 definition, Asset integrity can be defined as ***"the ability of an asset to perform its required function effectively and efficiently whilst protecting health, safety and the environment".*** In order to achieve high safety performance it is essential to apply the best approaches for each Asset integrity Management pillar shown on Fig. **10**, which are:

- **The Risk Management** means to define a risk target, hazard identification, incident and accident investigation, risk assessment, risk evaluation and risk mitigation, communicate the risk and prepare an emergency response plan. In order to identify hazards and assess the risk different qualitative and quantitative methods can be applied like PHA, FMEA, HAZID, HAZOP, FTA, ETA, SIL, LOPA, AQR and Bow Tie.

- The **Reliability and Maintenance** focus on operational availability, performance index achievement and begins in pre-feed phase to be assured in the design phase of RAM analysis, life time data analysis, accelerated testing, reliability growth analysis and DFMEA. In addition, a long operational phase, lifetime data analysis and RAM analysis are carried out to support decisions as well as RBI, RCM that will define maintenance and inspection policies to maintain asset availability and reliability a long operational phase by.

- The **Human Factors** are identified by human reliability analysis, which concerns all human performance factors considered by all critical activities that can lead in an accident or environmental impact. Based on such analysis, leadership, training and learning process are provided to avoid such human error influence on Asset integrity performance.

The asset integrity management start before design phase when benchmarking performance index as well as technology is defined. On design phase, the best reliability engineer methods (accelerated test, HALT, reliability growth analysis, DFMEA) are carried on in order to assure the asset reliability. In addition, the risk assessment is carried out in order to evaluate and mitigate risk on design phase as much as possible. On an operational phase, the reliability engineer methods (Life Time Data analysis, RAM, RBI, RCM and Human Reliability Analysis), risk Management and human factor are constantly assessed and revised in order to maintain a high Asset Integrity Performance.

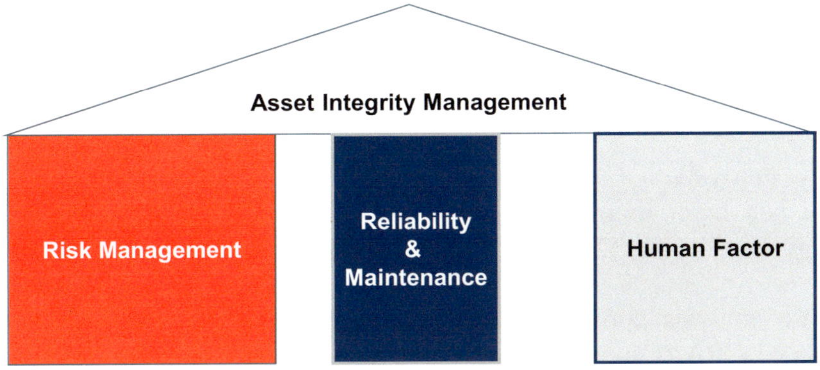

Figure 10: Asset Integrity Management.

In order to implement Asset integrity management the following steps are proposing:

⇒ First step: Define the asset to be assessed based on safety and environmental impact criticality;

⇒ Second step: Define Asset performance target;

⇒ Third step: Define Risk Analysis and reliability methods;

⇒ Fourth step: collect failure, incident, accident and repair data, inspections and maintenance policies;

⇒ Fifth step: Carry out the analysis defined in third phase by using the data collected in the fourth phase.

⇒ Sixth step: Implement actions, monitoring the performance index and implement improvement when necessary.

In the first phase, the Asset must be defined by client based on risk, safety performance or other criterion defined by the client.

The second phase is one of the most important because it will drive all asset integrity management process that is to define the representative index. Considering safety, societal risk, individual risk as well as operational availability, reliability, expected number of failures are the proposal index to be monitored a long asset integrity management life cycle.

In the third phase, risk analysis must be carried out in order to define hazards, assess risk, evaluate risk and implement mitigation actions. The risk analysis method more appropriated to be applied in integrity management is the Bow Tie Analysis. Such method enables to identify the root cause of incidents as well as the different consequences.

The Bow tie analysis is a quantitative risk analysis that has been in use since the 1970s, and has been incorporated into the Hazards Effects Management Plan methodology used by the Shell Oil Company on 90th.

The Bow tie analysis comprises FTA (Fault Tree Analysis), ETA (Event Tree Analysis) and LOPA (Layer Of protection Analysis) concepts and enable assess all combinations of events from incident causes to incident consequences considering the layer of protections that prevent accident and mitigate consequences. Such methodology can be used to assess different types of problems, but in safety terms is applied basically to assess and support accident analysis, Process hazard analysis and Risk Management. An example of Bow Tie is an incident of gas release from the pipeline as shows Fig. **11**. Other risk methods may support Bow tie analysis as for example Quantitative risk analysis to assess the consequence of accident as well as calculate the societal and individual risk.

On fourth phase the main event and equipment identified in the Bow Tie analysis must be assessed by reliability engineer methods like for example lifetime data analysis in order to define the failure occurrence profile for each unsafe failure as well as event. In order to implement such phase will be necessary collect historical data of events and equipment considered by Bowtie. In additions, the maintenance and inspection policies must be defined based on recognizing approaches like RBI and RCM based on reliability as well as specialist opinion. The RBI will support the inspection policies and RCM will take into account all preventive possibilities (predictive and based on equipment age time).

On fifth phase the bow tie will be updated with the failure rate functions defined in previous phase as well as the preventive actions, inspection and maintenance which mitigation risk. The final analysis will be updating Bow Tie analysis with all information a long asset life cycle that includes all inspections and maintenance policies in order to verify such impact on risk, number of failures, reliability and availability.

Consequently the bow tie will be dependent on time and once linked to failure rate function that will be updated over a long period time will generate a performance index to drive action to keep the asset integrity in high performance level. The Fig. **11** shows the time dependent bow tie.

In order to present the real Asset management application a case study will be demonstrated considering the methodology defined above. Mostly the main concern in Asset Integrity Management is related to safety. In this case study, the environmental impact will be highlighted.

Whenever The Asset Integrity Management is applied the first initial step is to define the asset and justify this choice. Indeed, based on the Asset Integrity Management concept, the main concern is about asset that in case of failure might cause a major accident or a significant environmental impact.

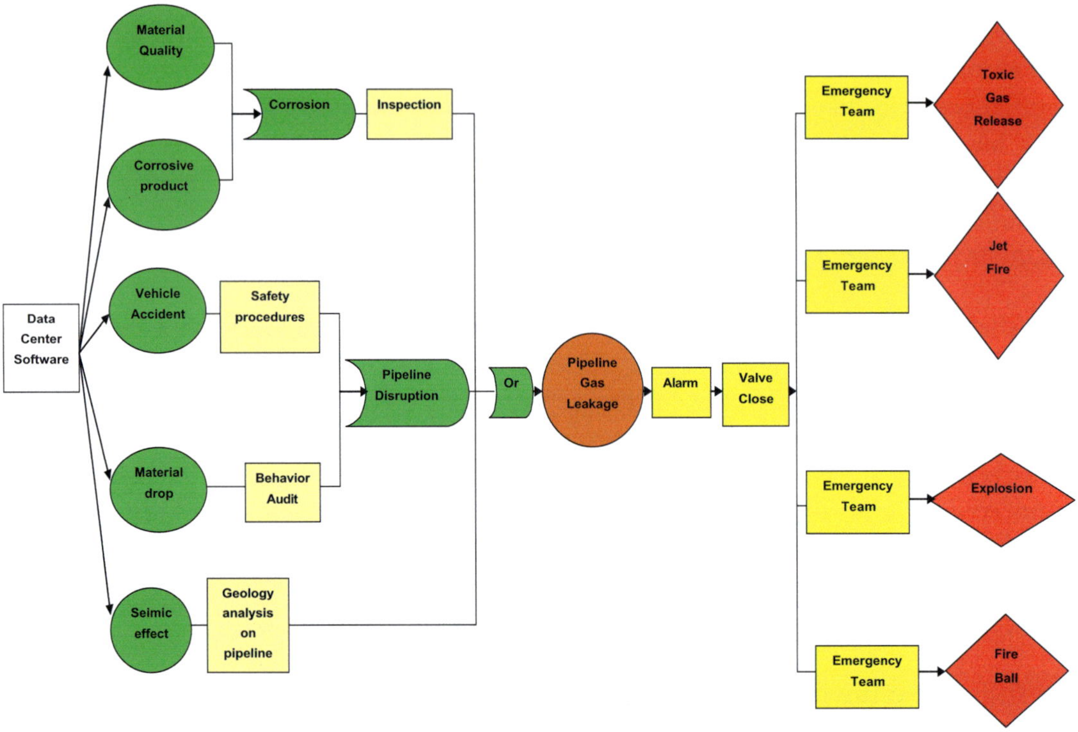

Figure 11: Pipeline leakage Bow tie time dependent (Source: Calixto. E, 2012).

The "first phase" of Asset Integrity Management is to define the asset to be assessed. This case study will focus on the Sulphur Recovery Unit plant, that in case of shutdown cause a direct impact on loss of production as well as an associated environment impact due the increasing level of sulphur components emission.

Despite all effort to avoid and control the sulphur emission, in some case, refineries those are located close to urban centres has restricted sulphur emission allowance. Consequently, in case of shutdown in Sulphur Recovery Plants it is necessary to shut down other refinery plants in order to avoid environmental impact caused by high levels of sugar component emission. In addition the loss of production is also associated with a bad consequence of such event.

The "Second phase" is to define the performance index for such asset integrity. In this case study based on company benchmarking as well as the environment target related to sulphur emission it is required that the sulphur recovery Unit achieve more than 99% of operational availability in each three years.

The "third phase" is to apply the quantitative methods to identify the hazards, assess the risk, verify the system performance and propose recommendation. In this case, the unwished event is associated with System outage. Therefore, the main concern in this case is to achieve high operational availability over time in such plant that starts in project phase by a RAM analysis, RCM and other reliability engineering methods. Once such methods are applied it is possible to define reliability requirements for critical equipment as well as preventive maintenance policies. The usual logic representation of a process plant in RAM analysis is a RBD (Reliability Block Diagram) which describe the effect of each equipment in system operational availability in case of failure as shows Fig. **12**. The pumps represented by the block 1.3 – B-01 A/B and the Filter represented by the block 1.0 – FT-01 A/B are parallel blocks. That means, there are one active equipment and other standby, the passive one. Such equipment causes, impact on system only in the case that both be unavailable at the same time.

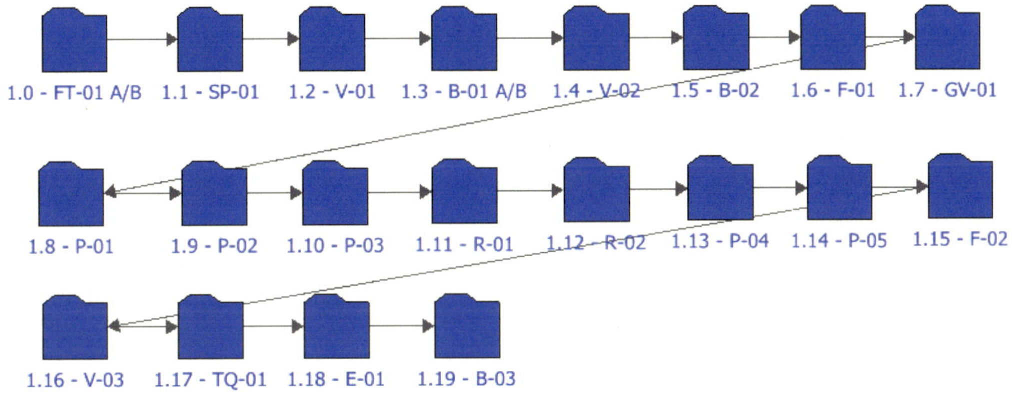

Figure 12: Sulfur Recovery Unit Plant RBD.

The Sulphur Recovery plant operational availability as a result of the RAM analysis show that system achieves initially is 99.5% in 3 years. The critical equipment is the Furnace which has 99.9 % of operational availability in 3 years. Such equipment is responsible for 40% of possible Sulphur Recovery Plant outage. In order to avoid such downtime the following actions are proposed:

⇒ Correct burner's material specification;

⇒ Operational procedure to avoid damage caused by high temperature during operation;

⇒ Preventive maintenance policy;

Once applying such recommendation is expected that the Sulphur Recovery Unit plant achieve 99.7 % of operational availability in 3 years.

Afterwards the project phase, once such recommendations is successfully achieving it is expected high performance on Sulphur Recovery Unit Plant. Indeed, there will be always the chance to have some outage due other equipment failures, despite low probability.

The "Fourth phase" is to carry out the Bow Tie Analysis concerning the holiest vision of all threats and consequences. Indeed, concerning the Sulphur Recovery Plant outage there are also other external events that might cause an impact on such plant and consequently outage like "Bad project definition", "Sabotage" and "Natural disaster". The holistic vision about the possible threat that might cause the Sulphur Recovery Unit Outage as well as their consequence is well represented by the bow tie diagram as shows Fig. **13**.

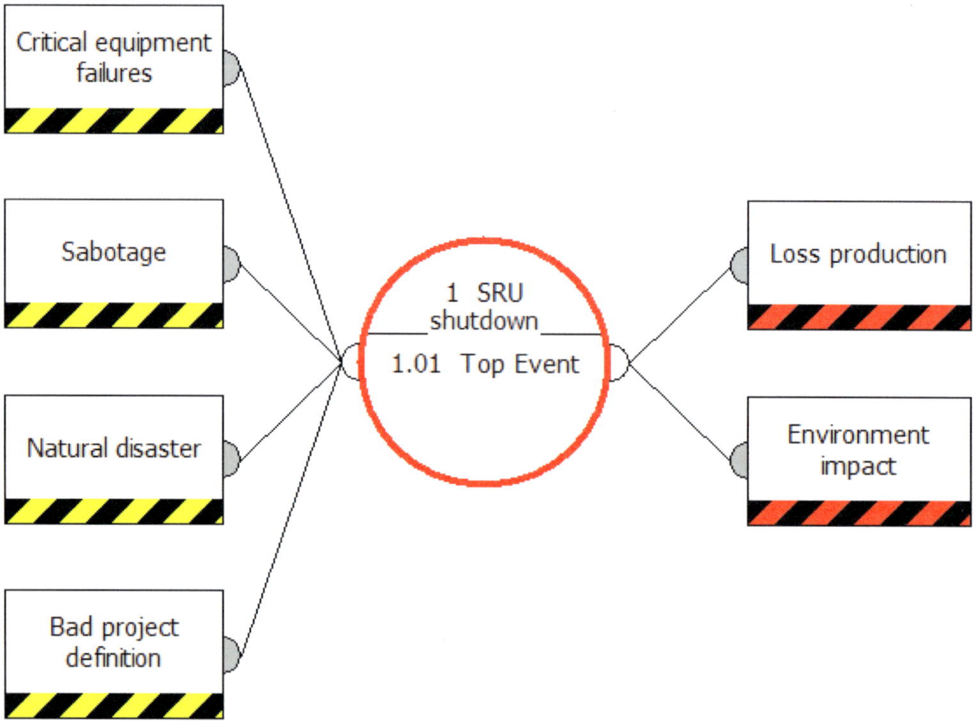

Figure 13: Sulfur Recovery Unit Bow Tie Diagram.

Despite the RAM analysis shows that it is possible to achieve high performance, which means 99.5% in 3 years, the external threats like Sabotage, Natural disaster and bad project definition might impact on such performance. In this case, it will be necessary to establish barriers to avoid that such treats take place as well as recovery action to avoid that once the Sulphur Recovery Unit shut down, the loss of production and environmental impact take place.

Considering the recovery action, once the Sulphur plant shut down there will not be too much to be done to avoid such consequence. Depends on equipment that lead to shut down of this plant there will be enough time to maintain the refinery plants in operation or not. In case of Shutdown in a furnace, for example, will be required around 120 hours to repair such equipment and consequently the whole refinery will be shut down.

Therefore, the barriers (the layer of protection) are very important to avoid that such treats trigger the TOP event, that means, cause Sulphur Recovery Plant outages. The Fig. **14** shows the complete Bow Tie with the main barrier that must be taken place to avoid the top event.

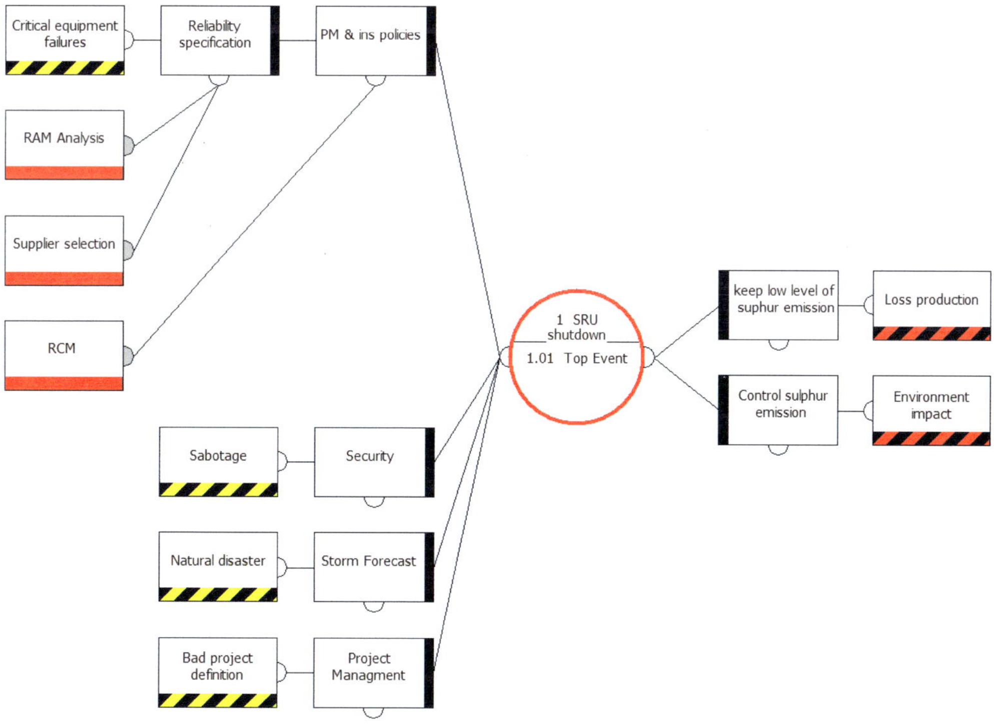

Figure 14: Sulfur Recovery Unit complete Bow Tie Diagram.

Considering the threat “Critical Equipment Failure” there are two barriers that must be taken place to avoid that such threat trigger a Top event that are “Reliability specification” and “Preventive maintenance and Inspection policies definition”. Those barriers are successful based on specific actions. In the first case, in order to define the reliability specification for critical equipment it is necessary to perform the RAM analysis and select the equipment supplier based on reliability requirement. In the second case, in order to establish the preventive maintenance and inspection policy it is necessary to perform the RCM analysis.

Considering the activity “RAM analysis” and the task “define reliability requirements for critical equipment” the final result is the reliability requirement definition for the critical equipment.

In addition, concerning the activity “Supplier selection” was defined two main tasks like “select supplier based on reliability requirement” and “critical equipment installation verification”. Those tasks have also subtasks as shows Table **18**.

Such similar steps must be carried out of the preventive and inspection policies defined on RCM. In addition, for the other threats like “Sabotage”, “Natural Disaster” and “Bad project definition” there will also be defined specific action to be implemented a long asset life cycle. Indeed, a good risk project management program is able to identify the risk and provide an action plan to mitigate such risk. The big challenge is to integrate all risk and monitoring the mitigate action on time.

Such risk management philosophy is also part of Asset Integrity management once the main objective is to maintain high performance of asset integrity by avoiding accident and environmental impact.

Table 18: Bow Tie Actions Plan

Bow Tie - Action Plan				
No.	**Code**	**Activity / Task**	**Who**	**When**
1	**1.1**	**RAM analysis / Define reliability requirement for critical equipment**	Reliability Engineer	Desing Phase
	1.1.1	Furnace (F-01) - 99.7% in 3 years and 0,04 number of failures	Reliability Engineer	Desing Phase
	1.1.2	Furnace (F-02) - 99.7% in 3 years and 0,04 number of failures	Reliability Engineer	Desing Phase
	1.1.3	Boiler (GV-01) - 100% in 3 years and 0 number of failures	Reliability Engineer	Desing Phase
2	**2.1**	**Supplier Selection / Select Supplier based on reliability requirement**		
	2.1.1	Require Lifetime data analysis or accelarated test from each supplier	Reliability Engineer	Desing Phase
	2.1.2	Compare different suplier analysis	Reliability Engineer	Desing Phase
	2.1.3	Chose supplier based on reliability assurance	Reliability Engineer	Desing Phase
	2.1.4	Specify warranty contract for 3 years	Reliability Engineer	Desing Phase
3	**3.1**	**Supplier Selection / Critical equipment installation verification**	Production Engineer	Executive Phase
	3.1.1	Verify critical task defined by supplier during Furnace (F-01) instalation	Production Engineer	Executive Phase
	3.1.2	Verify critical task defined by supplier during Furnace (F-02) instalation	Production Engineer	Executive Phase
	3.1.3	Verify critical task defined by supplier during boiler (GV-01) instalation	Production Engineer	Executive Phase

In order to monitor such action along the Asset Life cycle the risk considered by those two bow tie consequences (loss of production and environmental impact) must be assessed and such risk will associate the probability that the threats take place once the layers of protection are not effective, in other words, the preventive actions were not implemented on time or were not effective enough.

The Fig. **15** below shows the risk assessment of those two Bow ties final consequences (loss of production and environmental impact) based on the risk matrix concerning the consequence for people, environment, asset and reputation.

		Frequency Category					
		A (Extremely Remote)	B (Remote)	C (Little Frequent)	D (Frequent)	E (Very frequent)	F (Extremely frequent)
		At least 1 between from 1000 to 100.000 years	At least 1 between from 50 to 1000 years	At least 1 between from 30 to 50 years	At least 1 between from 5 to 30 years	At least 1 between from 1 to 5 years	At least 1 in 1 years
Severity Category	IV	M	NT	NT	NT	NT	NT
	III	M	M	NT	NT	NT	NT
	II	T	T	M	M	M	M
	I	T	T	T	M	M	M

Figure 15a: Risk Matrix.

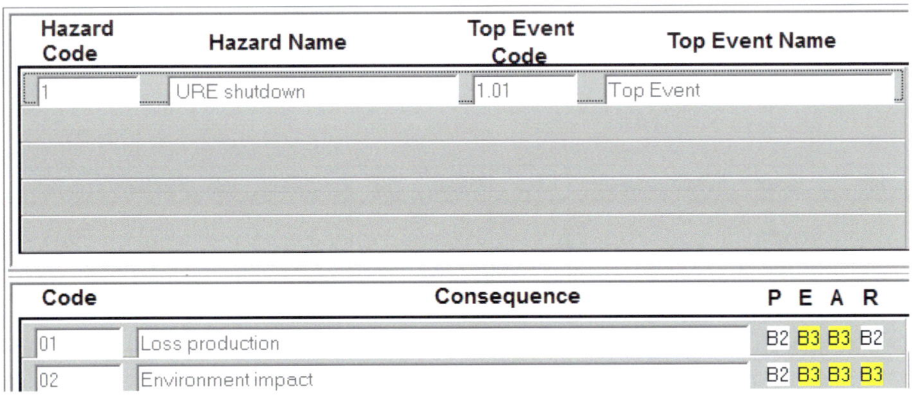

Figure 15b: Bow Tie Diagram risk assessment.

P = People

E = Environment

A = Asset

R = Reputation

The Risk level for the environment and asset are moderate that means additional actions are taken place only if is feasible enough, that means, such effort and investment bring the return expected related to risk mitigation. The Risk is calculated based on equation below:

$$Risk(Loss\ of\ production) = f_{Loss\ of\ Production} \times Consequence$$

$$Risk(Loss\ of\ production) = \left(f_{URE\ outage} \times P_{recovery}\right) \times Consequence$$

$$f(URE\ outage) = \sum_{n}^{1} f_{Threat_n} \times \left(\prod_{0}^{m} p_{(Barriers)_m}\right)$$

$$f(URE\ outage) = \{0.33 \times (0.1 \times 0.1)\} + \{0.1 \times (0.1)\} + \{0.5 \times (0.01)\} + \{0.33 \times (0.01)\} = 0.0193$$

$$f(Loss\ of\ production) = \{0.0193 \times (0.5)\} = 0.0097$$

The similar step was carried out for the risk assessment of Environmental impact. Finally, the Risk demonstrated in Table **19** regards the calculate frequency above and classify such frequency based on the risk matrix on Fig. **15** (0.097→B). Thereby, concerning the risk matrix consequence III the final risk is "B III".

Table 19: Bow Tie Risk Assessment

Threat	Freq	Barriers	Prob	Top Event	freq	Recovery	Prob	Consequence	Freq	Freq Matrix	Cons (Matrix)	Risk
critical equipment failure	0.33	Reliability Specification	0.1	URE outage	0.0193	Keep low level of sulphur emission	0.5	Loss of production	0.0097	B	III	B III
		PM and Insp policies	0.1									
Sabotage	0.1	Security	0.1									
Natural Disaster	0.5	Storm Forecast	0.01			Control Sulphur Emission	0.5	Enviroment Impact	0.0097	B	III	B III
Bad project Definition	0.1	Project Management	0.01									

Indeed, different approach to calculate the final risk can be applied. The consequence can also be expressed in monetary matter and consequently there will be a monetary risk measurement.

The importance of reliability and maintenance can be proved by Table **20** results where the risk level is intolerable if the reliability specification and PM and Inspection is not implemented. Such assumption is represented by 100% of failures in such layer of protections (1).

Table 20: Bow Tie Risk Assessment

Threat	Freq	Barriers	Prob	Top Event	freq	Recovery	Prob	Consequence	Freq	Freq Matrix	Cons (Matrix)	Risk
critical equipment failure	0.333333	Reliability Specification	1	URE outage	0.349	Keep low level of sulphur emission	0.5	Loss of production	0.175	E	III	E III
		PM and Insp policies	1									
Sabotage	0.1	Security	0.1									
Natural Disaster	0.5	Storm Forecast	0.01			Control Sulphur Emission	0.5	Enviroment Impact	0.175	E	III	E III
Bad project Definition	0.1	Project Management	0.01									

9.7. INTEGRATED MANAGEMENT SYSTEM

The Safety and occupational Health Management and the critical factors which influence on its performance were well described on itens above. The extension of such understanding is to establish the relation between safety management, quality management, environmental management and corporate and social responsibility management.

In order to establish such relation, the first step is to have a brief understanding about Quality management, Environmental Management and corporate and Social responsibility management.

The quality management became important in 80th decades where the standard ISO 9001 was established as a guide for organizations establish their Quality Management System (QMS). The main objective in implementing the ISO 9001 was to define systematic procedures for critical activities and process which take influence on product quality. Despite limitation like to be a generic standard, which does not address specific solution for different organizations in different industries as well as do not address efficiency steps to achieve efficacy and compliance with organizational process and products such standard support many organizations to achieve better performance. In fact, the next decade had a remarkable requirement from the society for the extension of the quality management that would comprise the environmental issues.

The environmental management, which focuses on the environmental aspect and impact related with organizational process was one of the most important issues discussed in 90th decades. Therefore, in addition to control process deviation which lead to product non compliance became also important to control process deviation which lead to environmental impact. In order to achieve environmental management process the ISO 14001 was established as the guideline which define technical terms as well as the importance to define the whole process environmental aspect and impact.

The end of 90's the safety management was the focus of attention in most of organizations all over the world due to the significant increase number of fatal accidents in different industries.

In order to establish a health & a safety management system (OH&SMS), the OHSAS 18001 was proposed as a guideline a was applied to many organizations all over the words. As is discussed in this chapter 8 the OH&SMS requires many other methods and approaches, but definitely, the OHSAS 18001 has been given a huge contribution to improve the safety performance in many organizations all over the world.

In the beginning of 2000 decades, the sustainability discussion became one important issue into an organizational context which the basic concept is that company must achieve economic success by preserving the environment from the effect of their enterprises as well as include the society in such economic benefits. Therefore, the corporate social responsibilities (CSR) became an important issue being guided for different standards like ISO 26000.

Indeed, all of such management dimension have in common their management framework which establishes by ISO standards like ISO 9001, ISO 14.001, OHASA 18001 and ISO 26000.

The main management principle is to establish a PDCA cycle with a defined requirement for each PDCA phase as shows Fig. **16**. Indeed, despite different focus, the database technology involved to support such management, the organizational resources and the framework can be commonly used for all of them despite in some cases requires different specialist to deal with different technical issues.

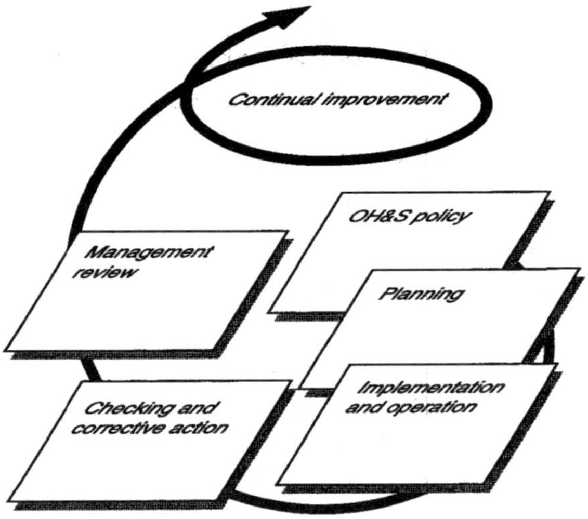

Figure 16: PDCA (Source: OHSAS 18001, 1999).

Considering the different ISO standards like ISO 14001, ISO 9001 and OHSAS 18001 it is possible to see the common aspects of such standards on Table **21**. That's shows the possibility to have a common management framework for different management systems considering the database technology, files, procedures, planning practices, internal audit practices and improvement actions practices.

Table 21: Common aspect of ISO 9001, ISO 14001 and OHSAS 18001

Item	ISO 9001	Item	ISO 14001	Item	OHSAS 18001
0	Introduction	0	Introduction	0	Introduction
1	Scope	1	Scope	1	Scope
2	Normative references	2	Normative references	2	Normative references
3	Terms and definitions	3	Terms and definitions	3	Terms and definitions
4	Quality management system	4	Environmental management system requirements	4	OH&S management system elements
4.1	General requirement	4.1	General requirement	4.1	General requirement
5.5.1	Responsibility and authority	4.4.1	Resources, roles, responsibility and authority	4.4.1	Resources, roles, responsibility and authority
5.1	Management commitment	4.2	Environmental policy	4.2	OH&S policy
5.4	Planning	4.3	Planning	4.3	Planning
5.2	Customer focus	4.3.1	Environmental	4.3.1	Hazard identification, risk

			aspects		assessment and determining controls
5.4.1	Quality objectives	4.3.3	Objectives, targets and program(s)	4.3.3	Objectives and programmer(s)
7	Product realization	4.4	Implementation and operation	4.4	Implementation and operation
8	Measurement, analysis and improvement	4.5	Checking	4.5	Checking
9.2.2	Internal audit	4.5.5	Internal audit	4.5.5	Internal audit
9.5.2	Corrective action	4.5.3	Nonconformity, corrective and preventive action	4.5.3	Nonconformity, corrective and preventive action
9.5.3	Preventive action				

Integration is the complete harmony and alignment of strategy and operations of an organization. It means that different departments and levels speak the same language and are tuned to the same wavelength (Garvin, 1991). In the literature, integration of management systems has been discussed as the merger of the quality management system (QMS), environmental management system (EMS), health & safety management system (OH&SMS), and corporate social responsibilities (CSR). Indeed, to survive and thrive in a period of global competition, organizations need to look at every aspect of their processes, including cost cutting, well-being of their employees, the working environment and the impact that organizational operations have on their neighbors and on the local community. Moreover, companies must address these issues while continuing to provide quality products and services. The concept of integrated management systems (IMS) has arisen from this need as shows the Fig. **17**.

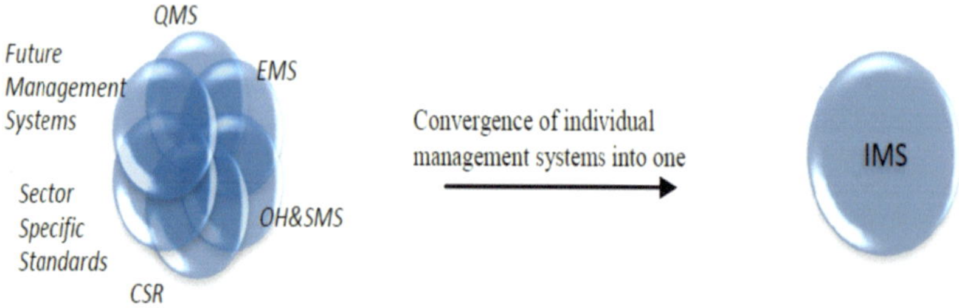

Figure 17: Integrated Management System (IMS) (Source: Asif, 2009).

The main questions that arise when integration is a possibility for organization which implement the ISO systems is about the main advantages of integration. Indeed, depends on organization maturity and efficiency in the implementation process of different ISO systems it is possible to get advantages like cost reduction in maintaining such ISO systems that are achieved basically by better resource allocation, operational system's efficiency and documentation reduction. The Table **22** shows the different researcher's opinion about ISO system integrations advantages.

Table 22: Integrated Management System advantages (Source: Asif, 2009)

Benefits from IMS implementation		
Benefit of IMS		**Supporting literature**
Documentation reduction	**Elimination of documentation duplication**	**Douglas & Glen, 2000; McDonald *et al.*, 2003; Zutshi & Sohal, 2005**
Customer's demand	**Pre requisite for business**	**McDonald *et al.*, 2003**
	Enhanced customer satisfaction	**Douglas & Glen, 2000; Zutshi & Sohal, 2005**
Cost reduction	**Cost reductions, *e.g.*, in manufacturing, operations, and insurance premiums**	**Jørgensen *et al.*, 2005; Wright, 2000, Douglas & Glen, 2000; Zeng *et al.*, 2007; Zutshi & Sohal, 2005; McDonald *et al.*, 2003**
Operational benefits	**Operational improvements**	**Fresner & Engelhardt, 2004; Holdsworth, 2003; Jørgensen *et al.*, 2005; McDonald *et al.*, 2003**
	Simplified systems	**Douglas & Glen, 2000; Zutshi & Sohal, 2005**
	Time saving	**Zutshi & Sohal, 2005**
	Better synergies between systems	**Rocha *et al.*, 2007**
	Unification of internal audits	**Salomone, 2008**
	Unification of training activities	**Salomone, 2008**
	Common framework for continual improvement	**McDonald *et al.*, 2003**
	Overall organizational performance improvement	**McDonald *et al.*, 2003**
Resources allocation and utilization	**Better allocation of resources**	**Zeng *et al.*, 2007**
	Saving of human resources	**Salomone, 2008**
	Better utilization of resources	**Rocha *et al.*, 2007**
Cultural change	**Teamwork promotion**	**Wright, 2000**

	Greater acceptance by employees	**Zutshi & Sohal, 2005**
	Cultural change (Learning organisation)	**Wright, 2000; Zutshi & Sohal, 2005**
Other benefits	**Strategic planning**	**Zutshi & Sohal, 2005**
	Holistic view	**Zutshi & Sohal, 2005**
	Enhanced interdepartmental communication	**Douglas & Glen, 2000; Wright, 2000; Zutshi & Sohal, 2005**
	Better definition of responsibilities	**Salomone, 2008**
	Means to sustainable development	**Fresner & Engelhardt, 2004; Rocha *et al.*, 2007**

REFERENCES

Bolden, R., Gosling, J., Marturano, A. and Dennison, P. (2003).A review of leadership theory and competency frameworks. University of Exerter.

Calixto, Eduardo. (2007). Integrated preliminary hazard analysis methodology considering the environment, safety and social issues: The platform risk analysis study. ESREL 2007, Stavanger.

Calixto, Eduardo. (2007). The safety integrity level as hazop risk consistence. The Brazilian risk analysis case study". ESREL 2007, Stavanger.

Calixto, Eduardo. (2009).Using Network Methodology to Define Emergency Response Team Location: The Brazilian Refinery Case Study". International Journal of Emergency Management, V6. Inderscience Publisher, 2009.

Calixto, Eduardo. (2008). Environmental Reliability as a Requirement for Defining Environmental Impact Limits in Critical Areas". ESREL 2008, Valencia.

Calixto, Eduardo. (2012).Process Risk Management based on Brazilian National Quality Award Methodology". Working on Safety conference, Poland, Sopot 2012.

Catherine L Wang & Pervaiz K Ahmed. (2002). A Review of the Concept of Organisational Learning. University of Wolverhampton 2002, ISSN 1363-6839

C.ERICSON. (1999). Fault tree Analysis-A history". 17 ° International System Safety Conference, 1999, EUA.

Choudhry RM, Fang D, Mohamed S. (2007). The nature of safety culture: a survey of the state-of-the-art. Saf Sci 2007; 45(10): 993-1012.

CROWL, DANIEL A; LOUVAR, JOSEPH F. (2002). Chemical Process Safety Fundamentals with Applications. Prentice Hall PTR Upper Saddle River, New Jersey 07458, 2002.

DE CICCO, F & FANTAZZINI, M. LUIZ. (2003). Tecnologias Consagradas de Gestão de Riscos; Série Risk Management; São Paulo; 2003.

Deal, T., Kenney, A., (1982). Corporate Culture: The Rites and Rituals of Corporate Life. Addison-Wesley, Reading MA.

Douglas, A., & Glen, D. (2000). Integrated management systems in small and medium enterprises. Total Quality Management, 11(4/5&6), 686-690.

DUARTE, M.(2002). Riscos industriais: Etapas para investigação e a prevenção de acidentes. Funenseg. Rio de janeiro, 2002.

Eisenhardt, K. M. (1989). Building Theories from Cases Study Research. Academy of Management Review, 14(4), 532-549.

Eisenhardt, K. M., & Graebner, M. E. (2007). Theory building from cases: Opportunities and challenges. Academy of Management Journal, 50(1), 25-32.

Eysenck, H. J. (1976) The learning theory model of neurosis – a new approach Behaviour Research and Therapy 14(4) pp. 251-267.

Elvik, R. (2000). Cost-benefit analysis of police enforcement. Working paper in EU project ESCAPE. WP1 29.2.2000SM/1116/2000. Institute of Transport Economics, Oslo.

Elvik (2007). Prospects for improving road safety in Norway. Report 897/2007. Institute of Transport Economics TØI, Oslo.
Erik J De Bruijn, Olaf A M Fisscher Corporate Motivation for Integrated Management System Implementation. Why do Firms Engage in Integration of Management Systems: A Literature Review & Research Agenda.
Fresner, J., & Engelhardt, G. (2004). Experiences with integrated management systems for two small companies in Austria. Journal of Cleaner Production, 12(06), 623-631.
Garvin, D. A. (1991). How the Baldrige Award really works. Harvard Business Review, 69, 80-93.
Glynn, M., Milliken, F. & Lant, T. (1992). Learning about organisational learning theory: an umbrella of organising processes Paper presented at The Academy of Management Meetings Las Vegas, Nevada.
Ginter, P. M. and White, D. D. (1982). A Social Learning Approach to Strategic Management: Toward a Theoretical Foundation. Academy of Management Review. April, 1982.
Hofstede, G. (1997). Cultures and Organisations. McGraw-Hill, New York.
Hopkins,Andrew. (2006). Studying organizational cultures and their effects on Safety Science Elservier. 10 May 2006.
Holdsworth, R. (2003). Practical applications approach to design, development and implementation of an integrated management system. Journal of Hazardous Materials, 104(1), 193-205.
Ikehara, H. T. (1999) Implications of Gestalt theory and practice for the learning organisation The Learning Organisation 6(2) pp. 63-69.
Jørgensen, T. H., Remmen, A., & Mellado, M. D. (2005). Integrated management systems-three different levels of integration. Journal of Cleaner Production, 14(08), 713-722.
Kerko P. (2001). Security management. Porvoo, Finland: WAS Bookwell Publications Ltd 2001.
Layer of Protection Analysis American Institute of Chemical Engineers Center for Chemical Process Safety. (CCPS)ISBN 0-8169-0811-7
Luthans, F. (1998) Organisational behaviour 8th Edition (Boston, MA: Irwin, McGraw-Hill).
MARZAL, E. M & SCHARPF, E. (2002).Safety integration Level selection. Systematics methods including layer of protection Analysis. The instrumentation, systems and Automation Society.2002
McDonald, M., Mors, T. A., & Phillips, A. (2003). Management system integration: Can it be done? Quality Progress, 36, 69-74
MORAES, G. ARAUJO. (2004).Elementos do Sistema de Gestão de segurança meio ambiente e saúde ocupacional. Gerenciamento Verde Consultoria Rio de Janeito: 2004
Oedewald P, Reiman T. (2006). Safety critical organizations. The specialfeatures. VTT Publications 593. Espoo, Finland: Otamedia Publications Ltd 2006.
OHSAS 18001. Especificação do Sistema de Gestão e Saúde Ocupacional. Health and Safety Assessment Series.BSI,1999.
Parker, D., Lawrie, M., Hudson, P. (2006). A framework for understanding the development of organizational safety culture. Safety Science 44 (6), 551–562.
Pavlov, I. P. (1927) Conditioned reflexes: an investigation of the physiological activity of the cerebral cortex (Oxford: Oxford University Press).
Pedler, M., Burgoyne, J. & Boydell, T. (1991) The learning company (London: McGraw-Hill).
Perls, F. S. (1973) The Gestalt approach and eyewitness to therapy (New York: Bantam Books).
PNQ – Prêmio Nacional de Qualidade. Disponivel em www.fnq.org.br. Acesso em 15 de novembro de 2011.
Ramos, D. Arezes, P. Afonso, P. (2012). Application of cbaohs model in the economic evaluation of risks and preventive measures.Working on Safety conference. Sopot-Poland. 2012.
Reason, J. (1997). Managing the Risks of Organisational Accidents. Ashgate, Aldershot.
Rocha, M., Searcy, C., & Karapetrovic, S. (2007). Integrating sustainable development into existing management systems. Total quality management, 18(1-2), 83-92.
Ruuhilehto, K & Vilppol, K. (2000). Safety culture and safety Promotion of the company. Tuk-Publication 1, VTT. Helsinki, Finland: Technical Research Center of Finland Publications 2000.
Salomone, R. (2008). Integrated management systems: experiences in Italian organizations. Journal of Cleaner Production, XX(2008), 1-21.
Schein, E. (1992). Organisational Culture and Leadership, 2nd ed. Jossey-Bass, San Francisco.
Simola A.(2005). Safety management superior piece of work. Doctoral dissertation. Oulu, Finland: Oulu University Publications 2005.
Wope. J.(1959). Psychotherapy by Reciprocal Inhibition.Stanford University Press, 1959.

Zutshi, A., & Sohal, A. S. (2005). Integrated management system: The experiences of three Australian organisations. Journal of Manufacturing Technology Management, 16(02), 211-232.

Subject Index

Eduardo Calixto

D

E

F

G

H

I

J

K

L

M

N

O

P

Q

R

S

T

U

V

W

Made in the USA
Las Vegas, NV
03 November 2026

57424108R00240